The Encultured Brain

The Encultured Brain

An Introduction to Neuroanthropology

edited by Daniel H. Lende and Greg Downey

The MIT Press
Cambridge, Massachusetts
London, England

First MIT Press paperback edition, 2015

This book was set in Stone Sans and Stone Serif by Toppan Best-set Premedia Limited.

Library of Congress Cataloging-in-Publication Data

The encultured brain : an introduction to neuroanthropology / edited by Daniel H. Lende and Greg Downey.
 p. cm.
Includes bibliographical references and index.
ISBN 978-0-262-01778-7 (hardcover : alk. paper)—978-0-262-52749-1 (pb. : alk. paper)
1. Neuroanthropology. 2. Anthropology. 3. Neurosciences. 4. Health—Social aspects.
5. Mental health—Social aspects. 6. Medicine—Humor. I. Lende, Daniel H., 1969–
II. Downey, Greg.
QP360.6.E53   2012
612.8—dc23
2011052298

Contents

Acknowledgments

Most academic books form the tip of an intellectual iceberg. So much of the exchange, discussion, and debate that produces what we see on the page is hidden, a vast collaboration below the surface. This volume in particular owes a great many debts to those who have helped this new field take shape. Here is just a glimpse of their hidden contributions. The ideas, words, and research represented in these pages emerged in part through a series of panels and conferences, including the American Anthropological Association annual meeting, the Society for Applied Anthropology annual meeting, the Critical Neurosciences workshop held in Montreal, and our stand-alone conference on "The Encultured Brain" in 2009. Along the way, we've amassed more intellectual debt than we can discharge.

We especially thank the Society for Psychological Anthropology (SPA) and Robert Lemelson for their support. The SPA made the original American Anthropological Association an invited session, and The Encultured Brain conference was supported by the Lemelson/Society for Psychological Anthropology Conference Fund, made possible by a generous donation from The Robert Lemelson Foundation. Many of the chapters in this volume took on their original form for this conference. Thanks to the leadership of people like Rob, Doug Hollan, and Ashley Maynard, this project moved along faster than many collected volumes often do.

The Encultured Brain conference was also supported by the Institute for Scholarship in the Liberal Arts, the Office of Research, the Kellogg Institute for International Studies, and the College of Arts and Letters at Notre Dame, and the Department of Anthropology at Macquarie University. For additional help on conference-related organizing, we also thank Marina Navia, Harriet Baldwin, Marie Blakey, and Chantelle Snyder. Jeff Granger of Granger Designs came through with a striking graphic design for the conference.

Both the editors and the contributors to this volume benefited extensively from the formal discussants and other colleagues who have participated in these various forums: Susan Blum, Ryan Brown, Joan Chiao, Lance Gravlee, Patricia Greenfield, Eric Lindland, Hal Odden, Naomi Quinn, Karl Rosengren, Robert Sapolsky, Rebecca Seligman, Claudia Strauss, Christina Toren, and Harvey Whitehouse.

Other people we must thank for their intellectual contributions and inspiration include Paul Mason, Juan Domínguez Duque, Carol Worthman, John Sutton, Vaughan Bell, Eugene Raikhel, Suparna Choudhury, Leslie Heywood, Laurence Kirmayer, Connie Cummings, Shinobu Kitayama, Brandon Kohrt, Jason DeCaro, Dan Hruschka, Jim Rilling, Thom McDade, Chris Kuzawa, Josh Reno, Nancy Campbell, Trevor Marchand, Tim Ingold, Anne Fausto-Sterling, Tanya Luhrman, Peter Taylor, Robert Turner, David Howes, Jim McKenna, Sue Sheridan, Agustín Fuentes, Chris Lynn, Jeff Snodgrass, John Hawks, David Dobbs, Jonah Lehrer, Mark Changizi, Seamus Decker, Mike Jindra, Darcia Narvaez, and Nicholas Malone, as well as the reviewers recruited by MIT Press.

Over the course of our relationship with MIT Press, we have been unfailingly impressed by their support of this volume, in particular Philip Laughlin and Katherine Almeida. We have been amazed by the speed, professionalism, and decisiveness of the Press at every level.

Finally, the Neuroanthropology weblog has formed part of the intellectual work, interdisciplinary outreach, and community building behind *The Encultured Brain*. We thank Jovan Maud for the initial inspiration, Vaughan Bell for crucial early encouragement, and Brian Mossop for great support as we made the transition from being an independent blog to forming part of PLoS Blogs. We want to express our gratitude to the Public Library of Science staff and our great colleagues at PLoS blogs. We also owe a great debt to the many colleagues and students who created guest posts for the Neuroanthropology blog, and assistants who helped behind the scenes, in particular Casey Dolezal and Naheed Ahmed. We are very grateful to all the people who linked to our posts; you are too many to name!

The rolling cloud of online discussion also includes too many names to mention, but some individuals are such repeat offenders that we have to single them out: Ryan Anderson, Jason Antrosio, Kate Clancy, Patrick Clarkin, Krystal D'Costa, Martijn de Koning, Max Forte, Kerim Friedman, Alex Golub, Dirk Hanson, Jason Baird Jackson, Lorenz Khazaleh, Barbara King, "Neurocritic," Julienne Rutherford, Mike Smith, and Ed Yong. Finally, thank you to the audience out there, from those who visited only once to our most regular visitors. You are the ones who helped make neuroanthropology what it is today.

Greg also wishes to recognize his colleagues at Macquarie University and in Australia, an outpost of neuroanthropological thought: not just John, Paul, and Juan, but also Monica Dalidowicz, John Evans, Kellie Williamson, Paul Keil, Roxi Tamba, Mark Wiggins, Victoria Burbank, Katie Gleason, and Alan Rumsey. Greg benefits for a very supportive and collegial department, especially under the capable guidance of Chris Houston: thanks to all our fellow anthropologists at Macquarie for your curiosity, camaraderie, and overall good spirits. We could not ask for a better place to work and think.

Daniel wants to thank a group of diverse colleagues who have provided support and sustenance at various stops along the interdisciplinary way: David Himmelgreen, Brent Weisman, Heide Castañeda, Ian Kuijt, Meredith Chesson, Vinay Kamat, Donna Murdock, Lara Deeb, Sarah Willen, John Wood, Bobby Paul, Claire Sterk, Irv DeVore, Jean-Paul Roussaux, and Neal Smith.

Daniel also thanks his wife Marina for her unfailing support and all her love over these many years. Greg is lucky that his wife Tonia puts up with his intellectual obsessions and helps to inspire his creativity.

Everyone who writes and edits know that they are strangely solitary activities that demand so much of everyone else around you. Thanks to everyone who has pitched in along the way.

I   On the Encultured Brain

1 The Encultured Brain: Development, Case Studies, and Methods

Daniel H. Lende and Greg Downey

On Origins

We met at the University of Notre Dame, our offices diagonal to each other, slotted into a back corner of the sixth floor of Flanner Hall. Our conversations roamed over the topics in which we shared interest, with our water cooler moments ranging from our time in Brazil and Colombia to extreme sports like free diving and cage fighting. Even with our different research topics—Greg interested in perception and training and skill, Daniel in addiction and desire and resilience—we found common themes: how individuals relate to sociocultural environments, the importance of neuroscience for illuminating that relationship and getting at what happens inside the person, and the use of ethnography to gain insight into how people actually experience and do things in the real world.

When we got really serious, speaking of theory and the future of anthropology, we found ourselves looking at the same problem from opposite sides. Greg had trained in anthropology at the University of Chicago, with its well-known strengths in cultural and historical theory, but its deep divisions between anthropology and psychological fields. The richness of cultural theory did not quite reach down to explain how perception differed so among capoiera practitioners in Brazil, who reported that their devotion to the Afro-Brazilian art affected their peripheral vision, balance, pain perception, even how they walked in everyday life. Daniel had trained in anthropology at Emory University, where a new program aimed to create a synthesis of biological and cultural anthropology. But that often meant a synthesis on biology's terms, a research focus on quantitative methods and biological markers and hypotheses derived from evolutionary theory that did not quite match with his three years' experience counseling troubled youths with drug problems in Colombia. Daniel saw the potential for a biocultural synthesis that took culture and ethnography more seriously; Greg saw the potential for drawing on neuroscience and psychology to strengthen sociocultural theory.

But our joint focus on neuroanthropology did not yet exist. We did not agree to a joint intellectual project, in fact, until after Greg left Notre Dame for a position at Macquarie University in Australia. At the next meeting of the American Anthropological Association, held in Washington, DC, (a looong trip from New South Wales), Greg proposed starting a weblog to continue to explore these ideas together, and to encourage the wider community to talk widely about this type of synthesis. As part of the blog Culture Matters at Macquarie, Greg had seen the galvanizing effect writing and communicating online could have, especially with our intellectual labor so often isolated and delayed by slow publication cycles. Daniel agreed, almost before Greg stopped talking. Daniel had worked in freelance journalism in Colombia and looked forward to the writing itself. That, combined with the chance to continue conversations with Greg and push a new field forward, made the blog an attractive idea, even if these online forums were not recognized as serious academic pursuits.

They agreed on the name "neuroanthropology," drawn from two graduate students in Australia, Juan Domínguez Duque and Paul Mason. The term captured the proposed area of synthesis in direct and pragmatic fashion; people would know instantly what we meant. The moniker also resonated with ideas that had swirled around Emory about the importance of neuroscience for understanding human behavior. Daniel's doctoral work had integrated neuroscience and anthropology, but without using the explicit name, "neuroanthropology." Later, among his old papers, he found an essay written by another Emory graduate student, Sally Seraphin, with the word "neuroanthropology" in the title. The term had been floating around for a few decades, advocated by sociologist Warren TenHouten (1976) and adopted by popular author and physician Oliver Sacks (1995) to describe his own distinctive portraits of life with neurological disorders, but the field had never really taken hold.

We created the Neuroanthropology weblog in December 2007 (www .neuroanthropology.net). That first month saw 1,267 visits, which we thought, at the time, was substantial traffic. People were coming to the site and seemed like they were actually interested in reading about this stuff! By March 2008, the visits had grown to over eight thousand, and we were finding psychologists, science writers and other commentators linking to our posts. People were definitely interested. We then put together a proposed panel for the next American Anthropological Association (AAA) annual meeting, this time held in San Francisco (while still long, definitely a shorter trip from Australia). The panel, "The Encultured Brain: Neuroanthropology and Interdisciplinary Engagement," became one of the Society for Psychological Anthropology's invited sessions. The session went very well, attracting high-quality papers and a packed room along with a brilliant discussion by Robert Sapolsky, Stanford neuroscientist, expert on endocrine processes, and MacArthur Fellowship winner. In November 2008, the same month as the AAA panel, the Neuroanthropology blog was averaging around 22,000 site visits a month based on a steady stream of posts by Daniel, Greg, and other contributors.

With the success of the conference panel and of Neuroanthropology. net, we decided the time was ripe for a dedicated conference on neuroanthropology, with the explicit goal that papers presented there would become the basis for this collection. With the support of the Lemelson/Society for Psychological Anthropology Conference Fund, as well as the University of Notre Dame and Macquarie University, the standalone conference, "The Encultured Brain: Building Interdisciplinary Collaborations for the Future of Neuroanthropology," took place on October 8 and 9, 2009, at Notre Dame.

This edited volume emerges from that conference. The conference statement, posted online at the start of the conference and read more than five thousand times, became the basis of chapter 2, "Neuroanthropology and the Encultured Brain." Joan Chiao's presentation demonstrated the affinities between her field, cultural neuroscience, and neuroanthropology, as chapter 2 discusses. Speed presentations, quick six-minute talks that delivered the core ideas behind someone's work, followed by extended coffee breaks to trade ideas in this exciting opportunity to discover what was happening across fields, yielded two more chapters: Casey Bouskill's work on humor and breast cancer and Ben Campbell's cross-cultural research on male vitality, embodiment, and testosterone. The second conference day, during which presentations from the original 2008 AAA panel became full-blown papers that we discussed at length, provided a critical mass of case studies to show neuroanthropology in action. Rachel Brezis, Cameron Hay, Katie MacKinnon and Agustín Fuentes, and Peter Stromberg all delivered talks, which we discussed with a number of guests.

Other conferences also proved important for this volume. An overflowing session, "The Globalized Brain: The Impact of Inequality and Exclusion," organized by Daniel Lende at the 2010 Society for Applied Anthropology meetings in Mérida, Mexico, brought two more chapters: work by Bill Dressler, Mauro C. Balieiro, and José Ernesto dos Santos on genetics, depression, and culture in Brazil, and chapter 10 by Erin Finley on post-traumatic stress disorder among US combat veterans returning from Iraq and Afghanistan. Lende's presentation on "The Cultural Brain" at the 2008 Critical Neuroscience Workshop in Montreal, Canada, prompted key insights through feedback from participants; the critical neurosciences approach, like cultural neuroscience, also became a key ally in this new engagement of the social sciences and the neurosciences.

The continued development of neuroanthropology's online presence has also contributed to the volume. The Neuroanthropology weblog become one of the founding members of the Public Library of Science's (PLoS) Blogs initiative in September 2010, moving to a new location and gaining the support of PLoS, a leader in open access publications. Our continued writing online helped us develop key ideas and explore specific topics, from the nature of embodiment to language as providing key insights into neuroanthropology. We continue to receive contacts from researchers around the world who are, in their own ways, trying to unravel the connections between cultural diversity, human variation, and the brain sciences.

Conversations that started in an office in South Bend, Indiana, have become the volume you hold in your hands. The trajectory was fed by more than shared coffee; the ideas were fed by having an academic conversation in the open, where unexpected collaborators joined in as we went. Working on a weblog together, complete with contributions from several of the authors in this volume, helped us to see the new field taking shape, both within our own minds and for a larger audience who increasingly made their presence felt. Conferences, with their formal demands to present research and more relaxed moments to discuss ideas, helped by creating an academic community with a substantive number of people involved. And now we are onto the next step—this volume on the encultured brain.

Overview of the Volume

The Encultured Brain is split into four sections. Part I, "On the Encultured Brain," contains this chapter; chapter 2, "Neuroanthropology and the Encultured Brain"; and chapters 3 and 4, which provide the biological background for doing neuroanthropology. This chapter, alongside the volume overview and our brief origin story, provides a brief guide to the ten case study chapters, followed by an examination of methods for doing neuroanthropological research. Chapter 2 describes the field of neuroanthropology, from early efforts at neuroanthropology by various scholars to how to understand the work of culture and patterns of human variation. This introductory chapter also examines some core ideas in neuroanthropology, from the importance of ethnographic fieldwork to approaches to culture that combine cultural neuroscience and anthropology.

Chapters 3 and 4 show how recent research and theory in primatology, evolutionary biology, and human evolution present us with a very different view of human biology than appears either in popular accounts or in older scholarship. As opposed to fixed accounts, in which genes were described as determining traits and discussions of human evolution focused on what set humans apart, the "new biology" emphasizes variability, plasticity, and interaction with the environment as well as continuity between the evolution of human and other life. Like neuroanthropology itself, this new biology looks at patterns of similarity and difference from a comparative approach, recognizing that both perceptions are equally valid.

Chapter 3, "Primate Social Cognition, Human Evolution, and Niche Construction: A Core Context for Neuroanthropology," uses that comparative approach to examine evolutionary dynamics through both primate and human evolution. Chapter 4, "Evolution and the Brain," discusses the emerging understanding of our baroque brains, built over evolutionary time by selective demands and developmental dynamics. This brain was built for social interaction and for culture, plastic in many parts, stereotypical and structured in others.

The rest of the book is split into two sections: case studies and the conclusion. The case studies are the core of the volume, the concrete demonstration of neuroanthropology in practice. The chapters address a range of different topics, with differing emphases on ideas from neuroscience and anthropology, and make use of a number of methods, from examining biomarkers to experience-near interviewing. This rich array shows neuroanthropology as a diverse field that uses a wide variety of theories and methods to permit empirical purchase on research problems related to the encultured brain.

Part II includes case studies that examine human capacities, skills, and variation. Specific chapters focus on knowledge and memory by health practitioners cross-culturally (chapter 5); the sense of balance as neuroanthropological rather than innate, formed as much by cultural practice as by biology (chapter 6); skill acquisition and the nature of habits, including how practitioners from different countries understand the same martial art (chapter 7); the role of humor in coping with treatment and recovery from breast cancer (chapter 8); and the role of testosterone and embodiment in understanding patterns of male vitality cross-culturally (chapter 9).

The case studies in part III examine human problems, pathologies, and variation. The five chapters provide discussions of a neuroanthropological model of post-traumatic stress disorder (chapter 10); how autism questions ideas about theory of mind and religion, and reveals novel aspects of cultural development (chapter 11); college smoking as illuminating the joint dynamics of the collective excitement of social life and a sense of lapses in individual agency (chapter 12); an examination of key components of addiction built from ethnography and neuroscience (chapter 13); and how culture and genetics work together to define different types of depression in Brazil (chapter 14).

Part IV provides the concluding chapter to *The Encultured Brain*. The conclusion reflects on what ties together the ten different case studies, charts new directions for research, and examines the wider implications of neuroanthropology.

Finally, the Neuroanthropology website, found at http://blogs.plos.org/neuroanthropology, offers a wide array of complimentary materials to this volume. Brief biographies of our authors can be found there, complete with links to additional works by them. A selection of popular posts online provide additional ways to explore the intersection of neuroscience and anthropology, ranging from "poverty poisons the brain" to human echolocation. Additional resources, such as a comprehensive bibliography on neuroanthropology, are also posted on this site, and we welcome you to join the conversation.

On the New Biology

The next three chapters each bring an important addition to a modern understanding of biology, based in emerging research and new lines of understanding rather than older assumptions and isolated disciplinary lines of research. One of the greatest challenges in any synthetic, cross-disciplinary research area is that each of the partners is on the move, and advances in contributing fields have implications for the shared intellectual labor. Only by keeping up with shifting ways of thinking in adjacent fields can we realize that old, well-worn arguments are no longer valid, that the old battle lines have not just shifted, they may no longer even be in conflict, and that new opportunities are afforded by innovative ways that we might collaborate. Too often, we believe, scholars make bold arguments to their like-minded peers against perspectives that are no longer held by the majority of theorists in other fields, except perhaps the most recalcitrant and battle-scarred in the old opposition.

Chapter 2, written by the two volume editors, provides an overview of neuroanthropology, relating "the new biology" to culture theory, ethnography, and critical analysis. By examining what we now know about neuroscience and brain function, and how emerging models contrast with older views of encapsulated cognitive modules and machine-like information processing, this chapter presents an overview of novel ways to draw on neurobiology and cognitive science when doing integrative research. In its discussion of neuroplasticity and the emerging field of cultural neuroscience, this chapter also demonstrates how cultural dynamics and social environments can literally help shape the architecture and functioning of the brain.

Chapter 3 was prepared by the primatologists Katherine C. MacKinnon and Agustín Fuentes. Drawing on both anthropology and evolutionary biology and grounded in a comprehensive understanding of primate evolution and behavior, MacKinnon and Fuentes show us that an understanding of primate social cognition and the role of niche construction in human evolution are central to understanding how the encultured brain has emerged. Primate-wide strengths in social cognition and the external dynamics of niches are part and parcel of how humans have become successful over evolutionary time, and provide central ideas for understanding how culture came to be such a dynamic force in human lives.

In chapter 4, the editors of this volume discuss the evolutionary emergence of the human brain, essential to understanding neuroanthropology. The dynamics we describe, however, are often quite different from those presented in many evolutionary accounts of human psychology, which either assume that the brain is a collection of mental modules, a cognitive Swiss Army knife designed during the Pleistocene, or is a black box focused on enhancing evolutionary success, an optimal machine for finding mates and fighting off competitors. The realities of brain evolution provide an empirical basis to understand how the brain is an inextricable mix of nature and nurture and biology and culture. After all, the brain has long evolved with exactly these mixed influences.

Synopses: Case Studies

In this section, we briefly review each of the case studies. These case studies form the heart of this volume, showing how neuroanthropology uses a variety of methods to tackle a wide range of theoretical and empirical problems. The test of any theoretical and analytical perspective is not its elegance in abstraction, but its usefulness in application. Part II, "Case

Studies on Human Capacities, Skills, and Variation," opens with chapter 5, "Memory and Medicine," by Cameron Hay, which asks: Where do healing, memorization, and knowledge meet? Cameron Hay contrasts the medical traditions of rural Indonesian healers and urban Californian doctors to demonstrate the interaction of memory and knowledge as a fundamental part of both traditions, but with quite different mechanisms. Using a comparative approach, Hay shows how traditions of knowledge are biocultural phenomena, formed jointly from the neurological processes of learning and memory and from the social expectations of healing and memory.

Chapter 6, "Balancing between Cultures: Equilibrium in Capoeira," by Greg Downey brings together multiple lines of evidence about the body's equilibrium system to argue that even nonconscious brain functions, like the vestibular system, are susceptible to enculturation. Downey's exploration of equilibrium began with his own participation in capoeira, a Brazilian art form that mixes martial arts and dance. He specifically contrasts the handstand techniques used in capoeira with those in Olympic gymnastics, but also develops a broader argument that the finely tuned equilibrium system, capable of balancing on the hands, is one example of how enculturation involves sensory learning and physiological adaptation in training, not just the acquisition of cultural information.

Katja Pettinen continues the discussion of bodily training in chapter 7 in her exploration, "From Habits of Doing to Habits of Feeling: Skill Acquisition in Taijutsu Practice." Examining the practice of Taijutsu, a noncompetitive martial art, both in Japan where it was created and in the United States, Pettinen reflects on how cultural traditions of bodily habits and sensory perception explain differences between practice in the two sites. Drawing on the traditional test of virtuoso ability, the *sakki*, in which a seated student must avoid a blow from behind, Pettinen suggests that different practical pedagogies are a result of diverging understandings of learning. The sakki, and the accompanying traditional pedagogy, arguably better respect certain dimensions of the neurology of proficiency than do forms of training imposed when the art was exported to the West.

In her research on the role of humor in resiliency among cancer patients in chapter 8, Kathryn Bouskill sensitively documents how breast cancer survivors' coping mechanisms, especially their gallows humor, demonstrate that stress is not merely a psychophysiological reaction. Bouskill draws on both evolutionary-biological and sociocultural accounts of humor to explain how laughing at the indignities of "cancer world," the day-to-day reality of treatment and recovery that survivors share, buffers individuals' ability to cope, even though the humor cannot be shared with a wider community. Humor has healing consequences because of its social effects as well as its neuropsychological correlates; neither can be understood in isolation.

In the final chapter of the section, "Embodiment and Male Vitality in Subsistence Societies," Benjamin Campbell seeks to combine biological and experiential approaches to embodiment by looking specifically at the links between hormones and subjective experience in men. Campbell examines how cross-cultural research on testosterone, biological function, and vitality can be bolstered by considering the cognitive processes that mediate our subjective sense of well-being. His research on testosterone among the Ariaal, pastoral nomads from northern Kenya, suggests that men have a different profile of testosterone levels over the life course due to diet, disease burden and activity levels in subsistence societies. Campbell proposes that the neuroanthropology of male vitality helps to explain a widespread cross-cultural pattern in which semen, the spine, and the brain are felt to be interrelated, a symbolic constellation referred to as *muelos*. Whereas an earlier generation of ethnologists dismissed a "nonsense and misinformation so ancient and pervasive," Campbell sees diverse subsistence societies recognizing a fact of human embodiment under socioecological stress.

In combination, the chapters in part II demonstrate how specific ethnographic studies of variation, especially in comparative perspective, illuminate the way that the human nervous system can be affected by cultural, social and environmental factors. Local forms of knowledge, including experts like medical practitioners and martial artists, often recognize in subjective ways how the brain and body are affected by different conditions, although they may use a language that incorporates elements that may appear mystical to outsiders, as the cases provided by Campbell, Pettinen, and Hay all demonstrate. At the same time, regimes of training, social networks of peers, and cultural practices build upon and extend human capacities in distinctive fashion, whether this is providing coping mechanisms that cross the boundary between the individual and group, as with Bouskill's cancer survivors, or creating distinctive neurobehavioral skill sets, as Downey explores. The overarching point is that humans build up their nervous systems in distinctive fashion because of the cultures in which they develop, and cultural practices make use of and extend these capacities; cross-cultural research can alert neuroanthropologists to locations of potential variation that we may not perceive in any single culture alone.

Part III provides "Case Studies of Human Problems, Pathologies and Variation," demonstrating how various brain dysfunctions illuminate the interaction of culture and the human nervous system. Mental and behavioral health problems are often treated as arising from a "damaged" or "malfunctioning" region of the brain; in fact, "disordered" functioning can often be as stable a dynamic as "normal" activity, and pathology can rarely be pinned down to a single location. Neuroanthropological research on psychological dysfunction feeds back into our understanding of all functioning because, on the one hand, cultural variation affects the way that psychological problems like trauma, depression, and addiction are lived, interacting in complex ways with our biological endowments and social worlds. On the other hand, careful study of breakdowns in functioning in particular settings can help us to more clearly see where our models of human psychology, including its disorders, oversimplify human potential and fail to recognize how many resources are built into our cognitive functioning. Variations within pathology may suggest new lines of research or intervention to address human problems and help us to address uneven patterns of human suffering.

In chapter 10, for example, Erin Finley provides us with a holistic model to understand post-traumatic stress disorder, or PTSD. Finley replaces core diagnostic features of PTSD—hyperarousal, reexperiencing, and avoidance— with a more experience-near model of stress, horror, dislocation, and grief, drawn from her work with military veterans of wars in Iraq and Afghanistan. Finley's model captures the unfolding trajectory of PTSD, how the experience of trauma over time involves a recalibration of the stress response, intrusive memories such as comrades lost, and interaction with the support structures and interpretive frameworks that veterans are encouraged to use as they struggle to recover. Finley's integrative approach permits her to delve deeply into human psychology and experience in ways that are informed by both neuroscience and anthropology, while remaining firmly focused on the people who actually develop and suffer from post-traumatic stress disorder.

In chapter 11, "Autism as a Case for Neuroanthropology: Delineating the Role of Theory of Mind in Religious Development," Rachel Brezis tests one of the dominant cognitive accounts of religion—that religious faith is a projection of human perception of other agents or "theory of mind"— against the belief that autism involves a deficiency in theory of mind. Using qualitative work with high-functioning Jewish autistic youth in Israel, Brezis tested the hypotheses that autistic individuals lacked an interpersonal understanding of God. Brezis found that not only did the majority of the autistic children have a sense of an agentive God, but they also seemed to draw upon religious narratives and their own talent for repeating prestructured narratives to compensate for deficiencies in autobiographical memories and the interpersonal coordination of the self. Rather than a deficits model of psychopathology, Brezis finds that her autistic subjects were using resources available in their symbolic niche to bolster their sense of self and their past; autistic individuals show that alternative routes can lead to faith, and that religion can serve diverse cognitive functions.

Peter Stromberg brings together contemporary neuroscience and classical social theory to make sense of college students' accounts of smoking in chapter 12, "Collective Excitement and Lapse in Agency: Fostering an Appetite for Cigarettes." Stromberg draws upon a longitudinal study of young men and women who get carried away with cigarette smoking in parties and other social settings but do not demonstrate signs of chemical dependency to discuss how our sense of autonomy and agency can be undermined. French sociologist Emile Durkheim employed the concept of "collective effervescence," the excitement felt in large crowds, such as at festivals and dramatic rituals, to illustrate how some psychological experiences were "social facts," inexplicable with reference only to the individual. Stromberg proposes that three processes form the neuroanthropological core of both effervescence and a sense of lost agency in college students' smoking: imitation and rhythmic entrainment, pretend play in socially sanctioned settings, and emotional arousal in groups. Together, these three processes drive changes in our sense of agency, and affect how we interpret what is happening in these collective moments. Given the local interpretation of smoking behavior—that cigarettes are "addictive" and thus undermine individual agency—students misread the reasons for their lapsed agency, their inability to stop themselves from smoking, attributing power to the cigarettes that they objectively and chemically do not possess.

Daniel Lende likewise problematizes simple accounts of addiction in chapter 13, "Addiction and Neuroanthropology," based on his research with young people struggling with drugs in Colombia. As Lende describes, drug addiction is often believed to be a problem of a badly "rewired" brain: pleasure circuits hijacked by narcotics, escalating their demands, and insufferable withdrawal when denied the chemical that greedy receptors have come to crave. By carefully listening to adolescents' accounts of "wanting more and more," inspired by daily rhythms, boredom, stress, or social conflict, Lende realized that much of addicts' suffering arose from the way that substance dependence forced them to choose between competing social worlds and life patterns; drug taking closed off the people and pursuits that had once been important to them. Chemically induced pleasure was less experientially potent than "wanting," a problem that suggested a competing neurological explanation for addiction: incentive salience, or the motivation to seek drugs. Lende recognized that, as the young people chose to inhabit a world consistent with drug seeking—meeting up with friends who were users, hiding from parents, cutting classes to go to places where they could find drugs—the salient cues for wanting narcotics multiplied and life without drugs seemed increasingly fraught and conflict filled. The spiral of addiction was not merely a neurological transformation, but a shift in habits, clothing, friends, hangouts, and other external factors that re-cued drug seeking behavior, drove addicts to take drugs, even when the young people sought to stay clean. Addiction is not simply in the brain, but in the way that the addict's brain and world support each other. Recognizing this recursive process helps us to better understand what prompts those who are addicted, against their own desire for a new life, to return to drug-seeking patterns of behavior.

In chapter 14, "Cultural Consonance, Consciousness, and Depression: Genetic Moderating Effects on the Psychological Mediators of Culture," William Dressler, Mauro Balieiro, and José Ernesto dos Santos offer a sophisticated, mixed methods examination of the interaction between culture, individual psychology, and genetics in susceptibility to depression. Based on a long-term research project in Riberão Preto, Brazil, Dressler, Balieiro, and Dos Santos seek to provide a quantifiable account of culture and its effect on individuals. The team specifically examined cultural consonance, the degree to which an individual's own beliefs and models of family, social support, lifestyle, and national identity match those pervasive in his or her community. Dressler and his colleagues posit that low consonance is stressful, as it is linked to higher blood pressure and greater subjectively perceived stress. Much of the effect of consonance on depression, however, is mediated by a gene polymorphism linked to serotonin function and emotional reactivity. Without the variant of the gene, individuals with low cultural consonance displayed the typical dysfunctional cognitive schemas associated with cognitive approaches to depression: they over-generalized negative experiences, highlighted failures, and personalized setbacks. With a variant of the gene (the "AA" subtype) linked to high reactivity, however, another pathway to depression led directly from lack of cultural consonance to depression, without the exacerbating effect of dysfunctional beliefs being necessary. These AA subtype individuals appeared more vulnerable to the negative effects of low cultural consonance, but the genetic variant augmented recovery when their consonance improved, such as a change in life situation to more closely approximate the cultural norm.

Field Research: Building on the Case Studies

Anthropology is marked by a rich, varied tradition of field research: from primatologists working in tropical rainforests and biological anthropologists unearthing human ancestors and sampling biological variation across societies, to archeologists performing digs and cultural and linguistic anthropologists studying with people around the globe today. To thrive, neuroanthropology will also need to be a field-driven endeavor. Drawing on the case studies included in this volume, we summarize here specific ways neuroanthropologists can approach field work. We also outline opportunities to expand and enrich research beyond what is represented in the case studies. These approaches represent a menu from which different combinations can be drawn to tackle distinctive types of problems, while working within the practical constraints imposed by field conditions, funding, and other types of institutional support.

Ethnography

All ten case studies draw on ethnography in one way or another, from providing context, interpreting quantitative research and helping to develop scales for survey work, to being the focal method of research. Participant observation was central to Pettinen's understanding of skill, enculturation, and physical practice. Interviews gave depth to Lende and Brezis' understanding of how addiction and autism work on the ground. Bouskill used linguistic analysis to understand how humor achieved its affiliative effects while also insulating cancer survivors from "cancer world." Participant observation and interviewing together helped Stromberg develop a new theoretical understanding of how collective effervescence and lapses in our sense of agency arise together. Interviewing and following experts gave insight into post-traumatic stress disorder, our sense of balance, and cross-cultural healing traditions that Finley, Downey, and Hay would not have achieved on their own.

Reviewing these chapters shows that fieldwork for neuroanthropology goes beyond the basic uses of ethnography: that is, participant observation to inform other parts of research, interviewing to get at what people say and do, behavioral and textual analysis to examine language, and consulting with expert informants to develop further insight. First, ethnography provides an absolutely essential framing for doing analysis of quantitative data. Campbell's distinction between Western and subsistence economies, with dramatically different socioecological conditions for people in either setting, informed his analysis of varying testosterone and vitality levels cross-culturally. Dressler's political economic view of consonance and Pettinen's historical view of mind-body philosophies similarly informed their interpretations. Second, for doing neuroanthropology, paying attention to the body is important. In bodily experience our sensory and motor systems interact with the world, the site where cultural experience can get under the skin, and the conceptual center for theories of embodiment and cognitive linguistics that are proving important in both cognitive science and anthropology.

Third, given the complexity of the problems that neuroanthropology generally tackles, breaking down ethnographic moments or contexts into specific parts makes great sense. Pettinen does this well when discussing the variety of ways bodies interact when learning martial arts strikes, thereby getting beyond the assumption of uniformity that lies behind ideas like "practice makes perfect." Similarly, Lende got empirical traction on addiction by not just focusing on pathological moments of craving or withdrawal, but by getting informants to describe a prototypical day of drug use step by step. Fourth, the world is full of "natural experiments" that permit us to do cross-cultural work on human variation in ways that we could not in a laboratory. Hay uses healing practices in two different societies to examine how knowledge traditions and memory meet. Just as Pettinen uses cross-cultural comparison with the same martial art, Downey draws on contrasting examples of expert balance to reveal the dynamism of an equilibrium system that can develop tremendous expertise in one arena, but not necessarily transfer that expertise to other athletic endeavors.

Fifth, ethnography can be used to test hypotheses. Brezis does this explicitly, utilizing ethnographic work with autistic individuals to test key proposals about theory of mind and religion. Lende also recounts how ethnographic study contradicted specific proposals about the workings of incentive salience. Finally, ethnography permits a community-based approach that complements theory-driven research. Bouskill recounts how women in the breast cancer support center proposed humor as a research problem, leading her into a rich arena of research. Downey emphasizes how capoiera masters' insistence that they perceived and acted in the world differently because of their training led to his engagement with cognitive science. Feedback from the community, from research problem formation to testing interpretations to applications of research, is a central part of neuroanthropology's ethnographic approach.

Biomarkers and Other Biological Measures

Chapters 9 and 14, by Dressler and colleagues and Campbell, actively incorporate biomarkers into their research. Data from different genotypes for serotonin receptor function proved central to Dressler, Balieiro, and dos Santos' analysis of the relationship between depression and cultural consonance; without this inclusion of individual biological variation, their results would have looked quite different. Similarly, Campbell shows how incorporating testosterone into embodiment research brings the physiological body into view. As he says, biological anthropologists have good ways to measure variation in hormonal responses and ecological conditions, opening up new ways to do research.

Biological anthropology also offers a range of other measures that can be incorporated into neuroanthropological research (Worthman & Costello, 2009). Cortisol and amylase, alongside simple measures of blood pressure, offer avenues to study the dynamics of stress in the field (DeCaro, 2008; Flinn, 2008; McDade, Williams, & Snodgrass, 2007). Anthropometrics, nutritional intake data, and daily activity levels provide central ways to examine how ecological conditions can affect the physical state of the body, and increasingly, to analyze how prenatal conditions, early growth, and other biocultural processes can affect brain development (DeCaro & Worthman, 2008; Himmelgreen, Romero-Daza, Vega, Cambronero et al., 2006; Kuzawa & Quinn, 2009). Another field-ready research methodology is psychophysiology, which can measure bodily reactions like heart rate and skin galvanic response. Rebecca Seligman has incorporated these measures into her neuroanthropological study of dissociation among Candomblé mediums in Brazil (Seligman, 2005, 2010).

Surveys, Scales, and Activity Measures

Campbell's and Dressler's teams both use surveys as a key way to test hypotheses and examine associations quantitatively in their work. The epidemiological approach has much to recommend it for neuroanthropology, since scales representing different hypotheses from evolutionary theory, culture theory, neuroscience, and the like can all be incorporated and then tested (Lende, 2007, 2005). Dressler and Lende also demonstrate how anthropologists can develop specific scales to measure culture and the role of dopamine in addiction. Continued development of scales will be an important part of the neuroanthropological endeavor; for example,

Campbell could develop scales that measure how the insula mediates the subjective experience of the body and vitality in ways suggested by a combination of neuroscience and ethnography. Measures that permit access to patterns of daily experience are a final approach to mention here. Jason DeCaro has developed portable ways to sample activities and experiences on a day-to-day basis through electronic devices, which he used in conjunction with cultural model and biomarker measures (DeCaro & Worthman, 2008, 2011).

Other Important Approaches

Four other methodological approaches deserve mention, as they will be important for the development of future research even if they are not an explicit part of the techniques utilized in the case studies presented here. First, experimental approaches have been used to good effect in field studies, examining, for example, cross-cultural patterns in reciprocity (Henrich, Boyd, Bowles, Camerer et al., 2001) and children's observational learning in Samoa (Odden & Rochat, 2004). Similarly, experimental design in primatology and psychology has led to important results that illuminate human evolution and neural function (MacKinnon & Fuentes, chapter 3, this volume; Chiao, Hariri, Harada, Mano et al., 2010). Experimental approaches to test the dynamics of balance or the role of dislocation in PTSD are potential avenues for further exploration among the case studies.

Critical approaches are also important for neuroanthropology. Questioning the underlying assumptions of research programs, analyzing the institutional forces that shape research and its implications, and being critical of the deployment of neuroscience and brain-based explanations in society are all part of a critical approach to neuroscience. At present, there is no "critical neuroanthropology" project. Such a project should draw on critical neuroscience and aim, not just to question, but to improve neuroscience research and its application. Adopting critical approaches to neuroscience from anthropology and the humanities, a critical neuroanthropology might help resist the inevitable neuroreductionism in the way brain-based ideas play out in popular discussions (Choudhury, Nagel, & Slaby, 2009; Choudhury & Slaby, 2011; Kraus, 2011; Martin, 2004).

Interpretive approaches, and humanities research more broadly, are another area where neuroanthropology will hopefully develop. Anthropology overlaps with the humanities in many ways, from the shared commitment to an interpretive approach to the writing at the core of ethnographic practice. A humanistic neuroanthropology will find creative ways to illustrate human variation around the world, while developing a grounded philosophy about fundamental human questions. Leslie Heywood, poet and humanist extraordinaire, writes about her own engagement with evolutionary theory: imagination through the humanities can be "an important tool for the creation of meaning," she notes, where interpretive approaches build not on a cumulative body of evidence, as in science, but create an intellectual conversation against which subsequent scholars define themselves (Heywood, Garcia, & Wilson, 2010). This type of intellectual project, from examining how creativity works to interrogating what it means to be human through grasping neural and ethnographic diversity at once, is another future development that beckons.

Finally, neuroimaging will inevitably form an important methodological resource for neuroanthropology. None of the case study chapters explicitly use neuroimaging in their research, though almost all draw on results from imaging studies, and some, like Campbell, propose hypothesis testable through neuroimaging. Cultural neuroscience has made extensive use of neuroimaging, an endeavor made easier by the development of neuroimaging facilities in urban areas around the globe (Chiao et al., 2010; Kitayama & Uskul, 2010). This diffusion of neuroimaging technology should permit neuroanthropologists to incorporate imaging into their methodological tool set. As Dressler and colleagues show with their use of biomarkers, neuroimaging should open up new theoretical models and provide the ability to test specific mediators in the integrated models neuroanthropologists develop.

Developing and Testing Models

Several chapters do develop specific models for the problems they are addressing. Finley's chapter on post-traumatic stress disorder, for example, outlines a six-factor model of PTSD, with stress, horror, dislocation, and grief at the core, and the cultural environment and cultural mediators as acting to bring those elements together. This model allows her to reinterpret key diagnostic criteria for PTSD, while proposing a cross-cultural approach to trauma more broadly. Other chapters propose similar models, from Hay's three-pronged approach to knowledge traditions to Campbell's tripartite model for male vitality in subsistence societies.

Campbell also makes a crucial proposal at the end of chapter 9: different studies can test separate elements of his account. Recognizing the value of partial validation is likely to be a valuable option for collaborative research. Neuroanthropologists can develop overarching models. Elements of those models can be addressed by individual studies, since technological limitations at present make it difficult to integrate all elements of research into one comprehensive design, especially given the practical challenges of field research settings in which anthropologists often work.

Conclusion

This overview suggests neuroanthropology is a field in rapid but early development. Our emphasis on the case studies in this volume is intended to show that a substantive interdisciplinary approach is workable. As we trace in chapter 2, we are hardly the first people to call for a merger of anthropology and brain science. From a theoretical perspective, the potential for synthesis is fairly obvious in light of emerging research on fundamental differences cross-culturally in brain function. International brain scientists will continue to study cultural differences in brains using neuroimaging and accompanying technologies. Significant gains will be made by bringing anthropological approaches as well as the discipline's traditional interest in human variation to bear on what laboratory research has shown. The case studies show that interdisciplinary field research matters for the empirical data it brings to questions both old and new, and illustrate how much we can gain by bringing anthropology and neuroscience together outside the laboratory. So much neuroanthropological variation is available for study throughout the world as long as we are willing to go find that variation where it lives.

References

Chiao, J. Y., Hariri, A. R., Harada, T., Mano, Y., Sadato, N., Parrish, T. B., et al. (2010). Theory and methods in cultural neuroscience. *Social Cognitive and Affective Neuroscience, 5*(2–3), 356–361.

Choudhury, S., Nagel, S. K., & Slaby, J. (2009). Critical neuroscience: Linking neuroscience and society through critical practice. *Biosocieties, 4*(1), 61–77.

Choudhury, S., & Slaby, J. (Eds.). (2011). *Critical neuroscience: A handbook of the cultural and social contexts of neuroscience*. Hoboken, NJ: Wiley-Blackwell.

DeCaro, J. A. (2008). Methodological considerations in the use of salivary α-amylase as a stress marker in field research. *American Journal of Human Biology, 20*(5), 617–619.

DeCaro, J. A., & Worthman, C. M. (2008). Culture and the socialization of child cardiovascular regulation at school entry in the US. *American Journal of Human Biology, 20*, 527–583.

DeCaro, J. A. and Worthman, C.M. (2011). Changing family routines at kindergarten entry predict biomarkers of parental stress. *International Journal of Behavioral Development*, *35*, 441–448.

Flinn, M. V. (2008). Why words can hurt us: Social relationships, stress, and health. In E. O. S. W. Trevathan & J. McKenna (Eds.), *Evolutionary medicine and health* (pp. 242–258). Oxford: Oxford University Press.

Henrich, J., Boyd, R., Bowles, S., Camerer, C., Fehr, E., Gintis, H., et al. (2001). In search of Homo Economicus: Behavioral experiments in 15 small-scale societies. *American Economic Review*, *91*(2), 73–78.

Heywood, L., Garcia, J., & Wilson, D. (2010). Mind the gap: Appropriate evolutionary perspectives toward the integration of the sciences and humanities. *Science & Education*, *19*(4–5), 505–522.

Himmelgreen, D., Romero-Daza, N., Vega, M., Cambronero, H. B., and Amador, E. (2006). "The tourist season goes down but not the prices:" Tourism and food insecurity in rural Costa Rica. *Ecology of Food and Nutrition*, *45*(4), 295–321.

Kitayama, S., & Uskul, A. K. (2010). Culture, mind, and the brain: Current evidence and future directions. *Annual Review of Psychology*, *62*, 419–449.

Kraus, C. (2011). Critical studies of the sexed brain: A critique of what and for whom? *Neuroethics*. Advance online publication.

Kuzawa, C. W., & Quinn, E. A. (2009). Developmental origins of adult function and health: Evolutionary hypotheses. *Annual Review of Anthropology*, *38*, 131–147.

Lende, D. H. (2005). Wanting and drug use: A biocultural approach to the analysis of addiction. *Ethos*, *33*(1), 100–124.

Lende, D. H. (2007). Evolution and modern behavioral problems. In E. O. Smith, W. Trevathan, & J. McKenna (Eds.), *Evolutionary medicine and health: New perspectives* (pp. 277–290). New York: Oxford University Press.

Martin, E. (2004). Talking back to neuro-reductionism. In H. Thomas & J. Ahmed (Eds.), *Cultural bodies: Ethnography and theory* (pp. 190–212). Malden, MA: Blackwell.

McDade, T., Williams, S., & Snodgrass, J. (2007). What a drop can do: Dried blood spots as a minimally invasive method for integrating biomarkers into population-based research. *Demography*, *44*(4), 899–925.

Odden, H. L., & Rochat, P. (2004). Observational learning and enculturation. *Educational and Child Psychology*, *21* (2):39–50.

Seligman, R. (2005). Distress, dissociation, and embodied experience: Reconsidering the pathways to mediumship and mental health. *Ethos*, *33*, 71–99.

Sacks, O. (1995). *An anthropologist on Mars: Seven paradoxical tales*. New York: Knopf.

Seligman, R. (2010). The unmaking and making of self: Embodied suffering and mind—body healing in Brazilian Candomblé. *Ethos, 38,* 297–320.

TenHouten, W. (1976). More on split brain research, culture, and cognition. *Current Anthropology, 17*(3), 503–506.

Worthman, C. M., & Costello, E. J. (2009). Tracking biocultural pathways in population health: The value of biomarkers. *Annals of Human Biology, 36*(3), 281–297.

2 Neuroanthropology and the Encultured Brain

Greg Downey and Daniel H. Lende

Our Encultured Brains

Our brain and nervous system are our most cultural organs. While virtually all parts of the human body—skeleton, muscles, joints, guts—bear the stamp of our behavioral variety, our nervous system is especially immature at birth, our brain disproportionately small in relation to its adult size and disproportionately susceptible to cultural sculpting. Compared to other mammals, our first year of life finds our brain developing as if in utero, immersed in language, social interaction, and the material world when other species are still shielded by their mother's body from this outside world. This immersion means that ideas about the self and methods of child rearing affect the environmental niche in which our nervous system unfolds, influencing gene expression and developmental processes to the cellular level. By recognizing that much of what makes humans distinctive lies both in the size and specialization and in the dynamic openness of the human nervous system, neuroanthropology places the brain at the center of discussions about human nature and culture.

Our ability to learn and remember, our sophisticated skills, our facility with symbolic systems, and our robust self-control all mean that the capacity for culture is, in large part, bought with neurological coin (see also Deacon, 1997; Tomasello, 2001). This dynamic infolding of an encultured nervous system happens over developmental time, through the capacity of individuals to internalize both experience and community-generated tools, and then to share thoughts, meanings, and accomplishments. In this way, activities, contexts, and experiences become central to forming what it means to be human and how humans are similar and different around the world.

This interactive approach means that humans' capacity for thought and meaning making emerges equally from social and individual sources, built of public symbol, evolutionary endowment, social scaffolding, and private neurological achievements. Rather than sets of genes or overarching systems of symbols as a basis for scholarly exploration and explanation, neuroanthropology examines how brain and culture interact, exploring the synthesis of nature and nurture while cutting through idealized views of biological mechanisms and cultural symbols. A central principle of neuroanthropology is that it is a mistake to designate a single cause or to apportion credit for specialized skills (individual or species-wide) to one factor for what is actually a complex set of processes.

The time is ripe for the mutual engagement of neuroscience and anthropology. Brain scientists are no longer content to treat cultural difference as simply a demographic variable, and anthropologists offer brain scientists more robust accounts of enculturation to explain observable differences in brain function. Anthropologists, no longer so afraid of "universalizing" or "psychologizing," can draw on the increasing evidence of how neuroplasticity plays a role in social and cultural dynamics; and through their understanding of biology, neuroscientists offer nuanced ways of enhancing the study of human variation and cultural dynamics. Finally, neuroanthropology itself offers broad benefits for each field: anthropologists can engage the brain sciences to enrich holistic anthropology, while cognitive scientists and neuroscientists can contribute to an enhanced cultural and social neuroscience.

Shifting the anthropological focus to the enculturation of the nervous system highlights a set of novel research questions, places anthropological theory on more confident footing, and creates opportunities for interdisciplinary collaboration (see Lende, 2008; Seligman & Brown, 2010). The project of neuroanthropology will be to examine different neural systems empirically, understand how neural capacities develop, and document which biological and environmental factors can shape their realization. As James Rilling writes:

The study of neuro-development also reveals which aspects of brain development are highly canalized and which are more labile, and in so doing, identifies mechanisms by which brain development can be affected by social and cultural environments to make people in one culture think and act differently from those in another. (2008, p. 3)

Neuroanthropology will advance as understanding of the brain grows more sophisticated, based on the recognition that theories about culture are at the same time theories about brain function and development.

Neuroanthropologists are after "brains in the wild," to paraphrase Edwin Hutchins (1995). This means aiming for laboratory research that is ecologically and ethnographically valid, and doing field research that draws on relevant insights from cognitive science and human development. Neuroanthropology should bring new questions to laboratories, as we will argue, but it should also road-test theories and insights in ecologically valid settings to make sure that they are not so fragile that they crumble outside the confines of carefully controlled psychological laboratories.

Laboratory study alone cannot offer a complete account of how the human brain works precisely because the nervous system is so crucially embedded in its environment, both in the immediate ecological sense in perceiving persons, but also in the long-term developmental sense. Although laboratory study is absolutely fundamental to our understanding of humans' cognitive peculiarity and potential, so too is wide-ranging field research, which locates and observes brains in action, in settings that may not be reducible to studies that can be run on prone subjects, lying still inside giant, clanging magnets. Moreover, the worldwide diversity in the cultural manipulation of the nervous system, especially the life-long cultivations of distinctive skills and cultural capacities, offers opportunities to explore the limits of our species' malleability.

Methodologically, neuroanthropology will require problem-driven resourcefulness and openness to mixed methods. No single psychological, ethnographic, or neuroimaging technique will be able to illuminate all aspects of the brain's variability. Neuroanthropology will make use of both established and emerging technologies in brain imaging; compelling anthropological questions will help to drive projects that are epistemologically sophisticated and to motivate technological refinements in methods.

Research on the brain is accelerating; our ability to manipulate neurobiological function in direct and indirect fashion is exploding through technology, pharmaceuticals, and the control and shaping of environmental conditions. Moreover, the brain has become a central part of how many societies explain the human condition, from pathologies to a lack of social success, and they sometimes use these explanations in ways that exacerbate social problems. Finally, across the globe, there are continued efforts to define human difference solely in biological terms or according to cultural differences. In both cases, this definition of difference often becomes a justification for doing nothing, since their "culture" or their "brains" are irremediably different. Neuroanthropology strikes at the heart of these efforts to use older ideas and research paradigms to overturn what science actually shows—that biology and culture jointly define us.

Early Efforts at Neuroanthropology

The realizations that the human brain is so culture capable, and that cultural learning is an intrinsic part of our species' evolutionary advantages, have led a number of scholars to propose that anthropology and the brain sciences should have strong relations. Reflecting on his own career, for example, anthropologist Victor Turner guardedly proposed an alliance of his home discipline with the brain sciences:

> My career focus mostly has been on the ritual process, a cultural phenomenon, more than on brain neuroanatomy or neurophysiology. But I am at least half convinced that there can be genuine dialogue between neurology and culturology, since both take into account the capacity of the upper brain for adaptability, resilience, learning, and symbolizing, in ways perhaps neglected by the ethologists pur sang, who seem to stop short in their thinking about ritualization at the more obviously genetically programmed behaviors of the lower brain. (1973, p. 104)

Likewise, early proponents of cognitive science were influenced by seminal anthropological theorists such as Gregory Bateson, Margaret Mead, and Claude Lévi-Strauss. The founders of cognitive science quite reasonably anticipated ongoing collaboration. For example, psychologist Howard Gardner (1985) suggested that anthropology would be among the six pillars of the "mind's new science" (see also Norman, 1980). Similarly, sociologist Warren TenHouten (1976) called for the development of "neuroanthropology," as have anthropologists Charles Laughlin and Eugene d'Aquili (1974; see also Laughlin, McManus, & D'Aquili, 1990). Anthropologists as diverse as Malcolm Crick (1982), Claudia Strauss and Naomi Quinn (1997), Maurice Bloch (1998), and Stephen Reyna (2002) have advocated dialogue with the brain sciences.

But the hesitation that we can read between the lines of Turner's prediction—the fact that he was only "at least half convinced," and not more confident—proved well-founded, as most attempts to bring together brain and cultural research have foundered and even more opportunities have been missed in the intervening decades. As Margaret Boden (2006, p. 515) describes in her magisterial history of cognitive science, anthropology has become the "missing discipline."

For example, the early work of Laughlin and d'Aquili on "biogenetic structuralism" (Laughlin & d'Aquili, 1974; Laughlin et al., 1990) argued that invariant cognitive structures such as those posited by theorists such as Lévi-Strauss and Noam Chomsky must arise from the genetically determined qualities of the human nervous system. These neurologically based frameworks ran into some of the same obstacles within the field that structuralism encountered, in particular a level of abstraction that often made the theoretical connection with everyday experience and practice difficult. Neurologist and author Oliver Sacks (1995, p. xx) has also described himself as a "neuroanthropologist." His work provides wonderful humanistic accounts of neurodiversity and a range of brain conditions, but these experience-centered accounts of patients leave aside central questions about diversity, acculturation, and the interaction of culture and neurobiology in pathology.

When anthropologists and other ethnographers have engaged with cognitive science, they have made remarkable contributions. Neuroscientists with anthropological inclinations have made similar important advances. But overall the traffic has been too little in both directions, and the contributions made have been piecemeal rather than systemic or sustained. The brain sciences need the research and insights that anthropologists have developed in order to seriously explore the wide variation in human cognitive and neural functioning. Anthropology must move beyond critique and engage with these fields in a constructive mode in order to answer basic questions about culture, inequality, and human difference. Together, we can help construct the frameworks that allow the best of diverse research on the brain and human nature to be shared across disciplinary lines.

Evolution, the Brain, and Neuroanthropology

The emphatically cultural human neurological condition is hardly a recent acquisition; over the evolutionary prehistory of our species, cultural and technological innovation, as well as social solidarity of a sophistication found in few other species, have allowed our infants to be born especially altricial, depending heavily on long-term support from parents and elders of many sorts. Without culture, social cooperation, and technology, the cost of such an incomplete nervous system would have been too high.

The developmental dependency of infants and children means that our complex hierarchies, even our societies' errors, cruelties, and injustices, end up shaping our neurological inheritance. The worlds we have built as families, cultures, societies, and civilizations affect the nervous system that each of our bodies builds during development. And this distinctive neurological potential is only possible because generations upon generations of humans have constructed worlds to support, protect, and teach each other. For example, the fine motor control and sequential memory necessary to make rudimentary tools requires social support and role models or teachers, as well as neurological refinement; but without early tools, these capacities likely would never have been selected so strongly in our species.

Although every animal's nervous system is open to the world, the human nervous system is especially adept at projecting mental constructs onto the world, transforming the environment into a sociocognitive niche that scaffolds and extends the brain's abilities. This niche is constructed through social relationships; accumulating material culture and technology; and the physical environments, ritual patterns, and symbolic constructs that shape behavior and ideas, create divisions, and pattern lives. Thus, our brains become encultured through reciprocal processes of externalization and internalization, where we use the material world to think and act, even as that world shapes our cognitive capacities, sensory systems, and response patterns.

The brain itself is baroque, fashioned over evolutionary time out of a host of modules and functional units that are still incompletely integrated. Different types of neurological activity do not obey the same rules, nor are they equally susceptible (or immune) to self-reflection and conscious thought. Some cognitive capacities are characterized by deeply ingrained stereotypical species-general responses; other functions are remarkably plastic, even susceptible to substantial revision and conscious redirection. No one, simple theory can explain how every system works, which means we should recognize that enculturation will vary even among the regions and networks within the brain. If an account of one system remains consistent with its functioning while defying expectations arising from other systems, this is as likely to be a product of the brain's heterogeneity as it is a reflection of differences in research methods or approaches.

Cognitive Science and Anthropology

Beginning in the 1960s, but emerging especially widely in the 1980s, empirical evidence of regional tissue specialization in brain architecture led some cognitive theorists to a strong revival of "nativist" tendencies: that is, the assumption that many brain capacities were largely innate and, by implication, that culture and environmental conditioning were superficial. Together with evidence of the domain specificity of some types of knowledge—the fact that information was sometimes not shared between different cognitive functions—this was interpreted as evidence of cognitive "modules" (see Fodor, 1983). Modules were conceptualized as innate cognitive structures similar to those Chomsky (1990) proposed for language. Later theorists expanded and modified Fodor's original argument to posit a much more extensive "massive modularity" of the human mind, a mental endowment of innate, special-purpose tools created by evolution (Barkow, Cosmides, & Tooby, 1992; Tooby & Cosmides, 1992).

Some anthropologists have productively engaged with the massive modularity thesis (Sperber, 1996; Sperber & Hirschfeld, 2004; cf. Whitehouse, 1996). But for many sociocultural anthropologists, the thesis simply makes too many unpalatable assumptions (as it does for many psychologists and cognitive theorists, as well, for that matter; see, for example, Cowie, 2002; Ellman, Bates, Johnson, Karmiloff-Smith, et al., 1997). For example, massive modularity seemed to imply—expressly in the work of some theorists—a whole series of assumptions anathema to understanding brain enculturation: nativism, or the idea that certain functions were preformed in the brain from birth; universality across all individuals and groups; inflexibility; and an adaptationist understanding of evolution, that is, the assumption that every human trait was adaptive (see Alcock, 2001).

Growing evidence of neuroplasticity and of regional "repurposing" has led brain scientists to reconsider the implications of data on tissue specialization (see Anderson, 2010; Pascual-Leone, Amedi, Fregni, & Merabet, 2005; Sanes & Donoghue, 2000). Although neuroanthropologists will continue to be interested in shared human capacities, such as the brain systems facilitating our species' sociality (see Adolphs, 2009) and the consequences of evolution (Gibson, 2002; Rilling, 2008), this work will be accompanied increasingly by research on the ways that cultural influences sculpt the brain's plasticity (see Chiao, 2009a; Park & Huang, 2010; Wexler, 2006).

It is abundantly clear that many neurological capacities, such as language or skills, do not appear without immersion in culture. Increasingly, neuroscientists recognize that differences in developmental environments, patterns of social interaction, formative symbolic niches, ongoing skill acquisition, first-language communities, emotional group dynamics, persistent motor patterns, child-rearing techniques, and the like have profound neurological consequences. Just as empirical research in genetics has led to a reassessment of the concept of "gene" to take into account more complex emergent properties of biological systems (Gerstein, Bruce, Rozowsky, Zheng et al., 2007), investigation of emergent brain mechanisms has highlighted contingency and variability (e.g., Edelman, 1987; Feldman, 2009; Mareschal, Johnson, Sioris, Spratling et al., 2007; Westermann et al., 2007), while neuroanthropology highlights how that immersion matters to the brain's construction and function.

The advent of noninvasive neuroimaging technologies has sparked a revolution in the brain sciences. With these innovations, data about both the architecture and functioning of the brain can be gained, not merely by inference from observed behavior or from postmortem autopsy, but also by directly detecting metabolic processes and structures through the skull of the living organism. In spite of the wide range of discoveries and rapid innovations that have resulted over the past three decades from neuroimaging, these insights have only recently touched anthropology, especially in biological anthropology (e.g., Deacon, 1997; Domínguez Duque, Turner, Lewis, & Egan, 2010) and archeology (Malafouris, 2009 Renfrew, Frith, & Malafouris, 2008). With some notable exceptions (Reyna, 2002; Vogeley & Roepstorff, 2009), cultural and social inquiry has carried on in anthropology with little acknowledgment that significant advances were occurring in research on the brain, some of which bore on crucial anthropological questions.

To many anthropologists, the brain sciences have seemed a risk to the field's integrity. Emily Martin, in a 1999 keynote address to the American Ethnological Society, cautioned:

With their research expensively underwritten by foundations, corporations, and the government, and their claim to provide reductive accounts of the social and cultural without (from my point of view) much information about social and cultural dimensions of existence, I see the neuroreductive cognitive sciences as the most dangerous kind of vortex—one close by and one whose power has the potential to suck in disciplines like anthropology, severely weakening them in the process. (Martin, 2000, p. 574)

The fear of reductionism in anthropology is an old one, shared by Alfred Kroeber (1917) and even Herbert Spencer in the 1850s, but currently this fear is most acutely felt among cultural anthropologists when biological concepts are invoked (see also Campbell & Garcia, 2009; Marchand, 2010, pp. S4-S5). Today, continued indifference, or even what sociologist Freese and colleagues (2003) term "biophobia," heightens the risk that social and cultural anthropology will be further marginalized in both public arenas and academic debate.

Because of simmering animosity between biological and cultural anthropologists, especially in the United States, and the isolation of social anthropology in Europe from fields like primatology, evolutionary theory, and physical anthropology, many social and cultural anthropologists have been reluctant to engage with the brain sciences. Writing and talking about the human brain has become associated with an intellectual workload distribution that pitted biological and cultural explanations against each other. To this day, these old habits persist, as if the only meaningful way to discuss the relationship of human biology and diversity were to apportion influence between them, choosing a side based upon how much weight a theorist wished to attribute to "one side" or the other.

Neuroanthropology demonstrates the necessity of theorizing culture and human experience in ways that are not ignorant of or wholly inconsistent with discoveries about human cognition from the brain sciences. Rather than broad-based concepts like habitus or cognitive structure, neuroanthropology focuses on how social and cultural phenomena actually achieve the impact they have on people in material terms. Rather than assuming structural inequality is basic to all societies, neuroanthropologists ask how inequality differentiates people and what we might do about that.

Fittingly, at the same time that Martin was issuing her warning, Robert LeVine (1999, p. 17) cautioned that continued disinterest in psychology threatened to undermine anthropological theorizing, especially as psychologists amassed remarkable data about human capacities. Continuing the tradition of influential work on cognition within anthropology (such as Boyer, 1994; D'Andrade, 1981, 1995; Hutchins, 1995; Shore, 1996; Sperber, 1996), neuroanthropology can help shape interdisciplinary cognitive science.

Variation Rather than Uniformity

One casualty of neuroanthropological exploration is likely to be the uneasy truce between cultural anthropology's respect for and emphasis of cultural diversity and the long-standing assertion in the field of the "psychic unity" of humanity. "Psychic unity," advocated by the "father" of American anthropology, Franz Boas (1911/1938), has long served as the "theoretical and moral bedrock" of the distinctive four-field approach to anthropology in the United States, according to Bradd Shore (1996, p. 15). Originally formulated by Adolf Bastian, mentor to Boas at the Ethnographic Museum of Berlin, the "psychic unity of mankind" was the commitment that, no matter what the race or background of a human being, all individuals operate with the same "elementary ideas" underlying the apparent diversity of their expression in "folk ideas" (see Koepping, 1983).

Understanding the profundity of neural enculturation plunges us empirically into the "psychic unity muddle" that Bradd Shore (1996) highlights. Social differences, including problematic categories like "race," can become biology because they shape the emerging nervous system (see Gravlee, 2009). We no longer need to argue, however, that it is possible for our "minds" to be different because our "brains" are all essentially similar. In fact, brains can be grown into different configurations, not because we are, at birth, fundamentally and irreducibly different, but because we become "en-brained" as a result of our own distinctive combination of genes, epigenetic influences, environmental factors, experiences, learning, and even understandings of ourselves.

At the same time, claims about a uniform human nature are another likely casualty of neuroanthropology. Indeed, the inherent variability of humans has made anthropologists squeamish about the term "human nature" for decades. When anthropologists reflect on the extraordinary variability of our species and the evolutionary time that humans have been living among and through cultures, most become suspicious of any overly glib assertions about what is essentially human, universal, homogeneous, or invariant. From a neuroanthropological perspective, the term seems to deny the very open-endedness and emergent qualities that are hallmarks of the sophisticated human nervous system.

Moreover, when assertions about human nature rest on a shallow research pool of subjects, disproportionately drawn from the wealthy, sedentary, highly educated, industrialized fractions of the world's population, we should not be surprised if anthropologists take familiar critical positions. As Henrich and colleagues (2010) point out, pervasive sampling bias distorts the types of variation explored by cross-cultural psychology, not just its sites, by producing subject pools that, from all available evidence, are themselves statistical outliers (see also Arnett, 2008; Choudhury, Nagel, & Slaby, 2009). Because this group of subjects is such a small fraction of our contemporary population and such a striking aberration in our evolutionary history (Henrich et al., 2010), it does not begin to reflect the real variability or even the common patterns that exist in the world.

Assertions of uniformity based on assumed concepts, whether race or gender or other posited "human types," are another casualty of neuroanthropology. Since Boas, anthropologists have repeatedly shown that sociocultural categories of people do not reflect empirical reality. As research has repeatedly shown (Gravlee, 2009; Fine, 2010), there is enormous diversity in gender, race, ethnicity, and other categories often defined as "natural" or assumed to have some inner essence that marks them as different from other, often "normal" people. Moreover, old ideas about ranking cultures or societies as inferior or superior, or along some progressive line of "civilization," are simply wrong. Yet these essentialist ideas often creep into scholars' work when they address cross-cultural differences or gender differences (Fine, 2010).

Overall, for neuroanthropologists, examining patterns of variation and looking at the processes that account for those patterns is the necessary empirical and conceptual approach. Assumptions of uniform difference, whatever their origin, are just that—assumptions. And these assumptions often reflect one's intellectual and social background, rather than documenting and explaining actual observed variation.

Cultural Neuroscience

Improving neuroimaging technology and growing infrastructure for cross-cultural research have permitted neuroscientists to demonstrate empirically that observable psychological differences often have neurological analogs (Chiao, 2009a; see reviews, Chiao, 2009b, 2010b; Han & Northoff, 2008). Growing appreciation of neural malleability and subtle research techniques have even revealed cultural differences in brain functioning not visible in psychological studies, so often based on explicit self-report (Fiske, 2002, p. 81–82). In the past decade, cultural differences in neural activity have been demonstrated in a range of activities, including making perceptual judgments (Hedden, Ketay, Aron, Markus et al., 2008); attentional control (Ketay, Aron, & Hedden, 2009); semantic tasks (Gutchess, Hedden, Ketay, Aron et al., 2010); music perception (Nan, Knosche, Zysset, & Friederici, 2008); amygdala responses to fearful faces (Chiao, Iidaka, Gordon, Nogawa et al., 2008a; Adams, Franklin, Rule, Freeman et al., 2010); self-construal and recognition (Lewis, Goto, & Kong, 2008; Sui et al., 2009); taste perception (McClure, Li, Tomlin, Cypert et al., 2004); and theory of mind (Kobayashi, Glover, & Temple, 2007; Kobayashi, Glover, & Temple, 2008). Although each case is distinct, the overarching pattern is that areas in which cross-cultural psychology has found observable cognitive divergences typically also demonstrate neural functional or structural divergence under close scrutiny (for reviews, see Chiao & Ambady, 2007; Fiske, 2009; Kitayama & Park, 2010).

For example, cross-cultural studies find consistent differences in the ability of Western and Asian subjects to judge relative and absolute sizes of objects and to recall information about focal objects and backgrounds from photographs and video. These studies support the argument that Western subjects perceive absolute size and focal objects more accurately, whereas Asian subjects recall context and relative size better (see Chiao, Li, & Harada, 2008b; Kitayama, Dufy, Kawamura, & Larsen, 2003; Masuda & Nisbett, 2006; Nisbett, Peng, Choi, & Norenzayan, 2001; Nisbett & Miyamoto, 2005). Goh and Park (2009) imaged matched subjects using carefully tested, identical magnet systems in Singapore and the United States. Their

experiments documented differences between the subject groups, especially older individuals, in the brain areas most active in figure-ground
recognition (see also Chee, Zheng, Goh, & Park, 2010; Gutchess, Welsh,
Boduroglu, & Park, 2006). Interestingly, some of the difference was miti-
gated by explicit instruction (Goh & Park, 2009, p. 106). This modulation
suggests that the neural disparities were due, not to inability, but to cognitive habits. In sum, behavioral disparities, such as patterns of eye movements when viewing faces or scenes (see Blais, Jack, Scheepers, Fiset et al.,
2008; Chua, Boland, & Nisbett, 2005), coincide with distinctive patterns
within networks that themselves show substantial cross-cultural architec-
tural consistency.

Likewise, Markus and Kitayama's (1991) classic paper on Western and
Asian self-construal provoked cultural neuroimaging examination (see also
Heine, 2001; Triandis, 1995). Markus and Kitayama argued that the Western
"self" is independent, whereas the Asian "self" is fundamentally relational
and interdependent. Markus and Kitayama had been inspired by anthropologists Shweder and Bourne (1984), who asked whether the "self" could
vary cross-culturally. Using functional magnetic resonance imaging (fMRI),
Zhu and colleagues (2007) found that the medial prefrontal cortex, especially active in Western subjects recalling traits of or making judgments
about themselves, was also active in Chinese subjects when making judgments about intimate relations (their mothers). They explain,

The relatively heavy emphasis on interpersonal connectedness in Chinese culture
has led to the development of neural unification of the self and intimate persons
such as mother, whereas the relative dominance of an independent self in Western
cultures results in neural separation between the self and others (even close others
such as mother). (2007, pp. 1314–1315)

Subsequent studies by Han and colleagues (2008, 2010) complicated the
picture by demonstrating that some religious backgrounds might accustom
individuals to distinctive forms of self-cognition. Devout Buddhists demonstrated patterns of neural activity in self-reflection that resembled those
more commonly seen in the evaluation of others, perhaps due to religious
advocacy of self-transcendence (see Han & Northoff, 2008 Kitayama &
Park, 2010; Sui et al., 2009; cf. Heatherton, Wyland, Macrae, Demos et al.,
2006).

The field of cultural neuroimaging has produced remarkable empirical
results, gratifying to anthropologists who have long argued that enculturation affects thought. Nevertheless, this emerging research area is dogged
by problems, some of which will be obvious to sociocultural theorists, such
as pervasive sampling and explanatory biases, with most of these studies

resketching familiar "East-West" comparisons. Cohen (2009, p. 194) writes: "A person reading these literatures could be excused for concluding that there is a very small number of cultural identities (North American vs. East or Southeast Asian), that vary principally on the dimensions of individualism-collectivism or independent-interdependent self-construal." For pragmatic reasons, the bias toward Asian-Western comparisons is understandable: neuroscience research demands specialized skills and expensive, often immobile imaging equipment. Nevertheless, cultural neuroimaging seems to lag behind cross-cultural psychology, which Lehman and colleagues (2004, p. 704) indicated was ready to "move beyond the east-west comparisons that have become so commonplace."

The empirical evidence generated by cross-cultural psychology itself undermines the strict division of Asians and Westerners(see Chiao, Harada, Komeda, Li et al., 2010a). As Oyserman and colleagues (2002) found in a meta-analysis of cross-cultural studies of "individualism" and "collectivism," close reading reveals that "Asians" and "Westerners" are hardly homogeneous groups and that the categories "individualist" and "collectivist" collapse heterogeneous, independent domains. Within psychology, reconsiderations of cross-cultural research have created an opportunity for greater cooperation with anthropologists. Fiske (2002, p. 79), for example, advocates theoretical and methodological approaches borrowed from anthropology (see also Rozin, 2009).

Moreover, given the pattern of East-West comparisons, the explanations of differences will strike sociocultural anthropologists as overly simplistic and, borrowing from Said (1978), as a kind of neural Orientalism. Such causal explanations for neurological differences refer to homogeneous, opposing civilizations, stretching back to ancient philosophers to highlight irreducible dichotomy between Asians and Westerners (see Nisbett, 2003). Nisbett and Masuda's (2003) influential account of Western "analytical" and Eastern "holistic" thought, for example, links these to "independent" and "interdependent" social life, tracing cultural isomorphism across causal attribution, logic, categorization, attention, perception, recognition of covariation, affordances, and perception of everyday life events. This East-West contrast assumes that a culture is characterized by a single pattern, and the assertion of internal homogeneity contrasts markedly with contemporary anthropological understandings of culture (see Roepstorff, Niewöhner, & Beck, 2010).

The problem, however, is not simply sampling bias or outmoded classificatory systems (see Bond, 2002), but also the way in which psychology conceptualizes culture. Anthropologist Domínguez Duque and colleagues

(2010, p. 143–144) argue that psychologists treat "culture" as a variable exogenous to research; subjects are assigned to groups based on an understanding of cultural differences as group identity. But culture itself, or enculturation as a process, is not a research subject (see Vogeley & Roepstorff, 2009). One task of neuroanthropology will be to aid interdisciplinary brain sciences to move beyond unproductive understandings of culture, including dichotomies between East and West. In the process, anthropologists can help to bring a more intriguing set of research questions to cultural neuroimaging.

On Culture

One of the problems with the "culture concept"—and anthropologists feel that there are many—is that people use the term to try to solve too many different problems, creating an intellectual muddle. The category of culture itself has become a kind of Ptolemaic category; when Ptolemy lumped the planets together with the moon and sun into a single category, "planets," he created a kind of incoherent meta-category that posed insoluble theoretical obstacles. In some ways, "culture" is the same sort of Ptolemaic construct, at once used to attempt to explain differences between groups, the inheritance within a population over time, and the gap between humans and other species.

From a neuroanthropological perspective, we agree with Bradd Shore (1996, p. 9) that the culture concept is more usefully refined than discarded, and believe that taking the nervous system as a perspective on theorizing culture is a helpful way to sort through the competing arguments. We will have to rethink the concept of "culture," to offer a more grounded and supple account of human diversity and the patterns of influence of environment on human development. Especially for research, we need to be specific about how the concept is being used. As Vogeley and Roepstorff (2009, p. 514) argue:

Culture, given its heterogeneity and fluidity, needs to be defined in an operationalised manner. ... Studying self-referential thinking, for example, would require characterizing the population from which the group of participants is recruited with respect to individualism versus collectivism; a stratification that is based on different mother languages in this case might be arbitrary and misleading.

This operationalization of "culture" will depend on assumptions made about the concept. Some anthropologists have argued that the profundity of our enculturation makes us a cultural animal, not a species to which culture is added or superficially acquired. Although many animals are capable of producing species-typical behaviors raised in isolation, for example, our socially rich developmental cradle is essential to the traits that we would consider most "human." As anthropologist Clifford Geertz has written:

Man's nervous system does not merely enable him to acquire culture, it positively demands that he do so if it is going to function at all. Rather than culture acting only to supplement, develop, and extend organically based capacities logically and genetically prior to it, it would seem to be an ingredient to those capacities themselves. A cultureless human being would probably turn out to be not an intrinsically talented, though unfulfilled ape, but a wholly mindless and consequently unworkable monstrosity. (1973, pp. 67–68)

For a long time, anthropologists have focused on culture as a system of symbolic associations, public signs, or shared meanings. But from the perspective of the nervous system, patterns of variation among different groups also includes significant nonconscious, non-symbolic traits, such as patterns of behavior, automatized response, skills, and perceptual biases. This neuroanthropological framing opens more space for considering why all types of cognition may not operate in identical fashion, and how non-cognitive forms of neural enculturation might influence thought and action. Given this type of functioning, neuroanthropologists will have to return to an older notion of "culture," one that considers capabilities, habits, and other forms of collective action, alongside meaning and symbols. While it can prove useful to speak principally of "culture" as shared representations, we also must recognize that "cultural variation" will include other sorts of patterned, shared conditionings of the nervous system.

The implication for neuroanthropology is obvious: forms of enculturation, social norms, training regimens, ritual, language, and patterns of experience shape how our brains work and are structured. But the predominant reason that culture becomes embodied, even though many anthropologists overlook it, is that neuroanatomy inherently makes experience material. Without material change in the brain, learning, memory, maturation, and even trauma could not happen. Neural systems adapt through long-term refinement and remodeling, which leads to what we see as deep enculturation. Through systematic change in the nervous system, the human body learns to orchestrate itself. Cultural concepts and meanings become neurological anatomy.

At the same, researchers must explore automization, endocrinology, emotion, perception, and other neural systems that contribute to patterns of variation but are not entirely susceptible to reflection. For example, practices of child rearing and early formative experiences are clearly influenced by cultural ideologies about how children should be nurtured, but many of the organic mechanisms through which these ideologies take hold of individuals and affect their long-term development may be unknown, even invisible to the individuals carrying out such practices (Worthman, 2009, 2010).

This broad view of culture highlights the challenge that a primarily ideational treatment of culture can bring. Ward Goodenough (1957, p. 168), for example, defined culture as "whatever it is one has to know or believe in order to operate in a manner acceptable to [a group's] members." Although he disagreed with Goodenough on many issues, Clifford Geertz (1973, p. 89) defined culture as "an historically transmitted pattern of meanings embodied in symbols, a system of inherited conceptions expressed in symbolic forms by means of which men communicate, perpetuate, and develop their knowledge about and attitudes toward life."

Long-running debates have raged about the definition and qualities of culture (see, for example, Kuper 1999), yet as Roy D'Andrade (2003) outlines, the treatment of culture as primarily an ideational system, however it is specifically termed—semantic structure, symbolic system, collection of schemata, cultural "imaginary"—is now firmly lodged in our field. Even when a less overtly cognitive structure is imputed to a cultural agent, such as a habitus or "embodied knowledge," too often this construct still behaves as information or an ideational structure. That is, when we discuss culture's effects on the senses, the body, or nonconscious dimensions of human beings, we still talk about "symbols" or "meaning" or "representation" or "information," assuming that nonconscious, even physiological effects of cultural difference can be treated the same as conscious, ideational effects (see Downey, 2010a).

An ideational culture concept has been a productive assumption for our discipline, and we still believe the conception is useful for some kinds of projects. But the ideational model has often harbored an unproductive opposition to "biology" that cannot be sustained at the level of neuroanthropological analysis. The metaphor, if not actively ameliorated, can lead sociocultural anthropology to an idealist "brain-mind," "hardware-software" or "organ-information" distinction (Reyna, 2002, p. 120). In contrast, brain sciences treat learning and neural "representation" as physical change in the organ (Ghose, 2004). A new working definition of culture will be necessary for neuroanthropological exploration, at least for the purposes of identifying cultural influences at the level of the nervous system. That is, our culture concept will need to be flexible enough to redeploy at the scale of the individual subject's brain, rather than simply assuming that culture can exist in the same way in both public expression and synaptic representation.

To paraphrase Naomi Goldblum (2001), neuroanthropologists will need to propose a "brain-shaped" culture—that is, a cultural theory consistent with neuroscientific evidence. Too often, cultural theory in anthropology takes a "product-shaped" approach to the psychological mechanisms underlying the production of culture. That is, theorists take mental products as analogous to the inner workings of human psychology; they treat product as if it were process. For example, because a human mind can produce a dictionary or systematic grammar—or, more accurately, a group of humans can produce and maintain a dictionary or grammar reference given sufficient time and training—does not mean that the inner mental workings of human vocabulary, word recall, sentence production, and symbolic processes are like consulting a dictionary or grammar. The example is intentionally ludicrous to highlight the improbability of this approach, but many other examples could hit closer to theoretical home.

Robert Turner (2001, p. 167) offers a neurocentric definition of culture as "relatively coherent and systematic biases in brain functional anatomy." Turner's definition assumes that the generative structure of cultural manifestations is the culturally inflected biological agent, not a structure within expression itself. In this approach, the generation of culture moves from the level of abstract structures to the nervous system. This shift in level of analysis is a hallmark of neuroanthropology and helps ground the power of culture in neural dynamics. In a complementary proposition, Super and Harkness (2002, p. 270) offer that culture "is usefully conceived ... as the organization of the developmental environment." Super and Harkness draw attention to the formative niche for the developing nervous system: the diverse medium that includes material culture, social interaction, language, symbols, technology, sensory experiences, behavior patterns, education, and other pervasive elements of the lived environment. As Fuentes (2009) comments, the use of niche construction for understanding the dynamics of inheritance is inherently constructive and integrative, as social, ecological, and biological dimensions are all part of a developing organism's niche (see also Li, 2003). This approach builds on evolutionary theories that incorporate non-genetic inheritance, such as dual inheritance theory (Durham, 1992; Richerson & Boyd, 2005; see also Lewontin, 2000; Jablonka & Lamb, 2005). At a developmental rather than evolutionary time scale in neuroanthropology, phenotypic adaptation is crucial to understanding enculturation, especially as we recognize that the human nervous system has likely been selected for versatility and under-determination (Sterelny, 2003). As Wheeler and Clark (2008, p. 3565) remind us, "much of what is most distinctive in human cognition [is] rooted in the reliable effects, on developmentally plastic brains, of immersion in a well-engineered, cumulatively constructed cognitive niche."

Rather than treating the nervous system as the medium for culture's propagation, neuroanthropology focuses on how the nervous system responds and adapts to social, material, and cognitive environments (see also Roepstorff et al., 2010). In this way, neuroanthropology can take a basic idea like Hebbian learning—"what fires together, wires together"—and examine how social and cultural processes shape the timing, exposure, and strength of activity, such that the coordinated action of brain systems emerges through cultural dynamics.

This approach can provide a powerful "person-centered" refocusing of our theoretical energies. The material environment, both natural and artificial, provides structure and information to the growing organism while being incorporated with its inherited biological legacy. This approach borrows from dynamic systems theory (Oyama, 2000; Oyama, Griffiths, & Gray, 2001); niche construction theory (Laland, Kendal, & Brown, 2007; Odling-Smee, Laland, & Feldman, 2003; Sterelny, 2007); and recognition of the "extended" human mind (Clark & Chalmers, 1998; Menary, 2010). As Ingold writes, culture and biology become inextricably entangled in the human individual:

If, as I have suggested, those specific ways of acting, perceiving and knowing that we have been accustomed to call cultural are incorporated, in the course of onto-genetic development, into the neurology, musculature and anatomy of the human organism, then they are equally facts of biology. Cultural differences, in short, are biological. (2001, p. 28; see also Ingold 2007)

Neuroanthropological research may suggest that some cultural theories are biologically implausible (such as Downey, 2010a; Turner, 2002). Nevertheless, one tenet of dynamic systems theory is that, at different scales of analysis, systems behave according to distinctive principles (Oyama et al., 2001). Dynamic systems have emergent properties at higher levels, thwarting the pernicious "neuroreductive" tendencies that Martin feared. As psychologists Ambady and Bharucha (2009, p. 345) describe, the combination of neurosciences and cultural research offers "the exciting opportunity to examine the mutual interplay of culture and biology across multiple levels of analysis, from genes and brain to mind and behavior, across the life span" (see also Li, 2003).

One way to incorporate neurological research and cultural theory is to treat culture, not as information, but rather as skill acquisition, an approach advocated by Ingold (2000, pp. 416–417; see also Marchand, 2007, 2010). Skill acquisition focuses on the process of enculturation, and examines the practices, settings, and interpretations that can drive the cultural patterning of the nervous system. Reconceptualizing enculturation as "enskillment" links to robust findings on neural architecture and functioning among highly skilled individuals: London taxi drivers (Maguire et al., 2000); musicians (Bengtsson, Nagy, Skare, Forsman et al., 2005; Gaser & Schlaug, 2003); jugglers (Draganski, Gaser, Busch, Schuierer et al., 2004; Draganski & May, 2008); and second-language learners (Golestani, Molko, Dehaene, LeBihan et al., 2007; Green, Crinion, & Price, 2007), for example. Even animal models show that tool use can affect cortical reorganization; macaques trained to use rakes to get food, for instance, evidence increasing cortex dedicated to visuo-tactile neurons (Frey, 2007, p. 369).

Asian-Western comparisons can be reinterpreted as highlighting the skill-like dimensions of culture. Tang and colleagues (2006), for example, interpret neurodynamic differences during mental arithmetic as resulting from habitual use of the abacus in primary school. Even when calculating without the tool, the training allows Asian subjects to use visual-spatial simulation for mental calculation, whereas Western subjects used verbal processing systems (see Cantlon & Brannon, 2006; Hanakawa, Honda, Okada, Fukuyama et al., 2003). Similarly, researchers have found that people from different backgrounds employ diverse cognitive resources when performing computer-based problem-solving tasks, which can be modified by explicit training (Güss, Tuason, & Gerhard, 2009; Strohschneider & Güss, 1998, 1999). This research could extend to sport, music, and other elite skill development for which we have evidence of stylistic difference that might indicate underlying neurological diversity (Downey, 2010b).

The Ethnographic Gaze

Rather than conceiving of subjectivity as a text to be interpreted and the brain as composed of hard-wired circuits or innate modules beholden to selfish genes and evolutionary algorithms, neuroanthropology posits that subjectivity and the brain meet in the things that people do and say and the ways we interact with one another and the environment. Neuroanthropology does not limit itself to psychology, which has a predominant focus on internal states, often separate from the body, physical activity, and the specifics of interaction with cultural environments. Moreover, neuroanthropology does not limit itself to Western notions of mind, self, or consciousness, which can dominate discussions in some academic settings.

Ethnography—the documentation and interpretation of lived experience and behavior in specific settings—is central to this approach. Laboratory research tests hypotheses, often in highly constrained designs and with limited ability to simulate sociocultural contexts. Ethnography offers brain researchers a way to test ideas from neuroscience against the reality of what people actually do, say, and experience. Rather than concluding from experimental or epidemiological evidence that "such-and-such process might reflect what happens in the real world," ethnography can add a crucial empirical dimension and test ideas using what actually happens in the real world. Ethnography can examine how neurological processes play out on the ground and how developmental settings, repeated activities, and social relationships shape how people interact with the world. Ethnography also offers crucial insight into the question "why?" by drawing on people's own insight and explanations for what they do, as well as providing empirical access to how sociocultural processes shape brain function, behavior, and meaning.

Neuroanthropology can also bring an ethnographic sensibility to brain research, including a willingness to take into consideration native theories of thought and individuals' accounts of their own experience. Thus, careful ethnographic research, in-depth interviews, and the analysis of indigenous worldviews will always be central to the neuroanthropological synthesis. As one example, neuroanthropologists can work as part of multidisciplinary projects, especially by helping to generate innovative, testable hypotheses about cultural diversity. Juan Domínguez Duque and colleagues (2010) have called for "neuroethnography," a multistage examination of human neurological variation in which ethnographic study generates questions for comparative imaging research (see Domínguez Duque, 2007; Seligman & Brown, 2010).

The inherent variety among different brain systems means that conscious reflection and experience-based accounts have a crucial relation to many of the phenomena we study. Consciousness itself is part of complex neural systems, adding degrees of self-regulation, restraint, learning, monitoring, cuing, and a host of other capacities. How people understand and experience their own thought is part and parcel of neural activities. At times individual conscious understanding and interpretation may be central to neural function, and at others times not relevant at all. Still, cultural interpretations of self, thought, emotion and relationships do play a role in shaping both understandings of experience and neural function itself. Given the limits of consciousness and the at times narrow focus and ideological force of cultural interpretations, neuroanthropological accounts may highlight the limits of conscious and cultural awareness and demonstrate the self-deceptions inherent in some kinds of neurological functioning. Most of our cultural and neural functioning is submerged, and accessible to consciousness through extraordinary effort and special techniques, if it is accessible at all. Thus, research techniques should focus on capturing both our conscious awareness of why we do what we do and the inherent processes that shape the flow and outcome of that doing.

For these reasons, subjects'-eye-view accounts are critical to neuroanthropology in a way that they might not be to other cognitive theorists. First, we recognize that theories about how the mind works or what it needs are themselves part of the developmental environment in which the brain is formed. Even if these ideas don't accurately represent actual neural function, they do influence the brain-culture system and can have an impact on the way the brain works even if that is in a way utterly unintended by those who hold the ideas. That is, whether or not indigenous theories of thought are accurate, they are part of the ecology of brain conditioning.

Cultural resources like subtle differences in language may support distinctive phenomenological insights into the human nervous system. That is, other cultures may notice things about the human nervous system that our own communities have not observed, thematized, or codified. For example, the cognitive neuroscience of highly skilled communities, or of specialists who refine certain brain functions, such as meditation, perceptual skills, or high-performance cognitive abilities in areas like mental calculation, recall, or spatial navigation, have demonstrated marked empirical differences in brain function in imaging studies. But something similar might happen as well in indigenous folk theories of thinking or other neural functions, and we lose a vital resource if we do not ask ourselves how ethnographic communities come to their own ideas about the mind and experience.

Areas for Research

Neuroanthropology will address a broad range of areas. Using social and cultural neuroscience in combination with psychological anthropology and cultural psychology, neuroanthropology builds in-depth analyses of mind, behavior, and self based on an understanding of both neurological function and ethnographic reality. This research creates robust analyses of specific neural-cultural phenomena, recognizing that each may demonstrate a distinctive dynamic; for example, neuroanthropological investigation reworks our understanding of human capacities like balance (often assumed to be something innate), studies how practices like meditation shape and piggyback upon neural functioning, and examines the interactive nature of pathologies like addiction and autism.

Neuroanthropological research can help us to evaluate conflicting explanations of pathological variation. Lende (2005), for example, used in-depth interviews and a specialized survey instrument with drug-taking Colombian youth to better understand their motivations. Lende argues from the ethnography that drug seeking—"wanting" and "craving"—is driven by a shift in the salience of drug-related stimuli, not simply by the felt effects of drugs. Lende's work uses close interpretive analysis of users' accounts to test competing neurological theories, an approach that flows in the opposite direction of the "neuroethnography" advocated by Domínguez Duque and colleagues (2010). Lende subjects neurologically based theories to ethnographic field testing rather than bringing field-based hypotheses to the laboratory. Both promising approaches, they require an openness to diverse quantitative and qualitative methods (see also Brown, Kuzara, Copeland, Costello et al., 2009).

Neuroanthropology has profound implications for our understanding of how societies become socially structured. Inequality works through the brain and body, involving mechanisms like stress, learning environments, the loss of neuroplasticity, the impact of toxins, educational opportunities (or their absence), and other factors that negatively shape development. Neuroanthropology can play a fundamental role in documenting these effects and in linking them to the social, political, and cultural factors that negatively impact the brain. Furthermore, social appeals to "hard-wired" differences remain a standard approach by people in positions of power to maintain racial, gender, sexual, and other inequalities; a deeper understanding of the complex origins and unfolding of key neural and physiological differences undermines accounts that assume these distinctions are inescapable.

Neuroanthropology points to new ways to think about how people become talented and ways to understand intelligence, resiliency, social relations, and other factors that shape success in life. At the same time, technological and pharmacological interventions are playing an increasing role in managing behavioral disorders, often with great profit for companies, while cognitive enhancement drugs, brain-computer interfaces, and neuro-engineering will surely be used in ways that create new separations between haves and have-nots. This focus on different forms of human capacity, from talent to technology, offers a rich area for research and application.

In societies across the globe, the brain now acts as a central metaphor, a substitute for self, a way to explain mental health, and a shorthand for why people are different. In reaction, critical approaches have looked at the interpretation and use of brain imagery, psychoactive pharmaceuticals, public presentations of neuroscience research, and related social phenomena (Choudhury & Slaby, 2011). Anthropologists will certainly find space for critical studies of the neurosciences as sites of knowledge production, in line with social studies of science or what Choudhury and colleagues (2009) call "critical neuroscience," turning anthropological analytical techniques on the neurosciences to better understand researchers' own cultural assumptions (e.g., Dumit, 2004; Joyce, 2008). We see neuroanthropology's role as a constructive contributor to integrative brain science, not just policing its borders or offering constant critical scrutiny. Critique has an important place, but without helping to produce better paradigms or suggestions for improvement, critique simply leaves conscientious researchers without positive alternatives to the practices that warrant criticism. Full engagement must include constructive proposals for improving both brain science and anthropological research.

The wider proliferation of imaging technology will, inevitably, demonstrate unrecognized areas of variation in brain development if neuroscientists are alerted to the possibility. The equipment involved, however, is still largely restricted to cities, although mobile devices can now deploy into remote areas (see Rilling, 2008, p. 27). The research is challenging for a host of reasons: subtle differences among equipment, difficulties matching populations, the low reliability of signals, and questions about statistical methods (Goh & Park, 2009, p. 99; Vul et al., 2009). More portable techniques are available, such as functional near-infrared spectroscopy, event-related optical signaling, diffuse optical imaging, and electroencephalography (EEG). Functional near-infrared spectroscopy, for example, uses light introduced to the scalp to monitor cortical activity, offering a portable, noninvasive, and cost-effective way to conduct imaging research (see Izzetoglu, Izzetoglu, Bunce, Ayaz et al., 2005; Izzetoglu, Bunce, Izzetoglu, Onaral et al., 2007). Although these technologies are limited (for example, allowing only outer cortical imaging), neuroanthropologists can help to accelerate their refinement by posing intriguing research questions that require their broader adoption.

Research Case Study: Sensory Learning

Anthropological studies of the senses have highlighted variety in sensory interpretation and experience (Classen, 1993, 1997; Geurts, 2002; Howes, 1991; Pink, 2009; Stoller, 1989). Alfred Gell (1995, p. 235), for example, noted how Umeda become particularly alert to sounds and smells in the forest where sight is limited by dense underbrush; Edmund Carpenter found that Inuit hunters have extraordinary visual ability and were able to spot distant people on low-contrast ice fields (1973, p. 36; cited in Ingold, 2001, p. 253). Ingold (2001, p. 250) argues that anthropology needs to take account, not merely of how senses might carry diverse meanings or be treated hierarchically, but also how the senses might be trained (see Downey, 2007; Grasseni, 2007, 2009).

Extensive psychological research demonstrates the plasticity of perceptual systems; that is, they become better adapted to the tasks for which they are consistently deployed (for reviews, see Ahissar, 2001; Edeline, 1998; Fahle & Poggio, 2002; Feldman, 2009; Ghose, 2004; Kellman & Garrigan, 2009; Seitz & Dinse, 2007). For example, Kuhl, Williams, Lacerda, Stevens et al. (1992) found that children's first language affected their ability to perceive speech sounds by six months of age, and psychologists have long known that all cultural groups are not equally susceptible to visual illusions (Segall et al., 1966; see also Henrich et al., 2010, pp. 64–65). In extraordinary circumstances, sensory adaptation can produce remarkable refinements; Gislén and colleagues (2003) found that "sea gypsies," the Moken of Southeast Asia, who forage by swimming, develop acute underwater vision by suppressing the automatic dilation reflex.

Neuropsychological research on sensory plasticity suggests that refinement can occur at many levels, from the conscious interpretation of sensations far "downstream" in perception, to patterns of sensation seeking (such as eye-fixation patterns in Chua et al., 2005; Goh, Tan, & Park, 2009), to better suppression of competing sensation (Ghose, 2004), even to quite "upstream" modification in the peripheral sensing organs (Sasaki et al., 2010; Zenger & Sagi, 2002). In contrast, some parts of the sensory systems appear much less susceptible to change (Zhang & Kourtzi, 2010). Some alterations to the peripheral nervous system occur through modulation by descending neural pathways that return from the brain to peripheral sense organs (on vision, see O'Regan & Nöe, 2001). These descending pathways mean that "higher" brain functions, such as emotions, motivations, attention, consciousness, and cultural interpretation, can bias sensory input at the neurological interface with the perceptual world.

A neuroanthropological approach recognizes that the senses serve as channels for deep enculturation, and are themselves liable to enculturation. The regularities of sense experience and early maturation of perceptual mechanisms can provide sufficiently uniform neurological conditioning that a sense-dependent trait appears almost universal even though that trait is neither innate nor genetically fixed. For example, research with the blind demonstrates that, in the absence of ocular stimulation, the visual cortex is reallocated to other sensory functions (Sadato et al., 1996). Research methods in neuroanthropology will include techniques borrowed from perceptual psychology and adapted for field use, such as increasingly versatile and field-ready eye-tracking technology and illusions. But awareness among anthropologists of diverse forms of perceptual plasticity will already allow ethnographers to generate testable hypotheses about the perceptual learning mechanisms engaged by enculturation.

Research Case Study: Embodiment
The anthropology of embodiment developed under the influence of phenomenology but has not engaged substantially with neuropsychology (see, for example, Csordas, 1990, 1992; Jackson, 1989; Strathern, 1996). The gap is particularly ironic given that Maurice Merleau-Ponty (1962), a central inspiration, read neuropsychology closely and was explicitly committed to locating body image in the central nervous system (see also Reyna, 2002, p. 185, fn.2). Campbell and Garcia (2009) have recently put forward a neuroanthropological exploration of bodily image, specifically exploring the physiology of malnourishment, and argue that "emotional embodiment" is a key area for future exploration (see also Damasio, 1994; Laughlin, 1997; Niedenthal, 2007; Seligman & Brown, 2010; Varela et al., 1991; Worthman, 2009). In fact, embodiment constitutes one of the broadest frontiers for future neuroanthropological exploration.

The sense of self is underwritten by both conscious and nonconscious embodiment, such as emotional and endocrine processes. For example, a series of papers examined cortisol and other stress hormones in relation to the "honor complex" among Southern males (Cohen, Nisbett, Bowdle, & Schwarz, 1996; D'Andrade, 2002; see also Kemper, 1990); the research suggests that a culturally distinctive form of masculinity involves heightened sensitivity to social slights and increased endocrine response. These findings parallel research on testosterone and challenges to status, including among nonhuman primates, which finds that the endocrine system is responsive to behavioral and emotional states (for reviews, see Archer, 2006; McAndrew, 2009; Sapolsky, 2004). Other research has explored populations with distinctive neuroendocrine and autonomic responses to basic human sensations, such as feeling compassion (Lutz, Dunne, & Davidson, 2007), pain (Grant et al., 2010; Kakigi, Nakata, Inui, Hiroe et al., 2005), or fear (Chiao et al., 2008a). These cases suggest that limbic systems can be affected by patterns of early experience, and are provoked (or suppressed) by symbolic situations, social contexts, or even learned techniques for self-manipulation, such as reappraisal (Barrett, Mesquita, Ochsner, & Gross, 2007; Davidson, Jackson, & Kalin, 2000; Ochsner & Gross, 2008; Worthman et al., 2010).

The neuroanthropological study of embodiment reveals that the body is not simply "good to think with," but that basic neurological functions influence, and are influenced by, conscious experience. Variability reveals how developmental trajectories can recruit phylogenetically ancient systems, skewing their responses, linking them to evolutionarily younger cognitive capacities, or cuing them with novel stimuli (Greenfield, Keller, Fuligni, & Maynard, 2003; Gogtay, Giedd, Lusk, Hayashi et al., 2004; Li, 2003). Carol Worthman (2009) discusses how systematic interaction between genes and environment, including maternal stress levels, affects serotonergic and other endocrine-related neurotransmission, partially explaining vulnerability to trauma, antisocial behavior, and depression.

How "experience gets under the skin" is an important emerging focus for research that examines the impact of inequality, racism, and trauma (Hertzman & Boyce, 2010; Gravlee, 2009). Ethnographic research on embodiment has increasingly embraced the emotional, bodily, and experiential dimensions of embodiment, providing a good complement to the emerging work in neuroscience and human development (Lende & Lachiondo, 2009; Lester, 1997; Tapias, 2006). Embodiment is also a crucial issue for neuroanthropological research in medical anthropology. For example, Seligman and Kirmayer (2008) have sought to integrate psychophysiological research on dissociation with anthropological explorations of altered states such as trance. Their rich account highlights the cortical inhibitory processes associated with dissociative states (hypnosis, trauma), but also demonstrates how cultural setting can lead dissociation to be classified as pathological or as a valorized skill, such that biological-social "looping" reinforces these states in non-pathological individuals (2008, p. 50). Anthropological inputs could be added to the emerging conversation between neurophysiology and developmental psychopathology, where, as Cicchetti and Cohen (2006, p. 3) discuss, researchers recognize that a "multiple-levels-of analysis approach" is necessary to understand the emergence of psychopathology (see also Hay, 2009; Kirmayer, 1989; Seligman & Kirmayer, 2008.

Conclusion

By placing the focus on the individual's nervous system and its relation to the world, neuroanthropology asks challenging questions of scale and depth for both neuroscientists and anthropologists, demanding both groups stretch beyond accustomed frames. For neuroscientists, seriously considering human diversity may require changes in research methods; in such basic processes as averaging and amalgamating imaging data, removing outlying data points (some of the most interesting individuals); and in finding test subjects. It can help cultural neuroimaging researchers to develop a much more sophisticated understanding about what results of comparative brain scanning of Asians and Western Europeans might mean, and why seeing doesn't always translate into cultural believing. Thus, neuroanthropology offers to neuroscientists more sophisticated ways of thinking about neural environment, based upon over a century of debate about the nature of cultural variation and how to conceptualize patterns of behavior.

The same thought and subtlety that go into understanding the relations among parts of the brain and body can be extended to consider how elements of the cultural and social environment are tied into specific brain functions, illuminating some of the specific ways that mind can become extended through cultural leveraging. That is, simply adding "culture" as a single population variable fails to really illuminate the dynamic, inconsistent processes through which neurological potential is channeled by specific cultural institutions or practices. Because the nervous system is embedded within the world, shot through with the environment down to its cellular structure, integrative models of its development must include interacting elements from both inside and outside of the skin.

Although brain scientists have reached out to other interlocutors, we believe that anthropology is an especially strong potential partner. The influence of culture, social interaction, and behavior patterns are immediate and susceptible to direct research, often more so than evolutionary theories about brain architecture origin. In addition, ethnographic research offers concrete evidence of how social and cultural dimensions of the environment might affect cognitive function, and illustrates the range of neuroplasticity in developmental outcomes well beyond what most experimental protocols consider. Anthropologists explore naturally occurring experiments in which the nervous system is developed, in diverging directions, over a lifetime.

For anthropologists, neuroanthropology entails a return to integrative research after decades in which many biological and cultural anthropologists have seen each other as the primary opposition. The anthropological study of the nervous system calls on anthropologists to make good on our promises of holism. Psychological anthropologists have called for a greater focus on elements of neuroanthropology—affect, memory, neural-based models of cognition, biocultural integration—but a wholesale shift requires anthropologists to maintain a simultaneous consideration of what may have previously been apportioned to different specialties in the field. The nervous system inherently spans boundaries between specialized knowledge of such areas as evolution, child development, physiology, perception, phenomenology, behavioral research, biology, and culture. Although some researchers might pull back from considering biology out of a fear of reductionism, the nervous system obstinately resists any simplistic explanation, throwing up counterexamples such as varying degrees of mental modularity, cognitive heterogeneity, and complex mixtures of neuroplasticity and innate endowments shaped by evolution.

Neuroanthropologists will have to keep abreast of new research techniques and findings, and be willing to modify, expand, or shed outright our theories if they are unsupported by data. Anthropology has tended to be a theoretically heterodox field, producing more than its fair share of paradigms for understanding human social life, so neuroanthropologists should have abundant resources on which to draw, as long as we are willing to range far and wide for our intellectual frameworks, including into the past paradigms of relevant fields.

Enough overarching theories have foundered on human neural heterogeneity to offer ample warning: neuroanthropological theory will have to be partial and incremental rather than overly generalizing and prematurely sweeping. That is, no single enculturation process affects all brain areas equally, so no single account of the relation between brain and culture is likely to prove compelling in all cases. We propose an evidence-based theoretical eclecticism, recognizing that some of our disagreements are likely to arise from the fact that we theorize from different case studies in neural acculturation.

The potential gains are enormous: a robust account of brains in the wild, an understanding of how we come to possess our distinctive capacities and the degree to which these might be malleable across our entire species. The applications of this sort of research are myriad in diverse areas such as education, cross-cultural communication, developmental psychology, design, therapy, and information technology, to name just a few. But the first step is the one taken here—by coming together, we can achieve significant advances in understanding how our very humanity relies on the intricate interplay of brain and culture.

References

Adams, R. B., Jr., Franklin, R. G., Jr., Rule, N. O., Freeman, J. B., Kveraga, K., Hadjikhani, N., et al. (2010). Culture, gaze and the neural processing of fear expressions. *Social Cognitive and Affective Neuroscience, 5*(2–3), 340–348.

Adolphs, R. (2009). The social brain: Neural basis of social knowledge. *Annual Review of Psychology, 60,* 693–716.

Ahissar, M. (2001). Perceptual training: A tool for both modifying the brain and exploring it. *Proceedings of the National Academy of Sciences of the United States of America, 98*(21), 11842–11843.

Alcock, J. (2001). *The triumph of sociobiology.* Oxford: Oxford University Press.

Ambady, N., & Bharucha, J. (2009). Culture and the brain. *Current Directions in Psychological Science, 18*(6), 342–345.

Anderson, M. L. (2010). Neural reuse: A fundamental organizational principle of the brain. *Behavioral and Brain Sciences, 33,* 245–313.

Archer, J. (2006). Testosterone and human aggression: An evaluation of the challenge hypothesis. *Neuroscience and Biobehavioral Reviews, 30,* 319–345.

Arnett, J. J. (2008). The neglected 95%: Why American psychology needs to become less American. *American Psychologist, 63*(7), 602–614.

Barkow, J. H., Cosmides, L., & Tooby, J. (Eds.). (1992). *The adapted mind: Evolutionary psychology and the generation of culture.* New York: Oxford University Press.

Barrett, L. F., Mesquita, B., Ochsner, K. N., & Gross, J. J. (2007). The experience of emotion. *Annual Review of Psychology, 58,* 373–403.

Bengtsson, S. L., Nagy, Z., Skare, S., Forsman, L., Forssberg, H., & Ullén, F. (2005). Extensive piano practicing has regionally specific effects on white matter development. *Nature Neuroscience, 8*(9), 1148–1150.

Blais, C., Jack, R. E., Scheepers, C., Fiset, D., & Caldara, R. (2008). Culture shapes how we look at faces. *PLoS ONE, 3*(8), e3022.

Bloch, M. E. F. (1998). *How we think they think: Anthropological approaches to cognition, memory, and literacy.* Boulder, CO: Westview Press.

Boas, F. (1938). *The mind of primitive man* (Rev. ed.). New York: Macmillan. (Original work published 1911)

Boden, M. A. (2006). The mystery of the missing discipline. In *Mind as machine: A history of cognitive science* (Vol. 1, 515–589). Oxford: Oxford University Press.

Bond, M. H. (2002). Reclaiming the individual from Hofstede's ecological analysis—A 20-year odyssey: Comment on Oyserman et al. (2002). *Psychological Bulletin, 128*(1), 73–77.

Boyer, P. (1994). *The naturalness of religious ideas: A cognitive theory of religion.* Berkeley, CA: University of California Press.

Brown, R. A., Kuzara, J., Copeland, W. E., Costello, E. J., Angold, A., & Worthman, C. M. (2009). Moving from ethnography to epidemiology: Lessons learned in Appalachia. *Annals of Human Biology, 36*(3), 248–260.

Campbell, B. C., & Garcia, J. R. 2009. Neuroanthropology: Evolution and emotional embodiment. *Frontiers in Evolutionary Neuroscience, 1*(4). Retrieved from: http://www.frontiersin.org/evolutionary_neuroscience/10.3389/neuro.18.004.2009/abstract

Cantlon, J. F., & Brannon, E. M. (2006). Adding up the effects of cultural experience on the brain. *Trends in Cognitive Sciences, 11*(1), 1–4.

Chee, M. W. L., Zheng, H., Goh, J. O. S., & Park, D. (2010). Brain structure in young and old East Asians and Westerners: Comparisons of structural volume and cortical thickness. *Journal of Cognitive Neuroscience, 23*(5), 1065–1079.

Chiao, J. Y. (Ed.). (2009a). *Cultural neuroscience: Cultural influence on brain function* (Vol. 178). Oxford, England: Elsevier.

Chiao, J. Y. (2009b). *Cultural neuroscience: A once and future discipline.* (Vol. 178, pp. 287–304). Oxford, England: Elsevier Press.

Chiao, J. Y., & Ambady, N. (2007). Cultural neuroscience: Parsing universality and diversity across levels of analysis. In S. Kitayama & D. Cohen (Eds.), *Handbook of cultural psychology* (pp. 237–254). New York: Guilford Press.

Chiao, J. Y., Harada, T., Komeda, H., Li, Z., Mano, Y., Saito, D., et al. (2010a). Dynamic cultural influences on neural representations of the self. *Journal of Cognitive Neuroscience, 22*(1), 1–11.

Chiao, J. Y., Hariri, A. R., Harada, T., Mano, Y., Sadato, N., Parrish, T. B., et al. (2010b). Theory and methods in cultural neuroscience. *Social Cognitive and Affective Neuroscience, 5*(2–3), 356–361.

Chiao, J. Y., Iidaka, T., Gordon, H. L., Nogawa, J., Bar, M., Aminoff, E., et al. (2008a). Cultural specificity in amygdala response to fear faces. *Journal of Cognitive Neuroscience, 20*(12), 2167–2174.

Chiao, J. Y., Li, Z., & Harada, T. (2008b). Cultural neuroscience of consciousness: From visual perception to self-awareness. *Journal of Consciousness Studies, 15,* 58–69.

Chomsky, N. (1990). *Modular approaches to the study of the mind.* San Diego, CA: San Diego State University.

Choudhury, S., Nagel, S. K., & Slaby, J. (2009). Critical neuroscience: Linking neuroscience and society through critical practice. *Biosocieties, 4,* 61–77.

Choudhury, S., & Slaby, J. (Eds.). (2011). *Critical neuroscience: A handbook of the cultural and social contexts of neuroscience.* Hoboken, NJ: Wiley-Blackwell.

Chua, H. F., Boland, J. E., & Nisbett, R. E. (2005). Cultural variation in eye movements during scene perception. *Proceedings of the National Academy of Sciences of the United States of America, 102*(35), 12629–12633.

Cicchetti, D., & Cohen, D. J. (2006). The developing brain and neural plasticity: Implications for normality, psychopathology, and resilience. In D. Cicchetti & D. J. Cohen (Eds.), *Developmental psychopathology* (2nd ed., Vol. 2, 1–64). New York: Wiley.

Clark, A., & Chalmers, D. (1998). The extended mind. *Analysis, 58*(1), 7–19.

Classen, C. (1993). *Worlds of sense: Exploring the senses in history and across cultures.* New York: Routledge.

Classen, C. (1997). Foundations for an anthropology of the senses. *International Social Science Journal, 153,* 401–412.

Cohen, A. B. (2009). Many forms of culture. *American Psychologist, 64,* 194–204.

Cohen, D., Nisbett, R. E., Bowdle, B. F., & Schwarz, N. (1996). Insult, aggression, and the Southern culture of honor: An "experimental ethnography". *Journal of Personality and Social Psychology, 70*(5), 945–960.

Cowie, F. (2002). *What's within?: Nativism reconsidered.* Oxford: Oxford University Press.

Crick, M. (1982). Anthropology of knowledge. *Annual Review of Anthropology, 11,* 287–313.

Csordas, T. (1990). Embodiment as a paradigm for anthropology. *Ethos, 18*(1), 5–47.

Csordas, T. (Ed.). (1992). *Embodiment and experience: The existential ground of self.* Cambridge, England: Cambridge University Press.

Damasio, A. (1994). *Descartes' error: Emotion, reason, and the human brain.* New York: Putnam Books.

D'Andrade, R. G. (1981). The cultural part of cognition. *Cognitive Science, 5,* 179–195.

D'Andrade, R. G. (1995). *The development of cognitive anthropology.* Cambridge: Cambridge University Press.

D'Andrade, R. G. (2002). Violence without honor in the American South. In T. Aase (Ed.), *Tournaments of power: Honor and revenge in the contemporary world* (pp. 61–77). London: Ashgate.

D'Andrade, R. G. (2003). A cognitivist's view of the units debate in cultural anthropology. *Cross-Cultural Research, 35*(2), 242–257.

Davidson, R. J., Jackson, D. C., & Kalin, N. H. (2000). Emotion, plasticity, context, and regulation: Perspectives from affective neuroscience. *Psychological Bulletin, 126*(6), 890–909.

Deacon, T. (1997). *The symbolic species: The co-evolution of language and the brain.* New York: W. W. Norton.

Domìnguez Duque, J. F. (2007). Neuroanthropology: The combined anthropological and neurobiological study of cultural activity. PhD dissertation, Melbourne University, Melbourne, Australia.

Domìnguez Duque, J. F., Turner, R., Lewis, E. D., & Egan, G. (2010). Neuroanthropology: A humanistic science for the study of the culture-brain nexus. *Social Cognitive and Affective Neuroscience, 5*(2–3), 138–147.

Downey, G. (2007). Seeing without knowing, learning with the eyes: Visuomotor "knowing" and the plasticity of perception. In M. Harris (Ed.), *Ways of knowing: New approaches in the anthropology of knowledge and learning* (pp. 222–241). New York, NY: Berghahn Books.

Downey, G. (2010a). "Practice without theory": A neuroanthropological perspective on embodied learning. *Journal of the Royal Anthropological Institute, 16*(s1), S22–S40.

Downey, G. (2010b). Cultural variation in elite athletes: Does elite cognitive-perceptual skill always converge? In W. Christensen, E. Schier, & J. Sutton (Eds.), *ASCS09: Proceedings of the Ninth Conference of the Australasian Society for Cognitive Science* (pp. 72–80). Sydney, Australia: Macquarie Centre for Cognitive Studies.

Draganski, B., Gaser, C., Busch, V., Schuierer, G., Bogdahn, U., & May, A. (2004). Neuroplasticity: Changes in grey matter induced by training. *Nature, 427*(6972), 311–312.

Draganski, B., & May, A. (2008). Training-induced structural changes in the adult human brain. *Behavioural Brain Research, 192,* 137–142.

Dumit, J. (2004). *Picturing personhood: Brain scans and biomedical identity*. Princeton, NJ: Princeton University Press.

Durham, W. H. (1992). *Coevolution: Genes, culture and human diversity*. Stanford, CA: Stanford University Press.

Edeline, J.-M. (1998). Learning-induced physiological plasticity in the thalamo-cortical sensory systems: A critical evaluation of receptive field plasticity, map changes and their potential mechanisms. *Progress in Neurobiology, 57*, 165–224.

Edelman, G. M. (1987). *Neural Darwinism: The theory of neuronal group selection*. New York: Basic Books.

Ellman, J. L., Bates, E. A., Johnson, M. H., Karmiloff-Smith, A., Parisi, D., & Plunkett, K. (1997). *Rethinking innateness: A connectionist perspective on development*. Cambridge, MA: MIT Press.

Fahle, M., & Poggio, T. (Eds.). (2002). *Perceptual learning*. Cambridge, MA: MIT Press.

Feldman, D. E. (2009). Synaptic mechanisms for plasticity in neocortex. *Annual Review of Neuroscience, 32*, 33–55.

Fine, C. (2010). *Delusions of gender: How our minds, society, and neurosexism create difference*. New York: Norton.

Fiske, A. P. (2002). Using individualism and collectivism to compare cultures—A critique of the validity and measurement of the constructs: Comment on Oyserman et al. (2002). *Psychological Bulletin, 128*(1), 78–88.

Fiske, S. T. (2009). Cultural processes. In G. G. Berntson & J. T. Cacioppo (Eds.), *Handbook of neuroscience for the behavioral sciences* (pp. 985–1001). New York: Wiley.

Fodor, J. (1983). *Modularity of mind: An essay on faculty psychology*. Cambridge, MA: MIT Press.

Freese, J., Li, J.-C. A., & Wade, L. D. (2003). The potential relevances of biology to social inquiry. *Annual Review of Sociology, 29*, 233–256.

Frey, S. H. (2007). What puts the how in where? Tool use and the divided visual streams hypothesis. *Cortex, 43*(3), 368–375.

Fuentes, A. (2009). A new synthesis: Resituating approaches to the evolution of human behaviour. *Anthropology Today, 25*(3), 12–17.

Gardner, H. (1985). *The mind's new science: A history of the cognitive revolution*. New York: Basic Books.

Gaser, C., & Schlaug, G. (2003). Brain structures differ between musicians and non-musicians. *Journal of Neuroscience, 23*, 9240–9245.

Geertz, C. (1973). *The interpretation of culture*. New York: Basic Books.

Gell, A. (1995). The language of the forest: Landscape and phonological iconism in Umeda. In E. Hirsch & M. O'Hanlon (Eds.), *The anthropology of landscape: Perspectives on place and space* (pp. 232–254). Oxford: Oxford University Press.

Gerstein, M. B., Bruce, C., Rozowsky, J. S., Zheng, D., Du, J., Korbel, J. O., et al. (2007). What is a gene, post-ENCODE? History and updated definition. *Genome Research, 17*(6), 669–681.

Geurts, K. L. (2002). *Culture and the senses: Bodily ways of knowing in an African community*. Berkeley, CA: University of California Press.

Ghose, G. M. (2004). Learning in mammalian sensory cortex. *Current Opinion in Neurobiology, 14*(4), 513–518.

Gibson, K. R. (2002). Evolution of human intelligence: The roles of brain size and mental construction. *Brain, Behavior and Evolution, 59*(1–2), 10–20.

Gislén, A., Dacke, M., Kröger, R. H. H., Abrahamsson, M., Nilsson, D.-E., & Warrant, E. J. (2003). Superior underwater vision in a human population of sea gypsies. *Current Biology, 13*(10), 833–836.

Gogtay, N., Giedd, J. N., Lusk, L., Hayashi, K. M., & Greenstein, D. Vaituzis, A. C., et al. (2004). Dynamic mapping of human cortical development during childhood through early adulthood. *Proceedings of the National Academy of Sciences of the United States of America 101*(21), 8174–8179.

Goh, J. O., & Park, D. C. (2009). Culture sculpts the perceptual brain. *Progress in Brain Research, 178*, 95–111.

Goh, J. O., Tan, J. C., & Park, D. C. (2009). Culture modulates eye-movements to visual novelty. *PLoS ONE, 4*(12), e8238.

Goldblum, N. (2001). *The brain-shaped mind: What the brain can tell us about the mind*. Cambridge, MA: Cambridge University Press.

Golestani, N., Molko, N., Dehaene, S., LeBihan, D., & Pallier, C. (2007). Brain structure predicts the learning of foreign speech sounds. *Cerebral Cortex, 17*(3), 575–582.

Goodenough, W. (1957). Cultural anthropology and linguistics. In P. Garvin (Ed.), *Report of the seventh annual round table meeting on linguistics and language study* (Monograph series on languages and linguistics, No. 9, pp. 167–173). Washington, DC: Georgetown University.

Grant, J. A., Courtemanche, J., Duerden, E. G., Duncan, G. H., & Pierre Rainville, P. (2010). Cortical thickness and pain sensitivity in Zen meditators. *Emotion, 10*(1), 43–53.

Grasseni, C. (Ed.). (2007). *Skilled visions: Between apprenticeship and standards*. New York: Berghahn Books.

Grasseni, C. (2009). *Developing skill, developing vision*. New York: Berghahn Books.

Gravlee, C. C. (2009). How race becomes biology: Embodiment of social inequality. *American Journal of Physical Anthropology, 139*(1), 47–57.

Green, D. W., Crinion, J., & Price, C. J. (2007). Exploring cross-linguistic vocabulary effects on brain structures using voxel-based morphometry. *Bilingualism: Language and Cognition, 10*(2), 189–199.

Greenfield, P. M., Keller, H., Fuligni, A., & Maynard, A. (2003). Cultural pathways through universal development. *Annual Review of Psychology, 54*, 461–490.

Güss, C. D., Tuason, M. T., & Gerhard, C. (2009). Cross-national comparisons of complex problem-solving strategies in two microworlds. *Cognitive Science, 34*(3), 489–520.

Gutchess, A. H., Hedden, T., Ketay, S., Aron, A., & Gabrieli, J. D. E. (2010). Neural differences in the processing of semantic relationships across cultures. *Social Cognitive and Affective Neuroscience, 5*(2–3), 254–263.

Gutchess, A. H., Welsh, R. C., Boduroglu, A., & Park, D. C. (2006). Cultural differences in neural function associated with object processing. *Cognitive, Affective & Behavioral Neuroscience, 6*(2), 102–109.

Han, S., Gu, X., Mao, L., Ge, J., Wang, G., & Ma, Y. (2010). Neural substrates of self-referential processing in Chinese Buddhists. *Social Cognitive and Affective Neuroscience, 5*(2–3), 332–339.

Han, S., Mao, L., Gu, X., Zhu, Y., Ge, J., & Ma, Y. (2008). Neural consequences of religious belief on self-referential processing. *Social Neuroscience, 3*(1), 1–15.

Han, S., & Northoff, G. (2008). Culture-sensitive neural substrates of human cognition: A transcultural neuroimaging approach. *Nature Reviews. Neuroscience, 9*(8), 646–654.

Hanakawa, T., Honda, M., Okada, T., Fukuyama, H., & Shibasaki, H. (2003). Neural correlates underlying mental calculation in abacus experts: Functional magnetic resonance imaging study. *NeuroImage, 19*(2), 296–307.

Hay, M. C. (2009). Anxiety, remembering, and agency: Biocultural insights for understanding illness. *Ethos, 37*(1), 1–31.

Heatherton, T. F., Wyland, C. L., Macrae, C. N., Demos, K. E., Denny, B. T., & Kelley, W. M. (2006). Medial prefrontal activity differentiates self from close others. *Social Cognitive and Affective Neuroscience, 1*(1), 18–25.

Hedden, T., Ketay, S., Aron, A., Markus, H. R., & Gabrieli, J. D. E. (2008). Cultural influences on neural substrates of attentional control. *Psychological Science, 19*(1), 12–17.

Heine, S. J. (2001). Self as cultural product: An examination of East Asian and North American selves. *Journal of Personality, 69*(6), 881–906.

Henrich, J., Heine, S. J., & Norenzayan, A. (2010). The weirdest people in the world? *Behavioral and Brain Sciences, 33*(1), 61–135.

Hertzman, C., & Boyce, T. (2010). How experience gets under the skin to create gradients in developmental health. *Annual Review of Public Health, 31*, 329–347.

Howes, D. (Ed.). (1991). *The varieties of sensory experience: A sourcebook in the anthropology of the senses*. Toronto, Canada: University of Toronto Press.

Hutchins, E. (1995). *Cognition in the wild*. Cambridge, MA: MIT Press.

Ingold, T. (2000). *The perception of the environment: Essays in livelihood, dwelling and skill*. London, England: Routledge.

Ingold, T. (2001). From complementarity to obviation: On dissolving the boundaries between social and biological anthropology, archaeology and psychology. In S. Oyama, P. E. Griffiths, & R. D. Gray (Eds.), *Cycles of contingency: Developmental systems and evolution* (pp. 255–279). Cambridge: Cambridge University Press.

Ingold, T. (2007). The trouble with "evolutionary biology". *Anthropology Today, 23*(2), 13–17.

Izzetoglu, M., Bunce, S. C., Izzetoglu, K., Onaral, B., & Pourrezaei, K. (2007, July/August). Functional brain imaging using near-infrared technology: Assessing cognitive activity in real-life situations. *IEEE Engineering in Medicine and Biology Magazine*, 38–46.

Izzetoglu, M., Izzetoglu, K., Bunce, S. C., Ayaz, H., Devaraj, A., Onaral, B., et al. (2005). Functional near-infrared neuroimaging. *IEEE Transactions on Neural Systems and Rehabilitation Engineering, 13*(2), 153–159.

Jablonka, E., & Lamb, M. J. (2005). *Evolution in four dimensions: Genetic, epigenetic, behavioral, and symbolic variation in the history of life*. Cambridge, MA: MIT Press.

Jackson, M. (1989). *Paths toward a clearing: Radical empiricism and ethnographic inquiry*. Bloomington, IN: Indiana University Press.

Joyce, K. A. (2008). *Magnetic appeal: MRI and the myth of transparency*. Ithaca, NY: Cornell University Press.

Kakigi, R., Nakata, H., Inui, K., Hiroe, N., Nagata, O., Honda, M., et al. (2005). Intracerebral pain processing in a yoga master who claims not to feel pain during meditation. *European Journal of Pain (London, England), 9*(5), 581–589.

Kellman, P. J., & Garrigan, P. (2009). Perceptual learning and human expertise. *Physics of Life Reviews, 6*(2), 53–84.

Kemper, T. D. (1990). *Social structure and testosterone: Explorations of the socio-bio-social chain.* New Brunswick, NJ: Rutgers University Press.

Ketay, S., Aron, A., & Hedden, T. (2009). Culture and attention: Evidence from brain and behavior. *Progress in Brain Research, 178,* 79–92.

Kirmayer, L. J. (1989). Cultural variations in the response to psychiatric disorders and emotional distress. *Social Science and Medicine, 29*(3), 327–339.

Kitayama, S., Dufy, S., Kawamura, T., & Larsen, J. T. (2003). Perceiving an object and its context in different cultures: A cultural look at new look. *Psychological Science, 14*(3), 201–206.

Kitayama, S., & Park, J. (2010). Cultural neuroscience of the self: Understanding the social grounding of the brain. *Social Cognitive and Affective Neuroscience, 5*(2–3), 111–129.

Kobayashi, C., Glover, G. H., & Temple, E. (2007). Children's and adults' neural bases of verbal and nonverbal "theory of mind". *Neuropsychologia, 45,* 1522–1532.

Kobayashi, C., Glover, G. H., & Temple, E. (2008). Switching language switches mind: Linguistic effects on developmental neural bases of "theory of mind". *Social Cognitive and Affective Neuroscience, 3,* 62–70.

Koepping, K.-P. (1983). *Adolf Bastian and the psychic unity of mankind: The foundations of anthropology in nineteenth century Germany.* St. Lucia: University of Queensland Press.

Kroeber, A. (1917). The superorganic. *American Anthropologist, 19,* 163–213.

Kuhl, P. K., Williams, K. A., Lacerda, F., Stevens, K. N., & Lindblom, B. (1992). Linguistic experience alters phonetic perception in infants by 6 months. *Science, 255*(5044), 606–608.

Kuper, A. (1999). *Culture: The anthropologists' account.* Cambridge, MA: Harvard University Press.

Laland, K. N., Kendal, J. R., & Brown, G. R. (2007). The niche construction perspective: Implications for evolution and human behaviour. *Journal of Evolutionary Psychology, 5*(1–4), 51–66.

Laughlin, C. D. (1997). Body, brain, and behavior: The neuroanthropology of the body image. *Anthropology of Consciousness, 8*(2–3), 49–68.

Laughlin, C. D., & D'Aquili, E. G. (1974). *Biogenetic structuralism.* New York: Columbia University Press.

Laughlin, C. D., McManus, J., & D'Aquili, E. G. (1990). *Brain, symbol and experience: Toward a neurophenomenology of human consciousness.* Boston, MA: New Science Library.

Lehman, D. R., Chiu, C., & Schaller, M. (2004). Culture and psychology. *Annual Review of Psychology, 55,* 689–714.

Lende, D. H. (2005). Wanting and drug use: A biocultural approach to the analysis of addiction. *Ethos, 33*(1), 100–124.

Lende, D. H. (2008, November). The encultured brain: Neuroanthropology and interdisciplinary engagement. Paper presented at the 2008 American Anthropological Association annual meeting, San Francisco.

Lende, D. H., & Lachiondo, A. (2009). Embodiment and breast cancer among African-American women. *Qualitative Health Research, 19*(2), 216–228.

Lester, R. J. (1997). The (dis)embodied self in anorexia nervosa. *Social Science and Medicine, 44*(4), 479–489.

LeVine, R. A. (1999). An agenda for psychological anthropology. *Ethos, 27*(1), 15–24.

Lewis, R. S., Goto, S. G., & Kong, L. L. (2008). Culture and context: East Asian American and European American differences in P3 event-related potentials and self-construal. *Personality and Social Psychology Bulletin, 34*(5), 623–634.

Lewontin, R. (2000). *The triple helix: Gene, organism and environment.* Cambridge, MA: Harvard University Press.

Li, S.-C. (2003). Biocultural orchestration of developmental plasticity across levels: The interplay of biology and culture in shaping the mind and behavior across the life span. *Psychological Bulletin, 129*(2), 171–194.

Lutz, A., Dunne, J. D., & Davidson, R. J. (2007). Meditation and the neuroscience of consciousness: An introduction. In P. D. Zelazo, M. Moscovitch, & E. Thompson (Eds.), *Cambridge handbook of consciousness* (pp. 499–551). Cambridge, MA: Cambridge University Press.

Maguire, E. A., Gadian, D. G., Johnsrude, I. S., Good, C. D., Ashburner, J., Frackowiak, R. S. J., & Frith, C. D. (2000). Navigation-related structural changes in the hippocampi of taxi drivers. *Proceedings of the National Academy of Sciences of the United States of America, 97*(8), 4398–4403.

Malafouris, L. (2009). Neuroarchaeology: Exploring the links between neural and cultural plasticity. *Progress in Brain Research, 178,* 251–259.

Marchand, T. H. J. (2007). Crafting knowledge: The role of "parsing and production" in the communication of skill-based knowledge among masons. In M. Harris (Ed.),

Ways of knowing: New approaches in the anthropology of knowledge and learning (pp. 173–193). Oxford: Berghahn Books.

Marchand, T. H. J. (2010). Making knowledge: Explorations of the indissoluble relation between minds, bodies, and environment. *Journal of the Royal Anthropological Institute, 16*, S1–S21.

Mareschal, D., Johnson, M. H., Sioris, S., Spratling, M. W., Thomas, M. S. C., & Westermann, G. (2007). *Neuroconstructivism: How the brain constructs cognition* (Vol. 1). New York: Oxford University Press.

Markus, H. R., & Kitayama, S. (1991). Culture and the self: Implications for cognition, emotion and motivation. *Psychological Review, 98*(2), 224–253.

Martin, E. (2000). Mind-body problems. *American Ethnologist, 27*(3), 569–590.

Masuda, T., & Nisbett, R. E. (2006). Culture and change blindness. *Cognitive Science, 30*(2), 381–399.

McAndrew, F. T. (2009). The interacting roles of testosterone and challenges to status in human male aggression. *Aggression and Violent Behavior, 14*(5), 330–335.

McClure, S. M., Li, J., Tomlin, D., Cypert, K. S., Montague, L., M., & Montague, P. R. (2004). Neural correlates of behavioral preference for culturally familiar drinks. *Neuron, 44*(2): 379–387.

Menary, R. (Ed.). (2010). *The extended mind.* Cambridge, MA: MIT Press.

Merleau-Ponty, M. (1962). *Phenomenology of perception* (C. Smith, Trans.). London: Routledge.

Nan, Y., Knosche, T. R., Zysset, S., & Friederici, A. D. (2008). Cross-cultural music phrase processing: An fMRI study. *Human Brain Mapping, 29*(3), 312–328.

Niedenthal, P. M. (2007). Embodying emotion. *Science, 316*(5827), 1002–1005.

Nisbett, R. E. (2003). *The geography of thought: How Asians and Westerners think differently ... and why.* New York: Free Press.

Nisbett, R. E., & Masuda, T. (2003). Culture and point of view. *Proceedings of the National Academy of Sciences of the United States of America, 100*(19), 11163–11170.

Nisbett, R. E., & Miyamoto, Y. (2005). The influence of culture: Holistic versus analytic perception. *Trends in Cognitive Sciences, 9*(10), 467–473.

Nisbett, R. E., Peng, K., Choi, I., & Norenzayan, A. (2001). Culture and systems of thought: Holistic versus analytic cognition. *Psychological Review, 108*(2), 291–310.

Norman, D. A. (1980). Twelve issues for cognitive science. *Cognitive Science, 4*(1), 1–32.

Ochsner, K. N., & Gross, J. J. (2008). Cognitive emotion regulation: Insights from social cognitive and affective neuroscience. *Current Directions in Psychological Science, 17*(2), 153–158.

Odling-Smee, F. J., Laland, K. N., & Feldman, M. W. (2003). *Niche construction: The neglected process in evolution* (Monographs in population biology, No. 37). Princeton, NJ: Princeton University Press.

O'Regan, J. K., & Nöe, A. (2001). A sensorimotor account of vision and visual consciousness. *Behavioral and Brain Sciences, 24*(5), 939–1031.

Oyama, S. (2000). *The ontogeny of information: Developmental systems and evolution* (2nd rev. ed.). Durham, NC: Duke University Press.

Oyama, S., Griffiths, P. E., & Gray, R. D. (Eds.). (2001). *Cycles of contingency: Developmental systems and evolution.* Cambridge, MA: MIT Press.

Oyserman, D., Coon, H. M., & Kemmelmeier, M. (2002). Rethinking individualism and collectivism: Evaluation of theoretical assumptions and meta-analysis. *Psychological Bulletin, 128*(1), 3–72.

Park, D. C., & Huang, C.-M. (2010). Culture wires the brain: A cognitive neuroscience perspective. *Perspectives on Psychological Science, 5*(4), 391–400.

Pascual-Leone, A., Amedi, A., Fregni, F., & Merabet, L. B. (2005). The plastic human brain cortex. *Annual Review of Neuroscience, 28*, 377–401.

Pink, S. (2009). *Doing sensory ethnography.* Los Angeles, CA: Sage Publications.

Renfrew, C., Frith, C., & Malafouris, L. (2008). The sapient mind: Archaeology meets neuroscience. *Philosophical Transactions of the Royal Society B: Biological Sciences, 363*(1499), 1935–1938.

Reyna, S. (2002). *Connections: Brain, mind, and culture in a social anthropology.* London: Routledge.

Richerson, P. J., & Boyd, R. (2005). *Not by genes alone: How culture transformed human evolution.* Chicago: University of Chicago Press.

Rilling, J. K. (2008). Neuroscientific approaches and applications within anthropology. *Yearbook of Physical Anthropology, 51*, 2–32.

Roepstorff, A., Niewöhner, J., & Beck, S. (2010). Enculturing brains through patterned practices. *Neural Networks, 23*(8–9), 1051–1059.

Rozin, P. (2009). What kind of empirical research should we publish, fund, and reward? A different perspective. *Perspectives on Psychological Science, 4*(4), 435–439.

Sacks, O. (1995). *An anthropologist on Mars: Seven paradoxical tales.* New York: Knopf.

Sadato, N., Pascual-Leone, A., Grafman, J., Ibañez, V., Deiber, M.-P., Dold, G., et al. (1996). Activation of the primary visual cortex by Braille reading in blind subjects. *Nature, 380*(6574), 526–528.

Said, E. W. (1978). *Orientalism*. New York: Pantheon Books.

Sanes, J. N., & Donoghue, J. P. (2000). Plasticity and primary motor cortex. *Annual Review of Neuroscience, 23*, 393–415.

Sapolsky, R. M. (2004). Social status and health in humans and other animals. *Annual Review of Anthropology, 33*, 393–418.

Sasaki, Y., Nanez, J. E., & Watanabe, T. (2010). Advances in visual perceptual learning and plasticity. *Nature Reviews. Neuroscience, 11*(1), 53–60.

Segall, M. H., Campbell, D. T., & Herskovits, M. J. (1966). *The Influence of Culture on Visual Perception*. Oxford: Bobbs-Merrill.

Seitz, A. R., & Dinse, H. R. (2007). A common framework for perceptual learning. *Current Opinion in Neurobiology, 17*(2), 148–153.

Seligman, R., & Brown, R. A. (2010). Theory and method at the intersection of anthropology and cultural neuroscience. *Social Cognitive and Affective Neuroscience, 5*(2–3), 130–137.

Seligman, R., & Kirmayer, L. J. (2008). Dissociative experience and cultural neuroscience: Narrative, metaphor and mechanism. *Culture, Medicine and Psychiatry, 32*(1), 31–64.

Shore, B. (1996). *Culture in mind: Cognition, culture and the problem of meaning*. New York: Oxford University Press.

Shweder, R. A., & Bourne, E. J. (1984). Does the concept of the person vary cross-culturally? In R. A. Shweder & R. A. LeVine (Eds.), *Culture Theory: Essays on Mind, Self, and Emotion* (pp. 158–199). New York: Cambridge University Press.

Sperber, D. (1996). *Explaining culture: A naturalistic approach*. Oxford: Blackwell.

Sperber, D., & Hirschfeld, L. A. (2004). The cognitive foundation of cultural stability and diversity. *Trends in Cognitive Sciences, 8*(1), 40–46.

Sterelny, K. (2003). *Thought in a hostile world: The evolution of human cognition*. Oxford: Blackwell.

Sterelny, K. (2007). Social intelligence, human intelligence and niche construction. *Philosophical Transactions of the Royal Society B, 362*(1480), 719–730.

Stoller, P. (1989). *The taste of ethnographic things: The senses in anthropology*. Philadelphia: University of Pennsylvania Press.

Strathern, A. (1996). *Body thoughts*. Ann Arbor: University of Michigan Press.

Strauss, C., & Quinn, N. (1997). *A cognitive theory of cultural meaning*. Cambridge: Cambridge University Press.

Strohschneider, S., & Güss, D. (1998). Planning and problem solving: Differences between Brazilian and German students. *Journal of Cross-Cultural Psychology, 29*(6), 695–716.

Strohschneider, S., & Güss, D. (1999). The fate of the MOROs: A cross-cultural exploration of strategies in complex and dynamic decision making. *International Journal of Psychology, 34*(4), 235–252.

Sui, J., Liu, C. H., & Han, S. (2009). Cultural difference in neural mechanisms of self-recognition. *Social Neuroscience, 4*(5), 402–411.

Super, C. M., & Harkness, S. (2002). Culture structures the environment for development. *Human Development, 45*, 270–274.

Tang, Y., Zhang, W., Chen, K., Feng, S., Ji, Y., Shen, J., et al. (2006). Arithmetic processing in the brain shaped by cultures. *Proceedings of the National Academy of Sciences of the United States of America, 103*(28), 10775–10780.

Tapias, M. (2006). Emotions and the intergenerational embodiment of social suffering in rural Bolivia. *Medical Anthropology Quarterly, 20*(3), 399–415.

TenHouten, W. (1976). More on split brain research, culture, and cognition. *Current Anthropology, 17*(3), 503–506.

Tomasello, M. (2001). *The cultural origins of human cognition*. Cambridge, MA: Harvard University Press.

Tooby, J., & Cosmides, L. (1992). The psychological foundations of culture. In J. H. Barkow, L. Cosmides, & J. Tooby (Eds.), *The adapted mind: Evolutionary psychology and the generation of culture* (pp. 19–136). New York: Oxford University Press.

Triandis, H. C. (1995). *Individualism and collectivism*. Boulder: Westview Press.

Turner, R. (2001). Culture and the human brain. *Anthropology and Humanism, 26*(2), 167–172.

Turner, S. P. (2002). *Brains/practices/relativism: Social theory after cognitive science*. Chicago: University of Chicago Press.

Turner, V. (1973). Body, brain and culture. In J. B. Ashbrook (Ed.), *Brain, culture and the human spirit: Essays from an emergent evolutionary perspective* (pp. 77–106). Lanham, MD: University Press of America.

Varela, F. J., Thompson, E., & Rosch, E. (1991). *The embodied mind: Cognitive science and human experience*. Cambridge, MA: MIT Press.

Vogeley, K., & Roepstorff, A. (2009). Contextualising culture and social cognition. *Trends in Cognitive Sciences, 13*(12), 511–516.

Vul, E., Harris, C., Winkielman, P., & Pashler, H. (2009). Puzzlingly high correlations in fMRI studies of emotion, personality, and social cognition. *Perspectives on Psychological Science, 4*(3), 274–290.

Westermann, G., Mareschal, D., Johnson, M. H., Sirois, S., Spratling, M. W., & Thomas, M. S. C. (2007). Neuroconstructivism. *Developmental Science, 10*(1), 75–83.

Wexler, B. E. (2006). *Brain and culture: Neurobiology, ideology and social change.* Cambridge, MA: MIT Press.

Wheeler, M., & Clark, A. (2008). Culture, embodiment and genes: Unravelling the triple helix. *Philosophical Transactions of the Royal Society B, 363*(1509), 3563–3575.

Whitehouse, H. (1996). Jungles and computers: Neuronal group selection and the epidemiology of representations. *Journal of the Royal Anthropological Institute* (N.S.), *1*(1), 99–116.

Worthman, C. (2010). The ecology of human development: Evolving models for cultural psychology. *Journal of Cross-Cultural Psychology, 41*(4), 546–562.

Worthman, C. (2009). Habits of the heart: Life history and the developmental neuroendocrinology of emotion. *American Journal of Human Biology, 21*(6), 772–778.

Worthman, C., Plotsky, P. M., Schechter, D. S., & Cummings, C. A. (2010). *Formative experiences: The interaction of caregiving, culture, and developmental psychobiology.* New York: Cambridge University Press.

Zenger, B., & Sagi, D. (2002). Plasticity of low-level visual networks. In M. Fahle & T. Poggio (Eds.), *Perceptual learning* (pp. 177–196). Cambridge, MA: MIT Press.

Zhang, J., & Kourtzi, Z. (2010). Learning-dependent plasticity with and without training in the human brain. *Proceedings of the National Academy of Sciences of the United States of America, 107*(30), 13503–13508.

Zhu, Y., Zhang, L., Fan, J., & Han, S. (2007). Neural basis of cultural influence on self-representation. *NeuroImage, 43*, 1310–1316.

3 Primate Social Cognition, Human Evolution, and Niche Construction: A Core Context for Neuroanthropology

Katherine C. MacKinnon and Agustín Fuentes

The taxonomic order to which we belong, Primates, has been mined extensively by anthropologists, biologists, psychologists, and philosophers, among others, in the search for insights into what binds humans together with our primate cousins and at the same time what makes us uniquely human. Broad evolutionary patterns in sociality and cognition are central among the areas where we look for answers. In this chapter, we will highlight the importance of primate studies to anthropology and the neurosciences, and provide examples from current work that demonstrate how such research can inform the integrative and interdisciplinary foci at the core of neuroanthropology. The primary framing questions we will consider here: a) How are humans both primates yet also unique, and how does taking a broader look across the Primate order provide a useful framework for understanding the role of an evolved social cognition? b) How can current thinking in evolutionary biology update our interpretation of primate strategies? c) How are sociality and cognitive functioning intertwined with evolutionary success in the nonhuman primates, as well as in our own genus?

To go about answering these questions, we will review some fundamental characteristics of the Primate order with an emphasis on how both ecological and social components are important for understanding variability, followed by a section on how current theories such as niche construction, and methodologies like social network analysis, can help contextualize the role of an increased cognitive capacity in primates. Finally, we examine cognition and sociality *vis à vis* characteristics of primate brains, learning in a social context, and brain-culture interactions to underscore our fundamental argument that knowledge gained from some exciting areas in modern primatology (both captive and field-based) continues to have broad cross-disciplinary relevance.

A Brief Background of Primatology

Primatology as a subfield of biological and physical anthropology in the North American tradition has its beginnings in early lab- and field-based work that examined our primate relatives in order to glean insights into human evolutionary patterns. Other primatological traditions (e.g., in Europe and Japan) have somewhat different intellectual origins centered in ethology or zoology; from the 1930s to the 1950s, Robert Yerkes, Clarence Carpenter, Solly Zuckerman, Kinji Imanishi, and Hans Kummer, among others, initiated studies of social behavior and ecology of the nonhuman primates, which helped to establish many of the frameworks used in modern primatology (Strier, 1994; Strum & Fedigan, 2000). However, North American primatology also has deep, tangled roots in anthropology and, as such, has a somewhat unique and varied history. Of particular note is Sherwood Washburn's call for a "new physical anthropology" (1951) to integrate laboratory and field work studies of comparative anatomy, functional morphology, and ecology and behavior, which synthesized theoretical and methodological practices and spawned generations of anthropological primatologists working under this new paradigm (e.g., see Kelley & Sussman, 2007, for a genealogy of field-based primatologists in North America). As Washburn stated in "The Promise of Primatology": "We are primates, products of the evolutionary process, and the promise of primatology is a better understanding of the peculiar creature we call man" (1973, p. 182). In the 1960s and 1970s, Louis Leakey famously sponsored Jane Goodall, Dian Fossey, and Birute Galdikas to conduct fieldwork on chimpanzees, mountain gorillas, and orangutans, respectively; their discoveries were promulgated by the National Geographic Society via its magazine and television documentaries, and glimpses into the lives of these fascinating species captured the public's imagination. Psychologists were also conducting captive studies in the areas of cognition and learning during this time; for example, Harry Harlow's experiments in the 1950s and 1960s on the significance of primate mother-infant attachment and social bonding. Paradigms shifts occurred, as they do, and subsequently affected how primatological questions were framed and contextualized. The 1970s and 1980s ushered in sociobiology and evolutionary psychology, the latter founded on the premise that the human brain contains cognitive components that evolved to solve the reproductive problems presumably faced by our hunter-gatherer ancestors. Neo-Darwinian theory has since been applied to explain questions of the evolutionary foundations of human morality, with extensions to the *inherent nature* of a range of characteristics, from human aggression and gender-typical behaviors to personality traits and self-esteem (e.g., Chisholm, 1999; Wright, 1994; Eagly & Wood, 1999; Neubauer, 1996; Zimmerman, 2000; Barkow, Cosmides, & Tooby, 1992; Caporael, 2001; Cartwright, 2000; Daly & Wilson, 1988, 1999; Ridley, 1993, 1997; Rossi, 1995).

Around the time that sociobiology and evolutionary psychology were emerging as dominant paradigms, there was a steady increase in the number of primate species studied under natural conditions around the world. This continued markedly over the last quarter of the twentieth century, and still continues in the twenty-first century, when many species are critically endangered or face extinction in our lifetimes. As Karen Strier discusses in "Myth of the Typical Primate" (1994), the strong diversity of studied species has significantly altered our views of primates and how they are best incorporated into questions relating to human evolution. We no longer focus exclusively on the great apes—or even on the savanna-living hominins—as superior examples for extrapolating the context of modern human origins, and we have begun to move away from categorizing "species-typical" primate social organizations (see Fuentes, 2011, for a review). As Strier emphasizes, there is not one generalized, overarching "primate pattern" in the natural world, but a *variety* of behavioral adaptations with some common themes (Strier, 1994).

Humans as General and Unique: The Primate Context

Who are we, as primates? There are several traits that, when taken together, are found in all primates within our biological taxonomic order; that is, there is no single, individual trait that stands alone as uniquely "primate," but rather, we share this cluster of biological features with everything from the largest gorillas to the smallest mouse lemurs. These include basic features such as prehensile hands and (except humans) feet, a reliance on visual and tactile sensory pathways (less so on olfaction), and extended periods of infant-dependency and development, as well as a significantly enlarged brain-to-body size ratio compared to other mammals. The role that vision has played in our evolutionary history appears to be significant. Primate brains show a disproportionate expansion of the visual system (Barton, 1998; Ross & Martin, 2007; van Essen, Anderson, & Felleman, 1992), and while integral to finding foods and avoiding predators, many studies converge on the fact that primate vision is also specifically tied to sociality, in that complex social signals (facial and gestural, in particular) are processed in an emotional context (Adolphs, 2001; Brothers, 1990; van Essen, Lewis, Drury, Hadjikhani et al., 2001, Young, Scannell, Burns, & Blakemore, 1994).

One of the oft-cited hallmarks of our order is the strong tendency toward sociality and group living. This extends to even the nocturnal Strepsirrhini species (e.g., lorises, galagos), which were incorrectly described as "solitary" for decades. Because of long-term fieldwork, we now know such species live in communities or dispersed groups, with higher levels of social interaction and connectedness than thought to exist previously (Bearder, 1987, 1999; Nekaris, 2003; Nekaris & Bearder, 2011). Indeed, without exception, all primates show a strong dependence on stable social relationships for emotional and physical well-being and, based on our more recent phylogenetic history, we can say that all hominins (humans and recent human ancestors) live, or have lived, in social networks that disseminate complex information about infant care, presence of predators, and peculiarities of the local habitat including resource (e.g., food) availability. In addition, for all primates, physical and emotional bonding and social attachment are crucial for the healthy development of the central nervous system, while the extended period of infant care necessitates greater support from, and communication with, the extended social group (both for the care of mom, as well as the infant; Hawkes, O'Connell, & Blurton-Jones, 2003). Thus, social bonding and the manipulation of those bonds are characteristic of all gregarious primate societies, and we view complex sociality as a core primate adaptation (Dolhinow, 1972; Kappeler & van Schaik, 2006; Silk, 2007).

The anthropoid brain, in particular, found in monkeys, apes, and humans, contains extensive neural networks dedicated to processing social interactions (Brothers, 1990; Adolphs, 2001). In part, this neural circuitry is likely related to the success of social plasticity as an adaptive mechanism in primates (see Bateson & Gluckman, 2011 and West-Eberhard, 2003 for discussion of plasticity and uses of the term). For example, capuchin monkeys discriminate against and thus likely recognize monkeys in photos shown to them (Pokorny & de Waal, 2009a, 2009b), and a recent study in humans has shown that our subconscious recognizes "gossip" (information disseminated through social networks) and that our eyes and mind pay particular attention when a facial photograph accompanies a bit of *negative* gossip (Anderson, Siegel, Bliss-Moreau, & Barrett, 2011). The authors posit that perhaps our brain and visual system selectively choose information that might help protect us from potentially dangerous individuals or situations.

Both ecological and social components are important for understanding variability in nonhuman primate and human adaptive strategies (see, e.g., Aiello & Wheeler, 1995; Aiello & Wells, 2002; Dunbar & Shultz, 2007; Leigh, 2004; Leigh & Blomquist, 2011; Walker, Gurven, Hill, Migliano et al., 2006). In an ecological context, finding and processing food is obviously one of the consistent challenges facing any individual (see Lambert, 2011, for a review). Primate strategies run the gamut from extreme dietary specificity—for example, bamboo lemurs (*Hapalemur*) who can bypass the toxic effects of the high amounts of cyanide in the bamboo they regularly exploit (Gould & Sauther, 2011), highly frugivorous spider monkeys (*Ateles*) who include over 80% ripe fruit in their diets year-round (di Fiore, Fink, & Campbell, 2011), and some marmosets (*Callithrix*) that are able to use gums and saps as a major source of food (Digby, Ferrari, & Saltzman, 2011)—to more generalized, unspecialized omnivores. The influence of food availability on diet and range use has been well-studied across the primate taxa, and the collective picture that emerges is one of extensive variability across the order (Chapman & Chapman, 1990; Chapman & Russo, 2011). Primate species are adept at dealing with shifting environmental challenges with different tactics: a social group might increase or decrease their daily path length, expand or contract their home ranges throughout the year, increase overall travel time to ensure more food sources are located, or decrease travel time to conserve energy (e.g., Chapman, 1987, 1988; Chapman & Fedigan, 1990; Chapman & Russo, 2011; Garber, 2011).

Such flexibility has been a component of primate evolutionary success in myriad habitats, but three aspects of foraging behavior highlight the role of cognitive functioning and behavioral plasticity in anthropoids: color vision and finding foods (Melin, Fedigan, Hiramatsu, et al., 2009; Melin, Fedigan, Young, & Kawamura, 2010); memory and the spatial mapping of resources (Garber, 2011; Garber & Brown, 2006; Garber & Jelinek, 2006); and communication about food sources as well as predators (Gouzoules & Gouzoules, 2011). The latter two are intricately tied to being able to deal with fluctuating resources and shifting environmental elements (e.g., Melin, Fedigan, Hiramatsu, & Kawamura, 2008).

Many primate species also occupy a broad geographic range and demonstrate success in differing environments. The genus *Macaca* (macaque monkeys) are found from Northern Asia and Micronesia to South and Southeast Asia and North Africa; their historical range is even larger. Across this massive geographic area, they also live in a diversity of habitats, including swamp forests, mountain ranges, secondary forests, plantations, temple complexes, villages, and major cities. Throughout this entire distribution there is also a remarkable flexibility in group demography, dietary foci, and foraging and ranging patterns (Gumert, 2011; Thierry, 2011). The genus *Cebus* (neotropical capuchin monkeys) occurs in highly seasonal tropical dry forest; primary rain forest, at varying elevations; and in mixed agricultural and forested areas with much anthropogenic disturbance and in close proximity to humans (Jack, 2011; Fragaszy, Visalberghi, & Fedigan, 2004). Accordingly, the genus shows a high level of behavioral plasticity and even intraspecific variation in areas that are ecologically very similar: in a study of cross-site differences in *C. capucinus* foraging behavior, long-term data sets from three Costa Rican field sites (in the same general geographic area) included 30% of common food items that were processed quite differently (Panger, Perry, Rose, Gros-Louis et al., 2002).

Examples of extensive social variation in primate societies abound as well. For example, in a study that involved a multi-researcher dataset in Costa Rica (19,000 hours, 13 social groups, and four study sites over a 13-year period), several social traditions were identified according to outlined criteria. It is hypothesized that capuchins use these group- or clique-specific social conventions to test the quality of their social relationships (Perry, Baker, Fedigan, Gros-Louis et al., 2003a; Perry, Panger, Rose, Baker et al., 2003b). In captive capuchins, Dindo, Whiten, & de Waal (2009) found results that indicated a "striking effect" of social conformity (i.e., social learning overriding individual learning) in behavioral techniques and foraging traditions consistent with field reports like the one just mentioned (i.e., Perry et al., 2003a, 2003b). Such socially mediated traditions by a highly encephalized primate like the capuchin seem to be convergent with chimpanzees and humans in terms of the extensiveness of cultural repertoires displayed (Dindo et al., 2009; de Waal & Bonnie, 2009; Whiten, Horner, & de Waal, 2005; Whiten, Spiteri, Horner, Bonnie, et al., 2007; Caldwell & Whiten, 2011; and see Pruetz & Bertolani, 2007; Pruetz & Lindshield, (2011); and Sanz, Call, & Morgan, (2009) for new insights on chimpanzee tool use complexity).

Summary of Ecological and Social Variability
Primatological inquiry over the past few decades has resulted in a dramatic expansion of data sets and knowledge about the biosocial characteristics of our order (see Campbell, Fuentes, MacKinnon, Stumpf, et al., 2011). In particular, complexity in behavior and grouping patterns has been observed within primate species across a wide range of habitats (e.g., Fuentes, 1999; Kirkpatrick, 1998; Kirkpatrick, Long, Zhong, & Xiao, 1998; Morgan & Sanz,

2003; Pruetz & Lindshield, 2011; Rylands, 1993; Strier, 1994; Treves & Chapman, 1996). Ecological pressures, the social landscape, and other elements in an individual's life history elicit responses governed by the parameters set by physiology, environment, and experience. The nature of such parameters allows for a wide range of potential expression, leading to variable or flexible behavior in individuals. As data sets increase, this flexibility becomes more visible at the level of the group, and it becomes increasingly evident that social organization is an emergent property (Allen & Starr, 1982; Fuentes, 2011) that has characteristics not readily reducible to the context-specific interactions between individual animals that produce it. We will examine this in the next section.

Advances in Evolutionary Theory and Frameworks: Niche Construction and Social Network Analysis

While the days of viewing evolutionary processes in a strictly adaptationist (and static) light seem to be fading, it is unfortunate to see that many of the exciting developments in modern, cutting-edge evolutionary biology not only do not trickle down to the common layperson's ideas about how evolution works, but also have not seemed to trickle across to other academic disciplines, and least of all to university textbooks. Current thinking now involves theoretical and methodological advancements in areas such as developmental systems theory, evolutionary developmental biology ("evo-devo"), multilevel selection, and niche construction, to name but a few. We will highlight niche construction here, as an example of a useful theoretical framework for contextualizing and explaining the evolution of social and cognitive complexity across our order; we will then incorporate social network analysis (SNA) as a holistic methodology applicable to our questions of an emergent social complexity.

Niche Construction

Niche construction is defined as modification of the functional relationship between organisms and their environment by actively changing one of the factors in that environment. As Day, Laland, and Odling-Smee write, "... through niche construction organisms not only influence the nature of their world, but also in part determine the selection pressures to which they and their descendants are exposed, and they do so in a non-random manner" (2003, p.80 ; and see Odling-Smee, Laland, & Feldman, 2003). In other words, organisms have significant effects on their environment, which can then affect those local populations. For example, the hypothesized responses to the energetic costs of an increasing brain size and an extended period of child rearing in the genus *Homo* ~1.5 million years ago included more cooperative interactions between group members, an associated increase in communicative complexity, increased effectiveness at avoiding predation (possibly with tools and multi-individual coordination of anti-predator behavior; see Hart & Sussman, 2008, and an expansion of the types and patterns of habitat exploited (Fuentes, 2009a; Fuentes, Wyczalkowski, & MacKinnon, 2010; Wells & Stock, 2007). As *Homo* became more costly for predators due to the above responses, predators shifted emphasis to more accessible prey, reducing the overall selective pressure of predation on the genus *Homo* (Fuentes et al., 2010). Given this, *Homo* experienced an increased opportunity for social interactions, range exploration, and testing of a variety of novel foraging opportunities, all of which demanded—and fed back into—an emerging higher cognitive functioning (see Hermann, Call, Hernandez-Lloreda, Hare et al., 2007).

At the populational level, we see that with an increase in cooperative interactions between members of *Homo* groups (and potentially between groups in local areas), foraging efficiency, predator avoidance, and care for offspring expand in effectiveness (as a form of positive feedback) facilitating the observed range and habitat expansion—as well as cultural expansion—in the period 1.8–1.0 million years ago. Therefore, human niche construction began to make use of the components of tool use, increased infant survivorship and better group health, and expanded information transference via more-complex communication patterns—all of which were tied to an evolving hominin cognition, which enabled success in myriad environments (see Laland, Odling-Smee, & Feldman, 2001). Such behavioral plasticity can be a prominent result of successful adaptation (West-Eberhard, 2003), and the positive feedback in populations that construct their niches can best be modeled in a nonlinear fashion (e.g., Fuentes, 2009b; Fuentes et al., 2010; Potts, 2004; and Tomasello, 1999).

For nonhuman primate examples, we can explore many aspects of social life in gregarious species living in large, stable social groups. For example, macaque species' dominance patterns range from strict linear ("despotic") systems to more relaxed ("egalitarian") ones; across species generally, while females rely heavily on alliances with female relatives (matrilines) to gain access to resources, males have relationships that can change rapidly, and there is a wide range of behavioral variation among individuals (Thierry, 2011). Macaques are a male-dispersing primate, so they cannot depend on kin to help them in conflicts—they must form associations with other males and females in order to resolve agonistic interactions and dominance disputes (Bercovitch & Huffman, 1999). Following Flack, Girvan, de Waal, and Krakauer (2006), there is the potential for social niche construction as individuals in macaque groups manipulate their social networks, altering selective landscapes (e.g., for adult males, associations with larger matrifocal groups might offer increased mating opportunities and a greater overall level of social interaction, which could lower stress from male-male competition; MacKinnon & Fuentes, 2011). Males also display high levels of prosocial behaviors with other males (grooming, playing); reciprocal altruism, mutualism, or a benefit of access to more social partners—or all three—might explain such interactions.

Across the *Cebus* species, males and females in the same social group are generally affiliative, with little physical aggression occurring between the sexes. The alpha female usually ranks directly below the alpha male, but can dominate over all other males; in several species, female coalitions form and can displace the alpha male in context-specific circumstances (Fragaszy et al., 2004). Kinship is an important factor in female-female relationships, although may not be as important a factor as in cercopithecines. While the overall characteristics of female social relationships are consistent across groups and species, male capuchins show extensive variation in their relationships with each other—ranging from despotic to highly cooperative and affiliative (Jack, 2003; Jack & Fedigan, 2004a, 2004b). Such diversity reflects behavioral plasticity in these primates and may be based on shifting ecological (resource abundance/defense) and social (kinship/familiarity) pressures (Janson, 1986).

As with macaques, interactions based on biological kin, as well as non-kin, intrasexual and intersexual coalition formation and competition are evident components of capuchin social systems (Perry, 1996, 1997, 1998). Because males typically transfer out, social relationships need to be reestablished within their new groups, however many times they switch groups in their lifetimes. Yet there is evidence that males may also exhibit extended networks of relationships among neighboring groups (see Jack & Fedigan, 2004a, 2004b). Female capuchins who stay in their birth group—and live upward of 20 to 30 years in the wild—have a dense constellation of intragroup social networks to keep track of and maintain over the course of their lifetimes. Each relationship has its own particular history of positive and negative reinforcements, possible kinship and relatedness factors, and ongoing dynamics. Thus, a form of social niche construction (see Flack et al., 2006) is likely in this genus, given their longevity and complex social landscapes.

Common chimpanzee (*Pan troglyodytes*) communities are characterized by a *fission-fusion* social organization, with individuals spending time in variable subgroups. These subgroups have different compositions across time and space, and so are a fluid aspect of chimpanzee sociality. Unlike macaques and capuchins, chimpanzees have female dispersal and male philopatry (although there is some female philopatry in at least a few eastern chimpanzee populations; Stumpf, 2011). Males are, on average, dominant over females, but both compete with others of their sex and establish hierarchical ranks (Stumpf, 2011). As in macaque societies, males can gain a high rank by forming coalitions and alliances with other males (often, but not always, of a similar age), and by using intimidating displays and other tactics (e.g., fighting), to manipulate other members of their community into ceding access to favored resources. Male competition for rank can result in serious injuries and, occasionally, death. Some males attain rank through extreme aggression and maintain that behavior once they are high ranking. Others appear to rely heavily on coalition partners, mutual grooming, and social bonding to achieve and maintain dominance status. For females, dominance is associated with substantially improved access to food sources and high infant survivorship (especially in East African populations). High-ranking females tend to have a large number of offspring in the group, and occasionally mother-daughter pairs act together to achieve or maintain high rank. Females do achieve dominance via aggressive displays and occasionally fighting, but they do not do so as frequently or intensely as males. In east African *P. troglodytes*, both males and females who are high ranking gain access to favored food sources and social partners and therefore have increased reproductive success. In West African forms, rank does not always result in increased reproductive success (Boesch, Hohmann, & Marchant, 2002; Stumpf, 2011).

These three examples of vastly different primate genera illustrate how social complexity as a niche-constructing mechanism is characterized by dynamic, fluid behavioral exchanges that are fundamental to primate sociality and group living. A highly evolved social cognition is required to monitor and remember complicated social networks, roles, rank, and histories.

As primates, we share an extended period of infant development and brain maturation, which allows for the acquisition of species-appropriate skill sets and knowledge. However, humans are unique in terms of the sheer amount of information that is spread via social mechanisms through space *and* time (e.g., spoken or written language characterized by abstraction and symbolism). In the relatively brief period of time in which the genus *Homo* has evolved, we have fundamentally altered and manipulated many environments, which has influenced our survival in terms of predator avoidance, dietary exploitation, cultural complexity, and increased infant survivorship and overall group health. The social-biological ecologies of human populations are modified by social behavior, which is in turn affected by the pressures of those same social-biological environments.

Integrated and holistic theories such as niche construction force us to think of sociality in a new light—not as an independent category, but as an interrelated aspect of a generated niche. It highlights how social living encompasses the cognitive developmental environment; social resources and competition for those resources; alteration of the selective landscape (while being continuous with the natural environment); material culture in humans and some of the apes; and an extension of cognitive capacities into distributed systems of multiple individuals.

Social Network Analysis

The components of a social niche, as illustrated by Hinde (1976), include many levels: at the basis there are the interactions between and among individuals (agonistic, affiliative), which lead to patterns of relationships and networks (kin-based, male-female, mother-infant), which then result in forming the overall social group structure and population characteristics; all feed back on one another with varying effects. One way to examine the emergent properties of the interactions among many individuals in a social group, beyond dyadic exchanges, is through social network analysis (SNA). This methodology—with eighteenth-century roots in mathematical graph theory, and later adopted by the social sciences in the twentieth century (Newman, 2010; Scott, 2000; Sueur, Jacobs, Amblard, Petit et al., 2011)—allows an examination of how individual behavior influences what happens at larger scales: the group and population levels. While elements of the conceptual frameworks (e.g., relational aspects of social group structure) and techniques (e.g., sociograms, social role theory) of SNA have been used in primatological investigations since the early 1960s (e.g., Hinde, 1976; Bernstein, 1964; Kummer, 1968; Sade, 1972; Fedigan, 1972; Bernstein & Sharpe, 1966; Berman, 1982; and see Brent, Lehmann, & Ramos-Fernández, 2011, for a review), there has been an increase in the use of more powerful analytic tools within the past ten years to address questions concerning the social behavior of many animal taxa, including primates (e.g., Croft, James, & Krause, 2008; Wey, Blumstein, Shen, & Jordan, 2008; Whitehead, 2008; Flack et al., 2006; Ramos-Fernández, Boyer, Aureli, & Vick, 2009; Sih, Hanser, & McHugh, 2009; Sueur & Petit, 2008).

Tiddi, Aurelil, Schino, and Voelkl (2011) used SNA to examine social relationships in wild *Cebus apella nigritus* and found that adult females' interactions with the alpha male showed much variation, with only some females having strong associations with him: those with high dominance ranks and high centrality in both proximity and grooming networks. In white-faced capuchins (*C. capucinus*), Crofoot, Rubenstein, Maiya, and Berger-Wolf (2011) found that dominance relationships do not follow a simple hierarchy, and their data support the notion that relationships between pairs of individuals may be less important than how individuals are connected within the group as a whole. They also found that high-ranking females were no more likely than lower-ranking females to participate in aggressive territorial defense of a group, suggesting that protecting the group does not require centrality in any social network (but see Perry, 1996).

In a study of baboon (*Papio anubis*) social complexity, SNA was used to compare multiple networks created from different behaviors such as grooming, displacements, aggression, etc. (Lehmann & Ross, 2011). The authors found that individual network positions were strongly affected by sex, and also that individuals central in one network tended to be central in other networks. However, individuals' network positions varied, depending on the behavior analyzed. They described the baboons' social environment as one best defined as a "multiplex network based on affiliative, aggressive, and sexual behavior" (Lehmann & Ross, 2011, p. 775). For desert baboons (*Papio ursinus*), King, Clark, & Cowlishaw (2011) used SNA to examine over 5,000 foraging events in a single troop and found that co-feeding was significantly correlated with grooming relationships but not genetic relatedness. Dominant individuals were central in the co-feeding network and frequently shared food patches with multiple troop mates.

Such group-level measures of social relationships help us to better understand the complex patterns of networks that individuals operate within, and highlight the social niche as one constructed from many linked levels of social organization. The socially constructed niche therefore feeds back into components of relationships (affiliative, agonistic, sexual) and thus affects overall health and fitness at the individual, group, and populational levels.

Cognition and Sociality: Brain-Culture Interaction and Behavioral Variation within the Social Context

The neocortex is necessary for many cognitive functions but sufficient for none. It is therefore misleading to view the neocortex as the "cognitive" part of the brain. (Barton, 2006, p. 231)

As mentioned earlier, primates display extended life-history variables relative to size, compared to other mammals (see Leigh & Blomquist, 2011; Sherrow & MacKinnon, 2011), and all species have a brain that is large in proportion to total body weight, with a particularly convoluted neocortex region (Le Gros Clark, 1971; Rilling & Insel, 1999). However, there is much variation in brain size within the order (~1.8 g mouse lemur to ~1,300 g modern human) and the significance of the cognitive implications of such variation remains poorly understood (Barton, 2006). While the neocortex is disproportionally expanded in primates compared to (some) mammals, and anthropoids have larger neocortices than prosimians (Barton & Harvey, 2000), this fact alone tells us little about cognitive ability. Complex cognitive processes are mediated by networks that link the neocortex with other structures—the functional systems of the brain cut across all major subdivisions (thalamus, cerebellum, medulla, neocortex; Barton, 2006). However, there are three primate genera in particular that are recognized to be the most "highly encephalized" given average body size—*Homo, Pan, Cebus*—and they have been the subject of many cognitive studies and explorations of evolutionary convergence (e.g., de Waal & Ferrari, 2010; de Waal, 2003, 2005).

Brain Growth during Development

An extended period of dependency and neurosocial development sets immature primates apart from other young mammals (Harvey & Clutton-Brock, 1985). Primates that live in social groups have an intricate web of social relationships to remember. Each relationship has its own particular history of ongoing social dynamics, including positive and negative reinforcements, possible kinship and familiarity factors, and differential reciprocity (see Dolhinow, 1994). An ability to simultaneously navigate through and manipulate such a complex landscape has become our adaptive advantage, and is the result of feedback systems between neurosocial networks and social and biotic environments woven together in our phylogenetic histories and ongoing evolution.

In primates, it is not just brain volume, but specifically neocortex volume, that correlates with increased sociality, previously defined as the size of the social group (Dunbar, 1992, 1998, 2002). We now recognize that it is not the mere *size* of a primate social group, but its level of social complexity that is correlated to larger neocortices in certain primate taxa (Dunbar & Shultz, 2007). Such complexities can be measured via social play, deception, coalition formations, altruistic acts, and other subtle social strategies (e.g., Burkart, Hrdy, & van Schaik, 2009; Byrne & Corp, 2004; Lewis, 2000; Kudo & Dunbar, 2001). Thus, when we examine brain development and the formation of the cerebral cortex, we see that expansion in phylogeny is paralleled by the emergence of more complex (social) functions.

Leigh (2004) found that primates are characterized by significant variation in patterns of brain growth, which generally fall along two lines of metabolic adaptations: 1) brain growth occurring during the prenatal period with heavy maternal investment (e.g., Old World monkeys), and 2) brain growth occurring during the postnatal period, with more investment from non-mother group members and the young primates themselves (e.g., New World monkeys and apes). Walker et al. (2006) have suggested that natural selection has decreased human growth rates so that there is more time for increased cognitive development (with lower "body-maintenance costs").

Primates are unique among mammals for the extended postnatal period in which myelination of pathways and neural circuitry formation take place and are then subsequently modified, particularly in the cerebral cortex. Complex "association systems" responsible for organizing intra-cortical circuits integrate information across functional domains from before birth—as in all mammals—yet continue to grow and reorganize into puberty (see Levitt, 2003, for review). After an initial organization of brain areas during the prenatal growth period, there is structural (and molecular) remodeling that occurs just prior to birth, which then continues into the neonatal, childhood, and pre- and even post-puberty stages (Levitt, 2003; Woo, Pucak, Kye, Matus et al., 1997, and see Gross, 2000, for a review of neurogenesis). Even complex functions such as emotional recognition can be greatly affected by early experience (Pollak & Kistler, 2002).

The three genera of extant primates that undergo the most rapid neurological change during the first few years of life are also the ones showing the highest degree of encephalization: *Homo*, *Pan*, and *Cebus*. All display a smaller proportion of the adult brain weight at birth compared to the rest of the primate taxa, and thus a period of extensive postnatal growth (see

Leigh, 2004). For example, postnatal brain growth in the chimpanzee (*Pan*) accounts for 65% to 75% of adult size (Vinicius, 2005), and the capuchin (*Cebus*) monkey brain at birth is 29 to 34 grams, or about 50% of the adult brain weight (Elias, 1977; Hartwig, 1996; Martin, 1983; Vinicius, 2005, and see Phillips & Sherwood, 2008). However, this growth is most dramatic in humans, with our total brain size increasing 100% in the first year of life and growing exponentially into adolescence (Gilmore, Lin, Corouge, Vetsa, et al., 2007; Giedd, 2004). Thus, among primates, humans have the least mature brain at birth, followed by a longer period of early rapid brain growth, taking place in an environment rich in social stimuli. For primates in general and humans specifically, the extended development of neurological networks appears to be a facet of the complex interaction of the individual with its social environment. This feedback system (social networks, extended development, social and ecological experience) simultaneously shapes the acquisition of behavioral patterns and the functional development of the brain. This, in turn, plays out in the lives of primates as they encounter, interact with, and potentially reconstruct their surroundings (social and structural).

In a recent study that is the first to follow the development of the chimpanzee brain and compare it to the human brain, researchers used MRI to examine three chimpanzees' brains from ages 6 months to 6 years (Sakai, Mikami, Tomonaga, Matsui, et al., 2011). Their results demonstrate that both infant humans and infant chimpanzees start out with an undeveloped prefrontal cortex, an area known to play an important role in cognitive functions like self-awareness and decision-making. This delay may provide both species an extended period to develop the knowledge and skills shaped by life experiences that are crucial for complex social interactions. Yet, despite this similarity as infants, the white matter in the prefrontal cortex of chimpanzees does not grow as rapidly as it does in humans, and this may contribute to our differing communication skills (language) and higher levels of intense social interaction (Sakai et al., 2011).

Mothers, Infants, and Social Learning

The formation and maintenance of the mother-infant relationship in mammals, and the mediation of these experiences by the brain and its neurochemistry, signals the importance of learning in a social context (Fleming, Tooth, Hassell, & Chan, 1999). This is especially marked in primates, as the neural development sequence is more complex and needs such a prolonged period of time (Clancy, Darlington, & Finlay, 2000, 2001). Parallels between the sensory, endocrine, and neural mechanisms of both mother and infant responses that underlie the early mother-infant relationship demonstrate an active, dynamic interplay built on feedback systems (and see Hrdy, 2009, McKenna, 1983).

In a comparison between chimpanzees (*Pan*) and capuchins (*Cebus*), Fragaszy and Bard (1997) found that both genera spend a similar proportion of their lives in a weaned but reproductively immature state. However, chimpanzees spend much more of their lives as nursing infants, and have far fewer offspring over their lifetimes. Among the Hominidae, humans have a longer period of infant dependency and are weaned far earlier than any of the great apes: bonobos and common chimpanzees wean around 5 years of age (Hiraiwa-Hasegawa, 1989; Kuroda, 1980; Pusey, 1983; Stumpf, 2011); mountain gorillas around 3–4 years (Robbins, Robbins, Gerald-Steklis, & Steklis, 2006; Robbins, 2011); and orangutans 6–7 years (Galdikas, 1981; Knott & Kahlenberg, 2011; van Noordwijk & van Schaik, 2005). Humans, on average, wean around 2.5 years (Kennedy, 2005). Kennedy (2005) argues that early weaning, when accompanied by more nutritious adult foods, is vital to the ontogeny of our larger brain; the human child's intellectual development, rather than its survival per se, is the primary focus of selection. The increased consumption of animal protein (facilitated by stone tool use) beginning around 2.6 million years ago propelled this phenomenon forward (Stanford & Bunn, 2001; Ungar, 2007; and see Antón, 2003, and Wrangham, 2009, for later *Homo*), but also brought members of the subtribe Hominina into closer competition with (and danger from) large powerful carnivores (Hart & Sussman, 2008). As we know from the fossil record, there was an increased mortality among young adults by predation during this time period in our evolution, and this likely necessitated myriad social responses via feedback loops, such as alloparenting and increased cooperation among and between social groups (Fuentes et al., 2010; Hawkes et al., 2003; Hrdy, 2009; Kennedy, 2005; and see Wall-Scheffler, Geiger, & Steudel-Numbers, 2007, for costs of infant carrying for a biped). With the longest periods of dependency and socially mediated learning among mammals (Deputte, 2000; Harvey & Clutton-Brock, 1985; Harvey, Martin, & Clutton-Brock, 1987), the attachments that young primates form with their mothers and members of the group are of primary importance for successful ontogeny (Hawkes et al., 2003; Hrdy, 2009; and see MacKinnon, 2011 for review).

All primates are shaped by the interplay between our physiology (i.e., central nervous system, endocrine system) and the social and biological environments in which we live. In particular, information crucial to our survival is disseminated through social networks, and those networks influence how we experience the world. For example, we know that the interactions between social effects (e.g., perception of status, dominance rank, etc) and physiology (e.g., endocrine and cardiovascular systems, the central nervous system; learning and memory) are well documented and are dynamic, experience-influenced, malleable systems (e.g., Sapolsky, 2003, 2005). We even see that what is assumed to be "species-typical" in terms of social structure or social characteristics can change in a very short period of time. Take the case of a highly despotic olive baboon (*Papio anubis*) society in Kenya, studied since 1978, that underwent a sort of extreme "social bottleneck" event when TB struck the troop in the mid-80s and the most aggressive and dominant males succumbed to the outbreak (for nonrandom behavioral reasons; see Sapolsky & Share, 2004). The cohort of unaggressive subordinate males that were left became part of the new founding population, and the troop has displayed a more egalitarian set of social behaviors to this day. New males transferring into this group quickly adopt the social norms displayed by its members (e.g., higher grooming rates, increased affiliation with females, a relaxed dominance hierarchy, and lower cortisol levels among low-ranking males), all of which are counter to what is typical for this type of baboon.

Thus, as primates we share a plasticity in behavioral and physiological responses, given ever-changing social landscapes: we can modify the affects of psychosocial stressors by the way our central nervous system interacts with and interprets extrasomatic events (see Sapolsky, 2003); we also see that a stable social setting facilitates transmission of group knowledge and behavioral norms (e.g., van Schaik, Deaner, & Merrill, 1999; Perry et al., 2003a, 2003b; Kummer, 1971; see Santorelli, Schaffner, Campbell, Notman et al., 2011, for community identity in a fission-fusion species, and see Caldwell & Whiten, 2011, for a review of social learning), a fact that is essential to the ongoing discussion of the evolution of human sociality and cognition.

The relationship between cognition and sociality appears across a wide array of topics such as cooperation, altruism, and social traditions (culture). However, attempts at demonstrating cooperative or even altruistic behavior in chimpanzees has often resulted in contradictory data from lab versus field studies: chimpanzees in lab experiments sometimes fail to show a willingness to help others, whereas chimpanzees observed in the wild routinely display such behaviors in the contexts of food-sharing, coalition formation during conflicts, helping to raise offspring, etc. Research design may be at fault, with earlier lab experiments being too complex for the chimps to really understand. A recent study (Horner, Carter, Suchak, de Waal, et al., 2011) suggests that chimpanzees may give help proactively simply because they understand others' need for assistance. Seven captive female chimps (trained to hand over colored tokens in exchange for food) were paired up but able to see each other in adjoining cages, and given tokens in two colors. If a chimp gave one color, she received some food while her partner got nothing. If she gave the other color, both got a food treat. Each female picked the color that gave them both food for a higher percentage of the trials (66.7%), and they showed no preference to help related individuals. However, Hamann, Warneken, Greenberg, and Tomasello (2011) found that their captive chimpanzees do not simply give up food so others can have it, but rather are prosocial when it is not too costly. By contrast, chimpanzees have been shown to readily help a human obtain a goal object even in costly situations (Warneken & Tomasello, 2006; Warneken, Hare, Melis, Hanus, et al., 2007). Mixed results in the captive chimpanzee literature point to the importance of context and the peculiarities of individuals, as well as the variance in research design (see Barnes et al., 2008).

Finally, capuchins appear to be able to sense fairness in reward or trade scenarios, and value equitable behavior in cooperative situations over rewards in certain tasks (Brosnan & de Waal, 2003, 2004a, 2004b; Brosnan, Freeman, & de Waal, 2006). They are also quite intent on grooming sick or injured individuals of varying rank, and are distressed when group members die (KCM, personal observation). Thus, they may be astute at empathizing with others' emotional states in varying social contexts (and see de Waal, 2008; de Waal, Leimgruber, & Greenberg, 2008; Preston & de Waal, 2002).

Discussion

Primates are characterized by a specific type of social intelligence (Dunbar & Shultz, 2007) such that "distinctive aspects of primate cognition evolved mainly in response to the especially challenging demands of a complex social life of constant competition and cooperation with others in the social group" (Hermann et al., 2007; Silk, 2007; and see Sussman & Garber, 2011), in addition to ecological pressures. There is a ratcheting up of this social complexity in anthropoids, which is increased in hominoids and exponentially enhanced in hominins (see figure 3.1).

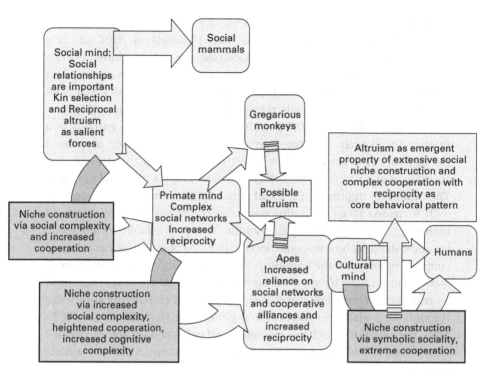

Figure 3.1
Source: MacKinnon and Fuentes (2011, p. 127).

Looking back through our evolutionary history, it appears that the basal sociality of complex gregarious mammals was expanded upon in primates, with primates then using their social networks, contexts, and enhanced cognition as tools to meet and modify the demands of the environment (the selective landscape). As local environments are modified, pressures may alter, changing the selective landscapes for those primate populations. Increased cognitive complexity in the hominoids facilitates a more rapid— or more intensive—utilization of social bonds and relationships as tools to meet ecological challenges. In some hominoid lineages, heightened levels of cooperation and reciprocity became core components of behavioral repertoires, constantly engaging the social and biotic ecologies, resulting in niche construction and concomitant shifting and modification of selective landscapes. Neurological complexity emerges both under direct selection and as a byproduct of the physiological and behavioral adaptations required to effectively negotiate social networks in which coalitions, multiparty social negotiations, and reciprocity are the primary avenues for social and reproductive success. Thus, social complexity *itself* acts as a niche-constructing mechanism to facilitate the intersections between individuals within a group, individuals with their environment, and conspecific groups within that local population. This feedback and feed-forward process became a central trend in hominin (human) evolution.

Dunbar and Schultz (2007) and Silk (2007) argue for the centrality of a primate social brain hypothesis in which social intelligence becomes a primary tool of adaptive success. We agree, and suggest that this be conceptualized as a process of neurological remodeling via the social interface. The expansion in primacy for social networks in the anthropoid primates, with increased focus on coalitionary and cooperative relationships and increasing reciprocity among the hominoids (apes), establishes a baseline for understanding how we could develop the level of cognitive-neurological complexity we see in the hominins. The "cultural intelligence" hypothesis argues that humans have a *species-specific* set of social-cognitive skills (that emerge early in ontogeny) for participating in and exchanging knowledge through particularly complex cultural groups (see Hermann et al., 2007). Thus, the extent to which our biological and cultural traits are intertwined and embedded in our species' evolutionary history is unique among primate taxa.

Primate social networks and complex social cognition have been central as niche-constructing strategies through time (MacKinnon & Fuentes, 2011). As the intersections of our biologies and environments run deep, looking at comparative examples from the order Primates can result in a constructive framework for understanding an evolved social cognition in neuroanthropology. Sociality and cognitive functioning display extensive plasticity within our phylogeny, resulting in nonlinear feedback loops among increasing brain size, social complexity, and evolutionary "success" via biosocial niche construction. By using a comparative framework, we can better contextualize the neurosocial traits humans share with other primates, as well as begin to tease apart the many ways in which our genus and species reflect a suite of specific evolutionary discontinuities with the other primates.

Human behavioral, dietary, and social variation exceeds that of all other living animals. Patterns of pairbonding and mating, intergroup associations and transfer, communication networks, intra- and inter-group alliances and coalitions, social hierarchies, politico-economic systems, and all manner of social structures vary around the world. Humans are among the slowest-developing primates, with extensive social learning integral to a convoluted developmental sequence involving symbolic and linguistic instruction. Manufacturing of material artifacts and technologies, along with massive manipulation of the environment, is ubiquitous in all human groups; social and biological-medical niche construction and alteration are ongoing, and now more recently and regularly at an accelerated rate across local, regional and global levels.

Conclusion

Studies of the nonhumans primates (lab and field) are fundamental to examining the evolution of social complexity and cognitive functioning in human societies. It is precisely what we share socially and cognitively, not where we differ, that can inform neuroanthropology. The other primates also create spheres of sociality—niches—that form a developmental milieu, partially buffering them from direct ecological forces and shifting the way that ecological (and social) selective pressures affect individuals. This relationship can create new selective forces both within intraspecific populations as well as between species in larger communities (e.g., modifying rates of predation, variable use of scattered resources). Thus, other primates have more general analogs to the social and cultural dimensions of human life, above and beyond the suite of evolved physical traits we share.

The current theories in evolutionary biology extend far beyond an outdated application of strict adaptationism with natural selection as the only mechanism. Sociality, at its core, is not just composed of altruistic acts based on shared kinship. Rather, sociality functions to meet the biological and developmental needs of offspring, shifts the way species-specific traits are inherited, and changes the selective equation by modifying the environment(s) in which groups of species live. As Fuentes (2009b, p. 17) illustrates, "humans can extend their net of caring, investing and bonding more widely than genetic kin, even beyond our species. If we remove the exclusivity of neo-Darwinian views of evolution, add the ideas of [developmental systems theory], niche construction, and social and symbolic inheritance and place them in the context of ethnographic knowledge, archaeological histories, contingency in human behavior and individual agency, we can derive better anthropological answers." Neuroanthropology contributes to this perspective in that a variety of lenses through which to examine human evolution is preferable to historically more narrow interpretations. This *constructivist evolutionary approach* (Fuentes, 2009b; Schultz, 2009), which seeks to incorporate modern evolutionary theory with anthropological methods, includes not only adapted (and phylogenetic) aspects of behavior and ecologies, but also those which may or may not have evolutionary impacts.

References

Adolphs, R. (2001). The neurobiology of social cognition. *Current Opinion in Neurobiology, 11*, 231–239.

Aiello, L. C., & Wells, J. C. K. (2002). Energetics and the evolution of the genus Homo. *Annual Review of Anthropology, 31*, 323–338.

Aiello, L. C., & Wheeler, P. (1995). The expensive-tissue hypothesis: The brain and the digestive system in human and primate evolution. *Current Anthropology, 36*(2), 199–221.

Allen, T. F. H., & Starr, T. B. (1982). *Hierarchy: Perspectives for ecological complexity.* Chicago: University of Chicago Press.

Anderson, E., Siegel, E. H., Bliss-Moreau, E., & Barrett, L. F. (2011). The visual impact of gossip. *Science, 332*(6036), 1446–1448.

Antón, S. C. (2003). A natural history of Homo erectus. *Yearbook of Physical Anthropology, 46*, 126–170.

Barkow, J., Cosmides, L., & Tooby, J. (Eds.). (1992). *The adapted mind: Evolutionary psychology and the generation of culture.* New York: Oxford University Press.

Barnes, J. L., Hill, T., Langer, M., Martinez, M., & Santos, L. R. (2008). Helping behaviour and regard for others in capuchin monkeys (*Cebus apella*). *Biology Letters 4*(6), 638–640.

Barton, R. A. (1998). Visual specialisation and brain evolution in primates. *Proceedings of the Royal Society B: Biological Sciences, 265*, 1933–1937.

Barton, R. A. (2006). Primate brain evolution: Integrating comparative, neurophysiological, and ethological data. *Evolutionary Anthropology, 15*, 224–236.

Barton, R. A., & Harvey, P. H. (2000). Mosaic evolution of brain structure in mammals. *Nature, 405*, 1055–1058.

Bateson, P., & Gluckman, P. (2011). *Plasticity, robustness, development and evolution.* Cambridge: Cambridge University Press.

Bearder, S. K. (1987). Lorises, bushbabies, and tarsiers: Diverse societies in solitary foragers. In B. B. Smuts, D. L. Cheney, R. M. Seyfarth, R. W. Wrangham, et al. (Eds.), *Primate societies* (pp. 11–24). Chicago: University of Chicago Press.

Bearder, S. K. (1999). Physical and social diversity among nocturnal primates: A new view based on long term research. *Primates, 40*(1), 267–282.

Bercovitch, F. B., & Huffman, M. A. (1999). The macaques. In P. Dolhinow & A. Fuentes (Eds.), *The nonhuman primates* (pp. 77–85). Mountain View, CA: Mayfield Publishing.

Berman, C. M. (1982). The ontogeny of social relationships with group companions among free-ranging infant rhesus monkeys I. Social networks and differentiation. *Animal Behaviour, 30,* 149–162.

Bernstein, I. S. (1964). Role of the dominant male rhesus monkey in response to external challenges to the group. *Journal of Comparative and Physiological Psychology, 57,* 404–406.

Bernstein, I. S., & Sharpe, L. G. (1966). Social roles in a rhesus monkey group. *Behaviour, 26,* 91–104.

Boesch, C., Hohmann, G., & Marchant, L. (Eds.). (2002). *Behavioral diversity in chimpanzees and bonobos.* New York: Cambridge University Press.

Brent, L. J. N., Lehmann, J., & Ramos-Fernández, G. (2011). Social network analysis in the study of nonhuman primates: A historical perspective. *American Journal of Primatology, 73*(8), 720–730.

Brosnan, S. F., & de Waal, F. B. M. (2003). Monkeys reject unequal pay. *Nature, 425*(6955), 297–299.

Brosnan, S. F., & de Waal, F. B. M. (2004a). A concept of value during experimental exchange in brown capuchin monkeys, Cebus apella. *Folia Primatologica, 75*(5), 317–330.

Brosnan, S. F., & de Waal, F. B. M. (2004b). Socially learned preferences for differentially rewarded tokens in the brown capuchin monkey (Cebus apella). *Journal of Comparative Psychology, 118*(2), 133–139.

Brosnan, S. F., Freeman, C., & de Waal, F. B. M. (2006). Partner's behavior, not reward distribution, determines success in an unequal cooperative task in capuchin monkeys. *American Journal of Primatology, 68*(7), 713–724.

Brothers, L. (1990). The social brain: A project for integrating primate behavior and neurophysiology in a new domain. *Concepts in Neuroscience, 1,* 27–51.

Burkart, J., Hrdy, S. B., & van Schaik, C. P. (2009). Cooperative breeding and human cognitive evolution. *Evolutionary Anthropology, 18,* 175–186.

Byrne, R. W., & Corp, N. (2004). Neocortex size predicts deception rate in primates. *Proceedings of the Royal Society B: Biological Sciences, 271,* 1693–1699.

Caldwell, C. A., & Whiten, A. (2011). Culture and cognition. In C. J. Campbell, A. Fuentes, K. C. MacKinnon, S. K. Bearder, & R. M. Stumpf (Eds.), *Primates in perspective* (2nd ed., pp. 652–662). New York: Oxford University Press.

Campbell, C. J., Fuentes, A., MacKinnon, K. C., Stumpf, R., & Bearder, S. K. (Eds.). (2011). *Primates in perspective* (2nd ed.). New York: Oxford University Press.

Caporael, L. R. (2001). Evolutionary psychology: Toward a unifying theory and a hybrid science. *Annual Review of Psychology, 52*, 607–628.

Cartwright, J. (2000). *Evolution and human behavior: Darwinian perspectives on human nature.* Cambridge, MA: MIT Press.

Chapman, C. A. (1987). Flexibility in diets of three species of Costa Rican primates. *Folia Primatologica, 49*, 90–105.

Chapman, C. A. (1988). Patterns of foraging and range use by three species of Neotropical primates. *Primates, 29*(2), 177–194.

Chapman, C. A., & Chapman, L. J. (1990). Dietary variability in primate populations. *Primates, 31*(1), 121–128.

Chapman, C., & Fedigan, L. M. (1990). Dietary differences between neighboring Cebus capucinus groups: Local traditions, food availability or response to food profitability? *Folia Primatologica, 54*, 177–186.

Chapman, C., & Russo, S. (2011). Primate seed dispersal. In C. J. Campbell, A. Fuentes, K. C. MacKinnon, S. K. Bearder, & R. M. Stumpf (Eds.), *Primates in perspective* (2nd ed., pp. 510–525. New York: Oxford University Press .

Chisholm, J. S. (1999). *Death, hope, and sex: Steps to an evolutionary ecology of mind and morality.* Cambridge: Cambridge University Press.

Clancy, B., Darlington, R. B., & Finlay, B. L. (2000). The course of human events: Predicting the timing of primate neural development. *Developmental Science, 3*, 57–66.

Clancy, B., Darlington, R. B., & Finlay, B. L. (2001). Translating developmental time across mammalian species. *Neuroscience, 105*(1), 7–17.

Crofoot, M. C., Rubenstein, D. I., Maiya, A. S., & Berger-Wolf, T. Y. (2011). Aggression, grooming and group-level cooperation in white-faced capuchins (I): Insights from social networks. *American Journal of Primatology, 73*(8), 821–833.

Croft, D. P., James, R., & Krause, J. (2008). *Exploring animal social networks.* Princeton, NJ: Princeton University Press.

Daly, M., & Wilson, M. (1988). *Homicide.* Hawthorne, NY: Aldine de Gruyter.

Daly, M., & Wilson, M. (1999). Human evolutionary psychology and animal behaviour. *Animal Behaviour, 57*, 509–519.

Day, R. L., Laland, K. N., & Odling-Smee, J. (2003). Rethinking adaptation: The niche-construction perspective. *Perspectives in Biology and Medicine, 46*(1), 80–95.

De Waal, F. B. M. (2003). Social roles, alternative strategies, personalities, and other sources of individual variation in monkeys and apes. *Journal of Research in Personality*, *36*, 541–542.

De Waal, F. B. M. (2005). A century of getting to know the chimpanzee. *Nature, 437*, 56–59.

De Waal, F. B. M. (2008). Putting the altruism back in altruism: The evolution of empathy. *Annual Review of Psychology, 59*, 279–300.

De Waal, F. B. M., & Bonnie, K. E. (2009). In tune with others: The social side of primate culture. In K. Laland & G. Galef (Eds.), *The question of animal culture* (pp. 19–39). Cambridge, MA: Harvard University Press.

De Waal, F. B. M., & Ferrari, P. F. (2010). Towards a bottom-up perspective on animal and human cognition. *Trends in Cognitive Sciences, 14*, 201–207.

De Waal, F. B. M., Leimgruber, K., & Greenberg, A. (2008). Giving is self-rewarding for monkeys. *Proceedings of the National Academy of Sciences of the United States of America, 105*, 13685–13689.

Deputte, B. L. (2000). Primate socialization revisited: Theoretical and practical issues in social ontogeny. In P. J. B. Slater, J. S. Rosenblatt, C. T. Snowdon, & T. J. Roper (Eds.), *Advances in the study of behavior* (Vol. 29, pp. 99–157). New York: Academic Press.

Di Fiore, A., Fink, A., & Campbell, C. J. (2011). The atelines. In C. J. Campbell, A. Fuentes, K. C. MacKinnon, R. Stumpf, & S. K. Bearder (Eds.), *Primates in perspective* (2nd ed., pp. 155–188). New York: Oxford University Press.

Digby, L., Ferrari, S., & Saltzman, W. (2011). The callitrichines. In C. J. Campbell, A. Fuentes, K. C. MacKinnon, R. Stumpf, & S. K. Bearder (Eds.), *Primates in perspective* (2nd ed., pp. 91–107). New York: Oxford University Press.

Dindo, M., Whiten, A. and de Waal, F. B. M. (2009). In-group conformity sustains different foraging traditions in capuchin monkeys (Cebus apella). *PLoS One, 4*(11), e7858.

Dolhinow, P. (Ed.). (1972). *Primate patterns*. New York: Holt, Rinehart and Winston.

Dolhinow, P. (1994). Social systems and the individual. *Evolutionary Anthropology, 3*(3), 73–74.

Dunbar, R. I. M. (1992). Neocortex size as a constraint on group size in primates. *Journal of Human Evolution, 22*(6), 469–493.

Dunbar, R. I. M. (1998). The social brain hypothesis. *Evolutionary Anthropology, 6*, 178–190.

Dunbar, R. I. M. (2002). Modelling primate behavioural ecology. *International Journal of Primatology, 23*, 785–819.

Dunbar, R. I. M., & Shultz, S. (2007). Understanding primate brain evolution. *Philosophical Transactions of the Royal Society, B 362*(1480), 649–658.

Eagly, A. H., & Wood, W. (1999).The origins of sex differences in human behavior: Evolved dispositions versus social roles. *American Psychologist, 54*(6), 408–423.

Elias, M. F. (1977). Relative maturity of cebus and squirrel monkeys at birth and during infancy. *Developmental Psychobiology, 10*, 519–528.

Fedigan, L. M. (1972). Roles and activities of male geladas (Theropithecus gelada). *Behaviour, 41*, 82–90.

Flack, J. C., Girvan, M., de Waal, F. B. M., & Krakauer, D. C. (2006). Policing stabilizes construction of social niches in primates. *Nature, 439*, 426–429.

Fleming, J., Tooth, L., Hassell, M., & Chan, W. (1999). Prediction of community integration and vocational outcome 2–5 years after traumatic brain injury rehabilitation in Australia. *Brain Injury, 13*, 417–431.

Fragaszy, D. M., & Bard, K. (1997). Comparison of life history in Pan and Cebus. *International Journal of Primatology, 18*, 683–701.

Fragaszy, D., Visalberghi, E., & Fedigan, L. M. (2004). *The complete capuchin monkey: The biology of the genus* Cebus. Cambridge: Cambridge University Press.

Fuentes, A. (1999). Variable social organizations: What can looking at primate groups tell us about the evolution of plasticity in primate societies? In P. Dolhinow & A. Fuentes (Eds.), *The nonhuman primates* (pp. 183–188). Mountain View, CA: Mayfield Publishing.

Fuentes, A. (2009a). *Evolution of human behavior.* New York: Oxford University Press.

Fuentes, A. (2009b). Re-situating anthropological approaches to the evolution of human behavior. *Anthropology Today, 25*(3), 12–17.

Fuentes, A. (2011). Social systems and socioecology: Understanding the evolution of primate behavior. In C. J. Campbell, A. Fuentes, K. C. MacKinnon, S. K. Bearder, & R. M. Stumpf (Eds.), *Primates in perspective* (pp. 500–511). New York: Oxford University Press.

Fuentes, A., Wyczalkowski, M., & MacKinnon, K. C. (2010). Niche construction through cooperation: A nonlinear dynamics contribution to modeling resilience and evolutionary history in the genus Homo. *Current Anthropology, 51*(3), 435–444.

Garber, P. A. (2011). Primate locomotor behavior and ecology. In C. J. Campbell, A. Fuentes, K. C. MacKinnon, R. Stumpf, & S. K. Bearder (Eds.), *Primates in perspective* (2nd ed., pp. 548–562). New York: Oxford University Press.

Garber, P. A., & Brown, E. (2006). Use of landmark cues to locate feeding sites in wild capuchin monkeys (*Cebus capucninus*): an experimental field study. In A. Estrada, P. A. Garber, M. Pavelka, & L. Luecke (Eds.), *New perspectives in the study of mesoamerican primates: Distribution, ecology, behavior and conservation* (pp. 311–332). New York: Kluwer.

Garber, P. A., & Jelinek, P. E. (2006). Travel patterns and spatial mapping in Nicaraguan mantled howler monkeys (*Alouatta palliata*). In A. Estrada, P. A. Garber, M. Pavelka, & L. Luecke (Eds.), *New perspectives in the study of mesoamerican primates: Distribution, ecology, behavior and conservation* (pp. 287–309). New York: Kluwer.

Galdikas, B. M. F. (1981). Orangutan reproduction in the wild. In C. E. Graham (Ed.), *Reproductive biology of the great apes* (pp. 281–300). New York: Academic Press.

Giedd, J. N. (2004). Structural magnetic resonance imaging of the adolescent brain. *Annals of the New York Academy of Sciences, 1021*(1), 77–85.

Gilmore, J. H., Lin, W., Corouge, I., Vetsa, Y. S. K., Smith, J. K., Kang, C., et al. (2007). Early postnatal development of corpus callosum and corticospinal white matter assessed with quantitative tractography. *American Journal of Neuroradiology, 28,* 1789–1795.

Gould, L., & Sauther, M. (2011). Lemuriformes. In C. J. Campbell, A. Fuentes, K. C. MacKinnon, R. Stumpf, & S. K. Bearder (Eds.), *Primates in perspective* (2nd ed., pp. 55–78). New York: Oxford University Press.

Gouzoules, H., & Gouzoules, S. (2011). The conundrum of communication. In C. J. Campbell, A. Fuentes, K. C. MacKinnon, R. Stumpf, & S. K. Bearder (Eds.), *Primates in perspective* (2nd ed., pp. 626–636). New York: Oxford University Press.

Gross, C. G. (2000). Neurogenesis in the adult brain: Death of a dogma. *Nature, 1,* 67–73.

Gumert, M. D. (2011). The common monkey of Southeast Asia: Long-tailed macaque populations, ethnophoresy, and their occurrence in human environments. In M. D. Gumert, A. Fuentes, & L. Jones-Enegl (Eds.), *Monkeys on the edge: Ecology and management of long-tailed macaques and their interface with humans* (pp. 3–45). Cambridge: Cambridge University Press.

Hamann, K., Warneken, F., Greenberg, J. R., & Tomasello, M. (2011). Collaboration encourages equal sharing in children but not in chimpanzees. *Nature*.

Harlow, H. F. (1959). Love in infant monkeys. *Scientific American, 200*(6), 68–74.

Harlow, H. F., & Harlow, M. K. (1962). Social deprivation in monkeys. *Scientific American, 207*(5), 136–146.

Hart, D., & Sussman, R. W. (2008). *Man the hunted: Primates, predators, and human evolution* (Expanded ed.,). New York: Westview Press.

Hartwig, W. C. (1996). Perinatal life history traits in New World monkeys. *American Journal of Primatology, 40,* 99–130.

Harvey, P. H., & Clutton-Brock, T. H. (1985). Life history variation in primates. *Evolution: International Journal of Organic Evolution, 39*(3), 559–581.

Harvey, P. H., Martin, P. D., & Clutton-Brock, T. H. (1987). Life histories in comparative perspective. In B. B. Smuts, D. L. Cheney, R. M. Seyfarth, R. W. Wrangham, & T. T. Struhsaker (Eds.), *Primate societies* (pp. 181–196). Chicago: University of Chicago Press.

Hawkes, K., O'Connell, J. F., & Blurton-Jones, N. G. (2003). Human life histories: Primate trade-offs, grandmothering socioecology, and the fossil record. In P. M. Kappeler & M. E. Pereira (Eds.), *Primate life histories and socioecology* (pp. 204–227). Chicago: University of Chicago Press.

Hermann, E., Call, J., Hernandez-Lloreda, M. V., Hare, B., & Tomasello, M. (2007). Humans have evolved specialized skills of social cognition: The cultural intelligence hypothesis. *Science, 317,* 1360–1366.

Hinde, R. A. (1976). *Interactions, relationships and social structure.* New York: McGraw-Hill Book Company.

Hiraiwa-Hasegawa, M. (1989). Sex differences in the behavioral development of chimpanzees at Mahale. In P. G. Heltne & L. A. Marquardt (Eds.), *Understanding chimpanzees* (pp. 104–115). Cambridge, MA: Harvard University Press.

Horner V., Carter J. D., Suchak M., & de Waal, F. B. M. (2011). Spontaneous prosocial choice by chimpanzees. *Proceedings of the National Academy of Sciences of the United States of America, 108,* 13847–13851.

Hrdy, S. B. (2009). *Mothers and others: The evolutionary origins of mutual understanding.* Cambridge, MA: Harvard University Press.

Jablonka, E., & Lamb, M. J. (2005). *Evolution in four dimensions: Genetic, epigenetic, behavioral, and symbolic variation in the history of life.* Cambridge, MA: MIT press.

Jack, K. (2003). Explaining variation in affiliative relationships among male white-faced capuchins (Cebus capucinus). *Folia Primatologica, 74*(1), 1–16.

Jack, K. (2011). The cebines. In C. J. Campbell, A. Fuentes, K. C. MacKinnon, R. Stumpf, & S. K. Bearder (Eds.), *Primates in perspective* (2nd ed., pp. 108–122). New York: Oxford University Press.

Jack, K., & Fedigan, L. M. (2004a). Male dispersal patterns in white-faced capuchins (Cebus capucinus). Part 1: Patterns and causes of natal emigration. *Animal Behaviour, 67*(4), 761–769.

Jack, K., & Fedigan, L. M. (2004b). Male dispersal patterns in white-faced capuchins (Cebus capucinus). Part 2: Patterns and causes of secondary dispersal. *Animal Behaviour, 67*(4), 771–782.

Janson, C. H. (1986). Capuchin counterpoint. *Natural History*, *2/86*, 45–53.

Kappeler, P. M., & van Schaik, C. (2006). *Cooperation in primates and humans: Mechanisms and evolution.* Berlin, Germany: Springer.

Kelley, E., & Sussman, R. W. (2007). Academic genealogy on the history of American field primatology. *American Journal of Physical Anthropology*, *132*, 406–425.

Kennedy, G. E. (2005). From the ape's dilemma to the weanling's dilemma: Early weaning and its evolutionary context. *Journal of Human Evolution*, *48*, 123–145.

King, A. J., Clark, F. E., & Cowlishaw, G. (2011). The dining etiquette of desert baboons: The roles of social bonds, kinship, and dominance in co-feeding networks. *American Journal of Primatology*, *73*(8), 768–774.

Kirkpatrick, R. C. (1998). Ecology and behavior in snub-nosed and douc langurs. In N. G. Jablonski (Ed.), *The natural history of the doucs and snub-nosed monkeys* (pp. 155–190). Singapore, China: World Scientific.

Kirkpatrick, R. C., Long, Y. C., Zhong, T., & Xiao, L. (1998). Social organization and range use in the Yunnan snub-nosed monkey Rhinopithecus bieti. *International Journal of Primatology*, *19*(1), 13–51.

Knott, C., & Kahlenberg, S. (2011). Orangutans. In C. J. Campbell, A. Fuentes, K. C. MacKinnon, R. Stumpf, & S. K. Bearder (Eds.), *Primates in perspective* (2nd ed., pp. 313–325). New York: Oxford University Press.

Kudo, H., & Dunbar, R. I. M. (2001). Neocortex size and social network size in primates. *Animal Behaviour*, *62*, 711–722.

Kummer, H. (1968). *Social organization of Hamadryas baboons: A field study.* Chicago: University of Chicago Press.

Kummer, H. (1971). *Primate societies: Group techniques of ecological adaptation.* Wheeling, IL: Harlan Davidson.

Kuroda, S. (1980). Social behavior of the pygmy chimpanzees. *Primates*, *21*, 181–197.

Laland, K. N., Odling-Smee, J., & Feldman, M. W. (2001). Cultural niche construction and human evolution. *Journal of Evolutionary Biology*, *14*(1), 22–33.

Lambert, J. (2011). Primate nutritional ecology. In C. J. Campbell, A. Fuentes, K. C. MacKinnon, R. Stumpf, & S. K. Bearder (Eds.), *Primates in perspective* (2nd ed., pp. 512–522). New York: Oxford University Press.

Le Gros Clark, W. E. (1971). *The antecedents of man.* Edinburgh, Scotland: Edinburgh University Press.

Leigh, S. R. (2004). Brain growth, life history, and cognition in primate and human evolution. *American Journal of Primatology*, *62*, 139–164.

Leigh, S. R., & Blomquist, G. E. (2011). Life history. In C. J. Campbell, A. Fuentes, K. C. MacKinnon, S. K. Bearder, & R. M. Stumpf (Eds.), *Primates in perspective* (pp. 418–428). New York: Oxford University Press.

Lehmann, J., & Ross, C. (2011). Baboon (Papio anubis) social complexity—A network approach. *American Journal of Primatology, 73*(8), 775–789.

Levitt, P. (2003). Structural and functional maturation of the developing primate brain. *Journal of Pediatrics, 143*(4): 35–45.

Lewis, K. P. (2000). A comparative study of primate play behaviour: Implications for the study of cognition. *Folia Primatologica, 71*, 417–421.

MacKinnon, K. C. (2011). Social beginnings: The tapestry of infant and adult interactions. In C. J. Campbell, A. Fuentes, K. C. MacKinnon, S. K. Bearder, & R. M. Stumpf (Eds.), *Primates in perspective* (pp. 440–455). New York: Oxford University Press.

MacKinnon, K. C. (in press). Ontogeny of social behavior in the genus *Cebus* and the application of an integrative framework for examining plasticity and complexity in evolution. In K. Clancy, K. Hinde, K., & J. Rutherford (Eds.), *Building babies: Proximate and ultimate perspectives of primate developmental trajectories*. New York: Springer.

MacKinnon, K. C., & Fuentes, A. (2011). Primates, niche construction, and social complexity: The roles of social cooperation and altruism. In R. W. Sussman & R. C. Cloninger (Eds.), *Origins of altruism and cooperation* (pp. 121–143). New York: Springer.

Martin, R. D. (1983). Human brain evolution in an ecological context (Fifty-second James Arthur lecture on the evolution of the human brain). New York: American Museum of Natural History.

McKenna, J. J. (1983). Primate aggression and evolution: An overview of sociobiological and anthropological perspectives. *Bulletin of the American Academy of Psychiatry and the Law, 2*(2), 105–130.

Melin, A., Fedigan, L. M., Hiramatsu, C., & Kawamura, S. (2008). Polymorphic color vision in white-faced capuchins (Cebus capucinus): Is there foraging niche divergence among phenotypes? *Behavioral Ecology and Sociobiology, 62*, 659–670.

Melin, A. D, Fedigan, L. M., Hiramatsu, C., Hiwatashi, T., Parr, N., & Kawamura, S. (2009). Fig foraging by dichromatic and trichromatic white-faced capuchin monkeys in a tropical dry forest. *International Journal of Primatology, 30*(6), 753–775.

Melin, A. D., Fedigan, L. M., Young, H. C., & Kawamura, S. (2010). Can color vision variation explain sex differences in invertebrate foraging by capuchin monkeys? *Current Zoology, 56*(3), 300–312.

Morgan, D., & Sanz, C. (2003). Naïve encounters with chimpanzees in the Goualougo Triangle, Republic of Congo. *International Journal of Primatology, 24*(2), 369–381.

Nekaris, K. A. I. (2003). Observations of mating, birthing and parental behaviour in three subspecies of slender loris (Loris tardigradus and Loris lydekkerianus) in India and Sri Lanka. *Folia Primatologica, 74*(5–6), 312–336.

Nekaris, A., & Bearder, S. K. (2011). The lorisiform primates of Asia and mainland Africa: Diversity shrouded in darkness. In C. J. Campbell, A. Fuentes, K. C. MacKinnon, S. K. Bearder, & R. M. Stumpf (Eds.), *Primates in perspective* (pp. 34–54). New York: Oxford University Press.

Neubauer, P. B. (1996). *Nature's thumbprint: The new genetics of personality.* New York: Columbia University Press.

Newman, M. E. J. (2010). *Networks: An introduction.* Oxford: Oxford University Press.

Odling-Smee, F. J., Laland, K. N., & Feldman, M. W. (2003). *Niche construction: The neglected process in evolution.* Princeton, NJ: Princeton University Press.

Panger, M., Perry, S., Rose, L. M., Gros-Louis, J., Vogel, E., MacKinnon, K. C., et al. (2002). Cross-site differences in foraging behavior of white-faced capuchins (Cebus capucinus). *American Journal of Physical Anthropology, 119,* 52–66.

Perry, S. (1996). Female-female social relationships in wild white-faced capuchin monkeys, Cebus capucinus. *American Journal of Primatology, 40*(2), 167–182.

Perry, S. (1997). Male-female social relationships in wild white-faced capuchins (Cebus capucinus). *Behaviour, 134,* 477–510.

Perry, S. (1998). Male-male social relationships in wild white-faced capuchins (Cebus capucinus). *Behaviour, 135,* 139–172.

Perry, S., Baker, M., Fedigan, L. M., Gros-Louis, J., Jack, K., MacKinnon, K. C., et al. (2003a). Social conventions in wild white-faced capuchins: Evidence for traditions in a neotropical primate. *Current Anthropology, 44*(2), 241–268.

Perry, S., Panger, M., Rose, L., Baker, M., Gros-Louis, J., Jack, K., et al. (2003b). Traditions in wild white-faced capuchin monkeys. In D. M. Fragaszy & S. Perry (Eds.), *The biology of traditions: Models and evidence* (pp. 391–425). Cambridge: Cambridge University Press.

Phillips, K. A., & Sherwood, C. C. (2008). Cortical development in brown capuchin monkeys: A structural MRI study. *NeuroImage, 43*(4), 657–664.

Pokorny, J.J. and de Waal, F. B. M. (2009a). Monkeys recognize the faces of group mates in photographs. *Proceedings of the National Academy of Sciences of the United States of America, 106,* 21539–21543.

Pokorny, J. J., & de Waal, F. B. M. (2009b). Face recognition in capuchin monkeys (Cebus apella). *Journal of Comparative Psychology, 123,* 151–160.

Pollak, S. D., & Kistler, D. J. (2002). Early experience is associated with the development of categorical representations for facial expressions of emotion. *Proceedings of the National Academy of Sciences of the United States of America, 99,* 9072–9076.

Potts, R. (2004). Sociality and the concept of culture in human origins. In R. W. Sussman & A. R. Chapman (Eds.), *The origins and nature of sociality* (pp. 249–269). New York: Aldine de Gruyter.

Preston, S. D., & de Waal, F. B. M. (2002). Empathy: Its ultimate and proximate bases. *Behavioral and Brain Sciences, 25,* 1–71.

Pruetz, J. D., & Bertolani, P. (2007). Savanna chimpanzees, *Pan troglodytes verus,* hunt with tools. *Current Biology, 17,* 412–417.

Pruetz, J. D., & Lindshield, S. (2011). Plant-food and tool transfer among savanna chimpanzees at Fongoli, Senegal. *Primates* Published online: 20 November 2011.

Pusey, A. E. (1983). Mother-offspring relationships in chimpanzees after weaning. *Animal Behaviour, 31,* 363–377.

Ramos-Fernández, G., Boyer, G., Aureli, F., & Vick, L. G. (2009). Association networks in spider monkeys (Ateles geoffroyi). *Behavioral Ecology and Sociobiology, 63,* 999–1013.

Ridley, M. (1993). *The red queen: Sex and the evolution of human nature.* New York: Macmillan Publishing Company.

Ridley, M. (1997). *The origins of virtue: Human instincts and the evolution of cooperation.* New York: Viking Press.

Rilling, J. K., & Insel, T. R. (1999). The primate neocortex in comparative perspective using magnetic resonance imaging. *Journal of Human Evolution, 37,* 191–223.

Robbins, M. (2011). Gorillas. In C. J. Campbell, A. Fuentes, K. C. MacKinnon, R. Stumpf, & S. K. Bearder (Eds.), *Primates in perspective* (2nd ed., pp. 326–339). New York: Oxford University Press.

Robbins, A. M., Robbins, M. M., Gerald-Steklis, N., & Steklis, H. D. (2006). Age-related patterns of reproductive success among female mountain gorillas. *American Journal of Physical Anthropology, 131*(4), 511–521.

Ross, C. F., & Martin, R. D. (2007). The role of vision in the origin and evolution of primates. In T. M. Preuss & J. Kaas (Eds.), *Evolution of nervous systems* (Vol. 5, pp. 59–78). Oxford: Elsevier.

Rossi, A. S. (1995). A plea for less attention to monkeys and apes, and more to human biology and evolutionary psychology. *Politics and the Life Sciences, 14*(2), 185–187.

Rylands, A. B. (Ed.). (1993). *Marmosets and tamarins: Systematics, behavior, and ecology.* Oxford: Oxford Scientific Publications.

Sade, D. S. (1972). Sociometrics of Macaca mulatta. I. Linkages and cliques in grooming matrices. *Folia Primatologica, 18,* 196–223.

Sakai, T., Mikami, A., Tomonaga, M., Matsui, M., Suzuki, J., Hamada, Y., et al. (2011). Differential prefrontal white matter development in chimpanzees and humans. *Current Biology, 21,* 1397–1402.

Santorelli, C. J., Schaffner, C. M., Campbell, C. J., Notman, H., Pavelka, M. S., Weghorst, J. A., et al. (2011). Traditions in spider monkeys are biased towards the social domain. *PLoS ONE, 6*(2), e16863.

Sanz, C., Call, J., & Morgan, D. (2009). Design complexity in termite-fishing tools of chimpanzees (*Pan troglodytes*). *Biology Letters, 5,* 293–296.

Sapolsky, R. M. (2003). Stress and plasticity in the limbic system. *Neurochemical Research, 28*(11), 1735–1742.

Sapolsky, R. M. (2005). The influence of social hierarchy on primate health. *Science, 308,* 648–652.

Sapolsky, R. M., & Share, L. J. (2004). A pacific culture among wild baboons: Its emergence and transmission. *PLoS Biology, 2*(4), 534–541.

Schultz, E. (2009). Resolving the anti-antievolutionism dilemma: A brief for relational evolutionary thinking in anthropology. *American Anthropologist, 11*(2), 224–237.

Scott, J. (2000). *Social network analysis: A handbook.* Beverly Hills, CA: Sage Publications.

Sih, A., Hanser, S. F., & McHugh, K. A. (2009). Social network theory: New insights and issues for behavioral ecologists. *Behavioral Ecology and Sociobiology, 63,* 975–988.

Silk, J. B. (2007). Social components of fitness in primate groups. *Science, 317*(5843), 1347–1351.

Sherrow, H. M., & MacKinnon, K. C. (2011). Juvenile and adolescent primates: The application of life history theory. In C. J. Campbell, A. Fuentes, K. C. MacKinnon, R. Stumpf, & S. K. Bearder (Eds.), *Primates in perspective* (2nd ed., pp. 455–464). New York: Oxford University Press.

Stanford, C. B., & Bunn, H. T. (2001). *Meat eating and human evolution.* New York: Oxford University Press.

Sueur, C., Jacobs, A., Amblard, F., Petit, O., & King, A. J. (2011). How can social network analysis improve the study of primate behavior? *American Journal of Primatology, 73,* 703–719.

Sueur, C., & Petit, O. (2008). Organization of group members at departure is driven by social structure in Macaca. *International Journal of Primatology, 29*, 1085–1098.

Strier, K. B. (1994). Myth of the typical primate. *Yearbook of Physical Anthropology, 37*(Suppl. S19), 233–271.

Strum, S. C., & Fedigan, L. M. (2000). Changing views of primate society: A situated North American perspective. In S. C. Strum & L. F. Fedigan (Eds.), *Primate encounters: Models of science, gender, and society* (pp. 3–49). Chicago: The University of Chicago Press.

Stumpf, R. (2011). Chimpanzees and bonobos. In C. J. Campbell, A. Fuentes, K. C. MacKinnon, R. Stumpf, & S. K. Bearder (Eds.), *Primates in perspective* (2nd ed., pp. 340–357). New York: Oxford University Press.

Sussman, R. W., & Garber, P. A. (2011). Cooperation, collective action, and competition in primate social interactions. In C. J. Campbell, A. Fuentes, K. C. MacKinnon, S. K. Bearder, & R. M. Stumpf (Eds.), *Primates in perspective* (pp. 587–599). New York: Oxford University Press.

Thierry, B. (2011). The macaques. In C. J. Campbell, A. Fuentes, K. C. MacKinnon, R. Stumpf, & S. K. Bearder (Eds.), *Primates in perspective* (2nd ed., pp. 229–241). New York: Oxford University Press.

Tiddi, B., Aurelil, F., Schino, G., & Voelkl, B. (2011). Social relationships between adult females and the alpha male in wild tufted capuchin monkeys. *American Journal of Primatology, 73*(8), 812–820.

Tomasello, M. (1999). The human adaptation for culture. *Annual Review of Anthropology, 28*, 509–529.

Treves, A., & Chapman, C. A. (1996). Conspecific threat, predation avoidance, and resource defense: Implications for grouping in langurs. *Behavioral Ecology and Sociobiology, 39*(1), 43–53.

Ungar, P. S. (2007). *Evolution of the human diet: The known, the unknown, and the unknowable*. New York: Oxford University Press.

Van Essen, D. C., Anderson, C. H., & Felleman, D. J. (1992). Information processing in the primate visual system: An integrated systems perspective. *Science, 255*, 419–423.

Van Essen, D. C., Lewis, J. W., Drury, H. A., Hadjikhani, N., Tootell, R. B. H., Bakircioglu, M., et al. (2001). Mapping visual cortex in monkeys and humans using surface-based atlases. *Vision Research, 41*, 1359–1378.

Van Noordwijk, M. A., & van Schaik, C. P. (2005). Development of ecological competence in Sumatran orangutans. *American Journal of Physical Anthropology, 127*, 79–94.

Van Schaik, C., Deaner, R., & Merrill, M. (1999). The conditions for tool use in primates: Implications for the evolution of material culture. *Journal of Human Evolution, 36,* 719–741.

Vinicius, L. (2005). Human encephalization and developmental timing. *Journal of Human Evolution, 49,* 762–776.

Walker, R., Gurven, M., Hill, K., Migliano, A., Chagnon, N., De Souza, R., et al. (2006). Growth rates and life histories in twenty-two small-scale societies. *American Journal of Human Biology, 18,* 295–311.

Wall-Scheffler, C. M., Geiger, K., & Steudel-Numbers, K. L. (2007). Infant carrying: The role of increased locomotory costs in early tool development. *American Journal of Physical Anthropology, 133*(2), 841–846.

Warneken, F., Hare, B., Melis, A. P., Hanus, D., & Tomasello, M. (2007). Spontaneous altruism by chimpanzees and young children. *PLoS Biology, 5*(7), e184.

Warneken, F., & Tomasello, M. (2006). Altruistic helping in human infants and young chimpanzees. *Science, 311*(5765), 1301–1303.

Washburn S. L. (1951). The New Physical Anthropology. *Transactions of the New York Academy of Sciences, 13,* 258–304.

Washburn, S. L. (1973). The promise of primatology. *American Journal of Physical Anthropology, 38,* 177–182.

Wells, J. C. K., & Stock, J. T. (2007). *The biology of the colonizing ape.* Yearbook of Physical Anthropology, 50, 191–222.

West-Eberhard, M. J. (2003). *Developmental plasticity and evolution.* New York: Oxford University Press.

Wey, T., Blumstein, D. T., Shen, W., & Jordan, F. (2008). Social network analysis of animal behaviour: A promising tool for the study of sociality. *Animal Behaviour, 75,* 333–344.

Whitehead, H. (2008). *Analyzing animal societies.* Chicago: University of Chicago Press.

Whiten, A., Horner, V., & de Waal, F. B. M. (2005). Conformity to cultural norms of tool use in chimpanzees. *Nature, 437,* 737–740.

Whiten, A., Spiteri, A., Horner, V., Bonnie, K. E., Lambeth, S. P., Schapiro, S. J., et al. (2007). Transmission of multiple traditions within and between chimpanzee groups. *Current Biology, 17,* 1038–1043.

Woo, T. U., Pucak, M. L., Kye, C. H., Matus, C. V., & Lewis, D. A. (1997). Peripubertal refinement of the intrinsic and association circuitry in the monkey prefrontal cortex. *Neuroscience, 80,* 1149–1158.

Wrangham, R. W. (2009). *Catching fire: How cooking made us human*. Basic Books.

Wright, R. (1994). *The moral animal: Evolutionary Psychology And Everyday Life*. New York: Pantheon Books.

Young, M. P., Scannell, J. W., Burns, G. A. P. C., & Blakemore, C. (1994). Analysis of connectivity: Neural systems in the cerebral cortex. *Reviews in the Neurosciences, 5,* 227–249.

Zimmerman, L. A. (2000). *The SE switch: Evolution and our self-esteem*, Orlando, FL: Rivercross.

4   Evolution and the Brain

Greg Downey and Daniel H. Lende

When Carl von Linne, the great Swedish naturalist, christened our species *Homo sapiens*, he recognized that among all of our species' strange traits—our odd bipedalism, relative hairlessness, unprecedented ability to speak, the absence of a penile spine, pronounced female breasts, exaggerated chins, extravagant tool use—our intelligence was what most distinguished us. Whereas most animals' skulls set aside a greater proportion of the architecture for food processing than thinking, our skulls show a shifted set of biological priorities. Our cranial vault has expanded at the expense of the jaw and its musculature, enlarged to the point of apparent precariousness. Alongside the skulls of our great ape cousins, and even some of our own hominin ancestors, we seem to have put our evolutionary eggs in one basket: grow the brain above all else.

Clearly, our brains help to make us an unusual, self-aware, cultural animal, and understanding the evolutionary pressures that produced this set of adaptations is incumbent upon neuroanthropology. Moreover, our distinctive brains have made us a remarkably successful animal; human biomass is estimated to be eight times the total of all wild terrestrial vertebrates (see Smil, 2002). With the recognition by Darwin that the similarities Linneus perceived among animals were actually the result of shared origins and common descent, however, the puzzle became to recognize how our distinctive suite of cognitive characteristics arose through evolutionary processes from a shared biological foundation. Not just why, but how, did evolution make us so different?

One of the challenges of neuroanthropology is to engage robustly and constructively with evolutionary psychology, recognizing that evolutionary theory and our knowledge of the fields that feed our understanding of human origins require constant updating and renewed theorizing. This chapter attempts to summarize some of the key recent developments that characterize a still-emerging understanding of how the human brain became distinct through evolutionary mechanisms, as well as how the "mental chasm" separating us from other great apes arose (Cosans, 1994), especially given that evolutionary processes tend to "tinker" rather than "redesign" wholesale existing structures (Jacob, 1977; Passingham, 2008).

This chapter will stress that an evolutionary account of the human brain needs to consider more than just adaptation. Increasingly, geneticists, comparative neuroscientists, paleoanthropologists, and evolutionary psychologists take into account a range of other factors, including constraints, genetic and developmental mechanisms that generate "evolvability," and the role of human actions in transforming selective pressures over our evolutionary history. Evolutionary explanations that privilege the brain's adaptive functions must be balanced by recognition of physiological and development constraint; that is, by an awareness of how biological mechanisms produce both organisms and specific organs (Alberch, 1982; Gould & Lewontin, 1979; Pilbeam & Gould, 1974). Without this balance, some theories of human cognitive evolution, as Terrence Deacon (1997, p. 338) suggests in diplomatic fashion, "take for granted assumptions that would be judged biologically implausible with respect to other organ systems (e.g. accretion of new structures, recapitulation, modular change)."

Theodosius Dobzhansky (1973) famously titled a short essay on the compatibility of science and theology, "Nothing makes sense in biology except in the light of evolution." While the aphorism is no doubt true, evolutionary theory also needs to be checked against biological and anthropological plausibility. As we increasingly understand the mechanisms that generate biological development, diversity, and adaptation, some accounts of evolution appear more or less credible, some "evolutionary explanations" little better than what the late Stephen Jay Gould called "just-so stories," flying in the face of biology and observed human differences. In the discussion of the evolution of the human mind and brain, theory has often run ahead of evidence, in part because the evidence for reconstructing this trajectory lies across widely divided scientific specialties. As Deacon (1990b, p. 629) elsewhere suggests, because knowledge about comparative neurology and paleoanthropology is not more fully integrated, for example, "speculative theories concerning brain evolution—especially human brain evolution—are widespread and often contain relatively little neuroanatomical or neurophysiological information."

Over the course of evolution, the most obvious change in our species' brain since the last common ancestor shared with chimpanzees is both an overall and a relative increase in brain size and changes in brain proportions (Holloway, 1995), so this review will begin there. More subtle changes, however, are equally important, and this review will seek to address them before inquiring into, not just brain function, but the consequences of evolutionary change as well.

Chapter Overview

This chapter focuses on the physiological dimensions of brain evolution, then places these changes within the theoretical frameworks that have been developed to explain the massive size and peculiar organization of the human brain. As we describe, size matters in considering the human brain; but absolute size is only a place to begin. Using a comparative framework across species and over phylogenetic (or evolutionary) time, we can begin to ascertain what the relative differences are among humans, primates, and other mammals, and how that relative difference in size emerged in our human ancestors. In humans, structural brain changes in evolution help us to discern the neural underpinnings of our cognitive capacity; our neocortex is disproportionately larger than other parts of the brain, while our olfactory bulbs are undersized. Moreover, we exhibit hemispheric asymmetry, exaggerated more than other great apes. Two explanations exist for these sorts of structural difference: the first is developmental dynamics, which posits that it is easier to size up brain areas that appear later during embryonic development, so the brain cortex can grow larger than brain stem; and the second is "mosaic" expansion, in which patterns of selection increase or decrease parts of the brain because they are involved in specific functional systems. Finally, connections matter, because brain regions do not function in isolation; neurons must communicate and function together. The human brain relies on connections both generally within regions and also in specific fashion; for example, neocortical connections to the viceromotor system handle the tongue, mouth, and larynx (important for the production of speech) and neocortical connections from prefrontal areas down into limbic areas increase the ability to regulate and control behavior, particularly in social settings.

Explanations for these patterns within the brain need to first reckon with how the brain actually grows and wires itself over developmental time, where gene expression, epigenetics, neuronal selection, and experience-dependent growth all help shape the brain's functionality from infancy through to adulthood and old age. This slow unfolding of human brains over the life course means that evolution can largely work through changing the developmental dynamics of the human brain, whether in internal fashion or through brain-body-environment interactions. We outline niche construction as a way to understand those brain-body interactions, and then present three main theories to explain human brain evolution. The social intelligence hypothesis focuses on the demands and opportunities that being a long-lived, highly social species brings and the role of competition with our own kind. The brain-development niche model incorporates ideas about cultural traditions, information transfer, and the development of technological, foraging, and social skills. Finally, sensory, motivational, and bodily changes emphasize that one way to shape the evolutionary and developmental function of the brain is not just changing the brain directly, but changing the body's sensory and other physiological systems and targeting basic motivational brain systems that can yield large differences in behavioral interactions and control. We suggest that all three approaches are important for thinking about human brain evolution, and that neuroanthropologists should draw on the specific set of patterns of brain evolution and selection pressures that best fits the analytical problem they are addressing.

Size Matters

As early as Aristotle, observers have noted that the human brain is disproportionately large, and wondered about the links with our species' distinctive cognitive capacities. Charles Darwin (1874/1890, p. 60), for example, hypothesized that human evolutionary enlargement was linked to greater intelligence than that found in other great apes: "As the various mental faculties gradually developed themselves the brain would almost certainly become larger. No one, I presume, doubts that the large proportion which the size of man's brain bears to his body, compared to the same proportion in the gorilla or orang, is closely connected with his mental powers." Our ability to measure more subtle neurological differences has thrown up other candidate explanations for human cognitive distinctiveness, but overall and relative brain size are still crucial considerations.

One thing is certain: brain tissue is metabolically expensive. Especially in humans, increased brain size can lead to complications in childbirth because skeletal changes linked to bipedalism include a narrowing of the pelvis (see Rosenberg & Trevathan, 1995, for a review). Bigger brains generate heat, have to pass signals over greater distances, and demand significant amounts of high-quality food, especially given that the human gut is disproportionately underdeveloped, perhaps as a direct result of metabolic trade-offs with an energy-greedy brain (see Falk, 1990; Ringo, 1991; Aiello & Wheeler, 1995).

Absolute and Relative Brain Size

Drawing on the postmortem data set produced by Heinz Stephan and colleagues (1981), who measured a dozen brain regions in numerous primates and other mammals, James Rilling (2006, p. 66) points out that the gap between average brain size in humans and in other great apes is yawning. Whereas human brains average around 1300 cubic centimeters, the average brain size among other great apes is between 300 and 400 cubic centimeters (Schoenemann, 2006, p. 383; see also Deacon, 1997; Jerison, 1973). Larger brains tend to be associated with greater social complexity and more sophisticated and varied food-seeking behavior across species, including both primates and parrots (see Barton & Dunbar, 1997; Byrne, 1997).

Georg Striedter (2006, p. 7) argues that absolute brain size is one crucial determinant of cognitive complexity, in part because larger brains may require structural and functional changes, such as greater specialization of brain regions, shifts in the proportional size of different parts, and systems of long-distance connections among increasingly separated regions (see also Striedter, 2005, pp. 126–131, 355–360).

But if absolute brain size alone explained intelligence, whales—with brains as much as six times larger than human counterparts—would be writing books on "neurocetaceaology," and elephants would be contemplating how their unusual cognitive capacities were linked to their large crania. In addition, if size alone explained intelligence, reptiles, birds, and mammals with the same size brains would be equally intelligent, which they are not (see Striedter, 2005, p. 11).

If humans' unusual cognitive abilities cannot simply be explained by the size of their brains, perhaps brain size relative to body size might explain why humans stand out; after all, the blue whale's mammoth brain only makes up 0.01% of its body's weight, whereas the human brain is 2% of our bodies' mass. Relative brain weight as an explanation for human cognitive distinctiveness presents a problem as well, however, since smaller animals have proportionally larger brains. The pocket and harvest mice, for example, have brains that comprise fully 10% of their bodies' mass (see Striedter, 2005, p. 93).

Comparative neurobiologists have long realized that plotting animals' brain sizes against their bodily mass in logarithmic coordinates produces a more linear cluster; the brain scales allometrically with body size. Given this relationship, neurobiologist Harry Jerison (1973) calculated "encephalization quotients," dividing a species' actual brain size by the mass predicted by this pattern of allometric scaling. In general, allometric relations between brain and body size do an excellent job of predicting most animals' brain size; according to Rilling (2006, p. 67), 94% of the variation in brain size among primates appears to be the result of the brain simply scaling along with the body.

Depending upon how one draws the "best-fit" line through the data, human encephalization quotients appear to be five to seven times higher than predicted for a mammal of our size (see Schoenemann, 2006, p. 381). Compared to another mammal of equal weight, human brains are freakishly large. Primates as an order have greater encephalization than other mammals, so the gap between humans and other animals diminishes when our brain-body ratio is compared only to the allometric plot for anthropoid apes (Deacon, 1990b). Shoenemann suggests the human encephalization quotient drops to 3.1 relative to other primates, but human brain size is still a marked outlier above the prediction for our body size (Schoenemann, 2006, p. 384; see also Passingham, 1973; Pilbeam & Gould, 1974; Rilling, 2006).

Phylogenetic Trajectory

Most paleoanthropologists now believe that the last common ancestor of humans and modern chimpanzees, who likely lived sometime between six and ten million years ago, had a brain close in size to that of modern chimpanzees. The initial difference between our hominin ancestor and the ancestor to chimpanzees was likely not brain size but lower body anatomy— specifically, bipedalism. By two million years ago, however, the genus *Homo* appeared, and a dramatic increase in brain size occurred. Although part of that growth was linked to the overall increased size of our genus— *Homo erectus* was bigger than most australopithecines—the increase in size was relative as well as absolute. After that, brain size grew slowly but steadily until again accelerating a bit more than a half-million years ago (see Holloway, Broadfield, & Yuan, 2004). In sum, our ancestors really started to look different from other apes around two million years ago, tripling brain size in a relatively short period of time (Ruff, Trinkaus, & Holliday, 1997).

Holloway argues that reorganization of the hominin brain preceded the marked expansion that distinguishes the brain size of our genus from that of other great apes. According to Holloway, even australopithecines, who had overall brain sizes in line with gorillas or chimpanzees, already evidenced a human-like brain pattern, likely linked to the cognitive demands being placed upon our ancestors by a new way of life (see Holloway et al., 2004). The reconstruction of extinct brain configurations is difficult, but

Holloway (2008) quite rightly points out that discussing brain evolution without taking notice of this data disregards the crucial information that skull endocasts can provide. Although comparisons with other living primates may help us to understand human distinctiveness, these comparisons are with extant cousins, not extinct ancestors, and all primates have undergone their own evolutionary changes since our last shared ancestors (see Aldridge, 2011; Oxnard, 2004).

Despite the marked behavioral and obvious morphological differences, genetic research initially discovered a surprisingly high degree of similarity between the human genome and that of chimpanzees. In fact, around a half-century ago, Morris Goodman (1963) argued on the basis of genetic similarity that humans and chimpanzees should both be in a single genus (presumably, *Homo*), an argument echoed in popular form by Jared Diamond (1992). Although the shared genus proposal is sufficiently radical that it has never been widely considered, genetic analysis of primate relations has pushed chimpanzees and humans closer together, especially in contrast to more distantly related great apes, such as the gorilla and orangutan (Marks, 2005, p. 49; but see Wildman, Uddin, Liu, Grossman, et al., 2003). The paradox is that quite subtle genetic difference is clearly foundational to extraordinary behavioral, physical, and ecological disparity (Marks, 2002; Oxnard, 2004).

To embryologists, the discovery of underlying genetic similarity likely came as little surprise, as they had long observed a remarkable degree of uniformity in initial development; across a wide variety of vertebrates, embryos go through an early stage that is highly, though not entirely, conserved (see Finlay & Darlington, 1995; Puelles & Rubenstein, 2003; Striedter, 2005). The recent "evo-devo" synthesis of evolutionary theory and developmental biology—of phylogenetic and ontogenetic approaches to organismal biology—has been especially persuasive in arguing that evolution does not simply work on finished genes or traits, but on the developmental programs that produce adults of a species as well (Carroll, 2003, 2005; Gerhart & Kirschner, 1997; Jablonka & Lamb, 2004; Kirschner & Gerhart, 2005; Wagner, 2005).

In primates, for example, changes in the adult brain size appear strongly linked to an extension of maturational processes, including disproportionally slow in utero body development, prolonged infancy, prolonged high brain-growth rates, delayed sexual maturity, and even greater overall longevity (see Cutler, 1976; Deacon, 1997). That is, adult brains of different species are built by growing an initial set of structures in diverse fashions, early precursors diversifying over developmental time, in part because of how they mature (e.g., Jiao, Medina, Veenman, Toledo, et al. 2000; see Striedter, 2005). Striedter (2006, p. 4) cautions that brain homology is not always the case, but argues strongly that ontogenetic emergence, or developmental processes, must be the primary target of evolutionary theory:

The more we learn about how brains evolve, the more it seems that evolution can tinker with any neuronal attribute, from embryonic origin to structural complexity and physiological function. A major challenge for the coming years is to reveal the details of that tinkering—to discover how evolution modified neural development to create diverse adult brains, and how those changes altered animal behavior.

Growing the extravagantly large and peculiar human brain over evolutionary time has meant tinkering with the developmental processes at the structural, connectional, cellular, and genetic levels.

Structure Matters

Many paleoanthropologists have argued that human cognitive distinctiveness is not merely, or even primarily, the result of overall brain enlargement, but rather of shifts in the organization of the brain and the relative size of different neural structures (see Jerison, 1973; Holloway, 1966). Some early comparative neurologists sought to find a unique human brain structure to explain the vast intellectual gulf between humans and other species (see Striedter, 2005, chapter 2). In preparation for the publication of *On the Origin of Species*, anatomist Richard Owen (1857), for example, asserted that only humans had several structures, including the hippocampus minor; T. H. Huxley (1863) demonstrated that these structures had corresponding parts, or "homologues," a word Owen himself had coined, in the chimpanzee brain. Even regions like Broca's and Wernicke's areas (Brodman's areas 44 and 45 and area 22, respectively), foundational to language ability, have homologs in other primate brains (see, for example, Petrides, Cadoret, & Mackey, 2005).

The majority of comparative neuroscientists have concluded that the primary differences between human and other primate brains are proportional rather than the result of evolution producing entirely new structures, although the increasing specialization permitted by brain expansion does suggest that new regions might emerge (see Preuss, 2001). The predominance of homologous regions across species in the brain, however, has important implications for the neurological underpinning of humans' distinctive cognitive abilities: the pattern suggests most new capacities have arisen from modifications to or repurposing of existing structures, what Ernst Mayr (1960) and other biologists call "pre-adaptations" (see also Anderson, 2010; Love, 2007).

Many researchers argue that humans are distinctive cognitively, in part, because changes in the relative size of different brain regions have shifted the neurological balance of power within the brain (see Jerison, 1973). Shifting regional proportions, in humans for example, might allow more complex problem solving and interaction or greater inhibition of instinctive behavior, or tip the balance among sensory modalities. Allometric scaling between species in relative brain size is also accompanied by disproportionate growth scaling rates among brain regions; as brains grow larger among related species, every part of the brain does not increase at the same rate, so species with large brains demonstrate predictable changes in the size of regions relative to each other.

In mammals, for example, the neocortex grows disproportionately, so that large-bodied animals tend to become heavily "neo-corticalized." Humans further exaggerate this trend, although how much is debated (see Holloway, 1979; Semendeferi, Lu, Schenker, & Damasio, 2002; Stephan et al., 1981). The ratio of neocortical gray matter to the size of the medulla in the brainstem of chimpanzees is 30:1; in humans, the ratio is 60:1, suggesting how disproportionate our neocortices are in relation to the rest of the brain (Striedter, 2005, p. 323). Rilling and Insel (1999) highlight changes, not just in the size, but also in cortical convolution (gyrification) and connectivity (proportion of white matter; see also Deacon, 1997). Rilling (2006) has argued convincingly that, among humans and our closest relatives, the great apes, the cerebellum has also been especially elaborated, possibly because the structure augments frontal lobe functioning.

In contrast, some other areas of the brain show decreased development. The olfactory bulb is smaller in humans than in chimpanzees, and simpler in configuration, and primates already have disproportionately small olfactory bulbs relative to other mammals. Holloway (1992) has found that the human primary visual cortex is less than half as large as expected from the size of this region in the chimpanzee brain, and that the posterior parietal (association) cortex has grown, suggesting an evolutionary trade-off between the areas (see also Holloway et al., 2004). The decreasing prevalence of visual areas is especially interesting in light of the way that primate brains, in general, show evidence of increased specialization for visual acuity (Striedter, 2005).

One other obvious way that the human brain has been reconfigured over evolutionary time is that asymmetries between the two hemispheres have become more pronounced, leading us to become a "lopsided ape," in the words of Corballis (1991). The pattern of hemispheric specialization in humans extends older evolutionary patterns, judging from widespread asymmetries in other primates (Sherwood, Subiaul, & Zawidzki, 2008, pp. 432–433). Specialization likely streamlined and sped up neural communication by restricting activity to one side of the brain (Ringo, Doty, Demeter, & Simard, 1994). The asymmetrical specialization, however, leaves our brains especially vulnerable to injury, as one side cannot easily compensate for damage to the opposite hemisphere.

How the Brain Grows Out of Balance
The mechanism that produces shifts in the relative size of brain regions is subject to debate. Barbara Finlay and Richard Darlington (1995) have influentially suggested that the relative enlargement of the neocortex occurs because of changes in the developmental processes that produce larger brains (see also Finlay, Darlington, & Nicastro, 2001; Rakic, 2009). Specifically, they detail how developmental stages in the embryo in large-brained mammals are prolonged so that more of the precursor cells that become neocortex are produced. As the neural developmental profile delays and immature phases elongate in larger-bodied animals, the last brain regions to emerge are the ones that grow the most, or, as they put it, "late equals large." The birth order of different regions correlates to especially exaggerated relative size. This developmental explanation suggests that most brain enlargement is "concerted," or the result of allometric scaling among brain regions, not evolutionary pressures acting specifically to grow one part or another (or, more accurately, perhaps selection targets one region but developmental mechanisms, unable to produce a more targeted effect, must grow them all).

In contrast, Robert Barton and Paul Harvey (2000) have championed the position that brain expansion has been "mosaic," disproportionately affecting regions that are linked together in specific functional systems. Barton (2006) argues that relative changes in the sizes of different brain regions reveal cognitive specialization in species. Specifically, he argues that primates have grown a functional system supporting greater visual acuity, and humans, especially, another system linking neocortex, diencephalon, and cerebellum together.

The debate between proponents of mosaic and concerted accounts of evolutionary change is fundamentally about the relative strength of biological constraints and natural selection, whether selective pressures can act with some precision to shape specific brain components or if the brain is constrained to grow in a predictable pattern. The implication of Finlay and Darlington's theory is that disproportionate growth in the neocortex is the result largely of a conserved sequence of neurogenesis and brain growth, rather than the neocortex being singled out for expansion by selective pressures. The brain is an integrated structure, and large parts are the predictable outcome of a large whole. In fact, a brain region might grow over evolutionary time as a side effect, or "spandrel," of other traits driving allometric scaling (see Gould, 1975; Gould & Lewontin, 1979; Riska & Atchley, 1985).

In contrast, the "mosaic" perspective suggests that disproportionate brain-region growth or decrease result from natural selection working on individual neurological systems (which may include several regions) as these systems vary independently. Georg Striedter (2005, pp. 149–159) cites a number of cases in his excellent review of brain evolution in which animals have grown distinctive and disproportionate "mosaic" linked systems of brain regions with suggestive parallels to their adaptive strategy, and where constraints on proportional growth seem to have been broken (for example, with the decreasing olfactory bulb in primates). Ultimately, however, Striedter argues that both mosaic and concerted growth patterns are likely to have been important in shaping species' brains, including humans, but that severe mosaic evolution is probably responsible for major brain changes, especially those that mark off the higher taxonomical differences, such as the order primates.

Connections Matter

As the human brain has reorganized during its evolution, significant changes have also occurred to patterns of connection among regions (see Holloway, 1966; Deacon, 1997). Size alone does not make a brain, or brain region, perform better, even in insects: specialization and interconnectivity are crucial (Chittka & Niven, 2009). Enlarging any part of the brain creates certain "design problems"; as the absolute distance between points in the brain grows, neuronal communication can get slower (see Kaas, 2000). Some neurons grow thicker axons to hasten conduction times, but too much thickening would make the brain grow crowded. Mammalian brains appear to counteract the problems of greater neocortex distances by increased folding (or "gyrification"), hemispheric specialization, and heightened regional specialization; from the fifteen or twenty distinct cortical fields in early mammals, humans likely have as many as 150 separate fields (Kaas, 2006, pp. R914).

Larger brains do not have bigger neurons, but rather more numerous neurons. As their number increases, each neuron is connected to a smaller proportion of the total (Kaas, 2000, 2006). Relative interconnectivity then drops, so areas of large brains become increasingly independent or modular, allowing them to specialize and diversify, becoming their own "small worlds" in the neural network (Watts & Strogatz, 1998). To partially counteract the increasing isolation, the total portion of the brain dedicated to axonal connections (white matter) increases more rapidly than the number of neurons (gray matter; see Zhang & Sejnowski, 2000). The increase of white connective matter, however, cannot keep pace with growing gray matter, so even though large brains are disproportionately composed of white matter, they are still less thoroughly connected (Ringo, 1991).

In addition, phylogenetically new connections can arise through "connectional invasion," when brains evolve links between regions that did not have strong axonal connection in ancestors. Terrence Deacon (1990b) proposed, as part of his "displacement hypothesis," that brain regions that grow disproportionately large over evolutionary time tend to send out connectional invasions, innervating regions for which there are no ancestral precedent. Striedter (2005, p. 237) describes "Deacon's rule" as the relation "large equals well-connected": that is, a hypertrophied brain region typically grows connections into brain regions not found in its ancestors. The enlarged primate neocortex, for example, includes a ventral premotor area, homologous to part of Broca's area in humans and specialized in arm and hand movements (Preuss, Stepniewska, & Kass, 1996). Georg Striedter (2005, p. 307) points out that this part of the neocortex has unusual descending direct connections to the spinal cord, "generally indicative of increased dexterity," but also possibly a channel for producing innovative, non-stereotypical hand and arm movements (see also Nudo & Masterton, 1990).

Perhaps the most interesting and best-documented case of connectional invasion, however, is the presence of well-developed projections directly from the human cortex to the brainstem nuclei underlying orofacial motor control (see Iwatsubo, Kuzuhara, Kanemitsu, Shimada et al., 1990; Kuypers, 1958). Deacon (1990a, p. 283) argues that the unusual innervation supports "unprecedented forebrain control of the human larynx. In this regard, we are not just divergent from other mammals but also from all other vertebrates—perhaps the only one with significant forebrain control of laryngeal muscles."

In comparison, laryngeal muscles are under automatic control by the viceromotor system in most animals, but the unusual neocortical invasion in humans may contribute to our ability, alone among land mammals, to learn to produce new sorts of sounds and novel combinations. Control over the muscles that move the face, tongue, mouth, and larynx, and allow inhalation and exhalation, has been split in humans between the automatic viceromotor system and the neocortex through a change in the pattern of brain intraconnection; projections from the neocortex to the brainstem exist in other immature mammals, but are only maintained and reinforced in humans and other primates to a lesser extent by the processes that "wire" our brains.

According to Edelman (2003), brain connectivity is essential for understanding human consciousness. The extensive presence of reentrant connections, often through "association" areas—those without direct external connections to sensory or motor systems, like the prefrontal cortex—grants our brains the more complex forms of self-awareness, evaluation, and other forms of cognitive integration. Increased neocorticalization, and the increase in the proportional strength of and widespread connectional invasion by the neocortex, likely allows greater top-down initiation and inhibition of lower-level functions, such as using visual areas to do mental simulations or seeking to consciously calm the autonomic nervous system, slow breathing, or disregard fear or unpleasant sensations. Neocortical abnormalities in individuals are sometimes found in individuals with anxiety disorders, suggesting that this self-regulatory mechanism can also go afoul (see Berkowitz, Coplan, Reddy, & Gorman, 2007). Reentrant connections would produce a brain better able to muster and redeploy its own neural resources, with greater versatility and ability to learn.

How to "Wet-Wire" a Big Brain

The computer metaphor, especially the idea that instincts or universally shared traits are necessarily "hard-wired" in the brain, is a pervasive misnomer. Brains, especially complex, slow-maturing brains like ours, are better understood as grown, or "wet-wired," with connections emerging through a competitive and selective process that Gerald Edelman has referred to as "neural Darwinism" (1987, 1993). With the shortfall of genetic information revealed by the sequencing of the genomes of complex animals (i.e., the relatively small number of protein-coding genes), many neurologists now believe that brain formation requires significant amounts of non-genetic information, including shaping by interactions within the brain itself, with the body, and with the environment.

Edelman argued, in simplified form, that genetic information underdetermines the structure of the brain; instead, early exuberant production of neural connections is followed by "neuronal group selection," a culling process in which active synaptic connections persevere while underused links gradually weaken and disappear (or become dormant). During embryonic development, the brain produces so many neurons that by adulthood, they contain only 20% to 80% of the number at the peak of neurogenesis, with the rest eliminated by natural processes of cell death (Oppenheim, 1985). Similarly, most neurons connect excessively, but undergo axonal "pruning" (see Stiles & Jernigan, 2010). In the motor system, spinal motor neurons compete to link an axon to muscle cells; those that fail, die. The resulting configuration perfectly matches the motor neurons to the muscles, even if a researcher alters the target number of cells either up or down (Deacon, 2000, p. 277). The postsynaptic cells provide some kind of growth factor that protects the successful neurons from culling.

The visual cortex, likewise, is structured in part by epigenesis and experience, including competition among different cortical regions and axonal connections, and retinal stimulation from patterned light. The gross architecture of the nervous system, including the way that neurons migrate in the embryonic brain and ocular nerves connect the retina to the cortex, reliably determines that the same parts of the brain will act as the primary visual processor because they receive the appropriate stimulation. But the structure is emergent from the selective process and competition within the nervous system; if the conditions of the selective process are shifted, the resulting architecture can vary.

In fact, a range of ingenious experiments has shown just how the selective process can lead the brain to unusual configurations if conditions vary. For example, in the absence of auditory stimulus, the primary auditory cortex can be induced to interpret visual afferents; Mriganka Sur and his team (1988), in one experiment, surgically redirected ocular nerves to the auditory cortex in extremely immature ferret newborns. The auditory cortex successfully interpreted the ocular information in a visual fashion (see also von Melchner, Pallas, & Sur, 2000). The mechanisms that provoke axons to grow, signal them where to go, tell them to sprout collaterals, and stabilize connections are conserved across a range of vertebrates and invertebrates (Chisholm & Tessier-Lavigne, 1999). In fact, Deacon (1997, 2000) discusses "chimera" transplants, in which immature brain cells from one species are transplanted into the embryonic brains of another species, and yet local signaling proteins successfully instruct the formation of the transplanted material to develop into functioning brain matter.

This process of local signaling and neuronal "selection" means that the body of the organism also instructs the maturing nervous system, allowing the brain to adapt to a wide range of conditions, including injuries, disabilities, and even structures with no evolutionary precedent. Law and Constantine-Paton (1981), for instance, transplanted a third eye onto frogs and found that the innervation from the additional organ produced a striped pattern of ocular dominance on the visual area of the tectum, a pattern normally not seen in frogs but found in animals with overlapping fields of vision (like the overlap produced by the third eye). Processes of neuronal selection mean that variation in the demands placed upon the nervous system can be compensated for without a corresponding mutation in a preexisting genetic "blueprint" for the brain; functions will be coordinated by the process of neural and axonal overproduction and culling (see Deacon, 1997, p. 349). In humans, we can see similar neural adaptation manifest where blindness deprives the visual cortex of stimulation; in some remarkable cases, humans can even learn to echolocate, in part because neural selection can reassign the visual cortex to extract spatial information from echoes (see Merabet & Pascual-Leone, 2010; Thaler, Arnott, & Goodale, 2011).

In most mammals, neurogenesis runs its course quite early; whereas fish, birds, and reptiles continue to grow their brains throughout their lifetimes, the mammalian brain tends to be disproportionately large early in development and then grow much less, stopping entirely once the animal has matured. In contrast, primate brains continue to develop longer, producing larger adult brain sizes, a pattern that appears strongly linked to an extension and retardation of maturational processes (see Cutler, 1976).

Much of humans' relative brain expansion is achieved by prolonging the earlier periods of rapid growth, expanding three times in size after birth while most primates' brains only double in size (see Striedter, 2005, p. 319). Ironically, our brains continue to grow even though they pass many crucial developmental landmarks prior to birth that suggest our brains are fairly mature at partition (Clancy, Darlington, & Finlay, 2001). This continued rapid growth makes human brains especially responsive to the environment and susceptible to enculturation even though we may be neurologically "mature" at birth. Preuss (2011) argues that the gap between human and other brains is, in part, the result of genetic modifications to synapse formation and energy metabolism that make the human brain use more energy than expected for a brain of its size. Preuss suggests that our brains are "running hot," with greater plasticity and activity levels.[1] Just as is the case with the individual (Li, 2003), evolution has orchestrated changes across multiple levels of brain development, from the overall size to the cellular structure.

Not a Brain Alone

Richardson (2000) and others have cautioned that the human brain alone does not make intelligence; we should not simply focus on the individual in isolation in thinking about the emergence of human distinctiveness. Just as all of the human brain's distinctiveness is not carried in our genes, all of the differences in human cognition are not produced by the brain alone. The way in which brains develop, especially the extended periods of maturation for our children, makes the human brain evolutionarily primed for learning, domestication, socializing, up-skilling, and enculturation, but also especially altricial or dependent upon caregivers, necessitating that we invest in protecting our children. The anatomical, connective, and molecular peculiarities of our species' neural endowment help us to understand both why evolution selected and exaggerated these properties and how this evolutionary legacy affects our adaptability.

The niche construction model highlights how an organism's own activities can affect its ecological niche to such a degree that selective pressures on that organism change (see Lewontin, 2000; Odling-Smee, 1988; Odling-Smee, Laland, & Feldman, 2003; Turner, 2000b; MacKinnon & Fuentes, chapter 3, this volume). In a simple evolutionary approach, the environment is described as causing adaptation in organisms through selection. In the niche construction approach, members of a species can alter the environment, and thus how selective pressures act on them. Niche modifying creates a feedback loop from behavior to environment to selection in ways that are not generally represented in most evolutionary scenarios. The niche construction model appears to be an extremely robust way of thinking about brain evolution for neuroanthropology because it offers a way to bring together thinking about how patterns of environmental change represent simultaneously adaptation and selective pressure, as subsequent generations face the cumulative effects of altered developmental and evolutionary landscapes. For the neuroanthropological discussion of human brain evolution, understanding cultural evolution as ongoing niche modification offers one way to reconcile what are sometimes seen as competing explanations for our ancestors' unusual evolutionary trajectory, highlighting the complex ways that cognitive, technological, and social changes were simultaneously selective challenges as well as adaptations to rapidly shifting conditions.

Attempts to bring together discussion of biological and cultural evolution are currently quite vigorous. Beginning with the work of Cavalli-Sforza and Feldman (1981), Durham (1991), and Boyd and Richerson (1985), many evolutionary theorists have begun to model biology-culture relations in humans as "dual inheritance" or "gene-culture co-evolution," to capture the way that cultural and biological inheritance mutually shape each other (see also Danchin, Charmantier, Champagne, Mesoudi et al., 2011; Laland, Odling-Smee, & Myles, 2010; Richerson, Boyd, & Henrich, 2010). Other proponents have suggested considering a "triple helix" model of gene, organism, and environment (Lewontin, 1983, 2000), or modeling evolution in "four dimensions": genetic, epigenetic, behavioral, and symbolic (Jablonka, 2001; Jablonka & Lamb, 2004). Each model describes what is essentially a similar recognition in slightly different terms, so we are sympathetic to each approach, which contrast importantly with the assumption by some evolutionary theorists that organismic development is irrelevant and that genes are the sole vehicle for inheritance (Maynard Smith, 2000; see also Griffiths & Gray, 1994; Oyama, 2000; Oyama, Griffiths, & Gray, 2000.

Odling-Smee (2007) suggests that the gains from talking about three channels (or two or four) may be outweighed by the complications of designation; for example, dividing culture from ecology in the abstract sounds straightforward. However, in the complex built environments and anthropogenic "natural" settings that humans have inhabited for a very long time, using fire, hydraulic alterations, game management, selective culling of trees and plants, animal husbandry of varying intensities, and other techniques to subsist, the line between the "environment" and "culture" can blur. We agree with Kendal and colleagues (2011, p. 786) that "niche" is a neutral explanatory device that does not draw a strong distinction between external selective pressures and endogenous pressures created by the organisms themselves; this is an especially useful ambivalence when trying to understand the complex evolutionary dynamics accompanying human brain development (see also Laland et al., 2010 MacKinnon & Fuentes, chapter 3, this volume). Our developmental niche is not simply the speech, handling, child-rearing techniques, and social support of our elders, but also their disease profile, built environment, tool types, epigenetic influences, prior effect on predators, and other elements. Like the evo-devo discussion of genetic inheritance and expression, the niche construction model encourages us to think on both evolutionary and developmental scales simultaneously.

But the bigger question about human brain evolution is not how to talk about it, but why it happened. The relatively consistent allometric relationship between body and brain size across mammals, according to Deacon, has "long been thought to reflect some overarching evolutionary economy of intelligence or metabolism that holds across diverse adaptations and sizes" (1997, p. 339). The human departure begs some explanation, some suggestion of the evolutionary benefits of increased intelligence that might offset the metabolic and other costs of greater encephalization. The problem has often been that we have no shortage of theories—in fact, as Barbara Finlay (2007, p. 294) writes, we have "too many, and all are probably correct, at least in part." For that reason, we will focus on two of the more robust and inclusive theories—social intelligence and cultural niche creation—and point out a third, less well-elaborated theory—perceptual and emotional self-regulation—that coincides with many of the neurological changes we have discussed, and with the goals of neuroanthropology.

Social Intelligence
One of the most influential explanations put forward for the explosive growth in hominin encephalization is the "Machiavellian intelligence" or, less colorfully, the "social intelligence" hypothesis (see Byrne, 1997; Byrne & Whiten, 1989; Dunbar, 1998). The theory built upon early suggestions by Humphrey (1976) that increased intelligence might be more useful for social problem-solving than environmental adaptation, and the realization among many primatologists that our distant cousins led complex, strategically demanding lives (see Byrne, 1997). The social intelligence hypothesis helped explain why a wide range of candidate explanations—such as tool use, complex foraging, and cooperative hunting—had not produced similar brain growth in species that we increasingly realized shared these behaviors. Social selective forces also explained extravagant intellectual abilities, capacities like "theory of mind," which did not seem to directly impact adaptation (see Flinn, Geary, & Ward, 2005, pp. 11–12). The widespread use of "Machiavellian" to describe apparently cooperative behavior was symptomatic of a deeper set of assumptions that, since selection could only be individualistic, a deeper selfish benefit must underwrite any apparently prosocial behavior. The approach highlighted, however, that successful deception and strategizing in social environments required subtle cognitive skills to anticipate what others knew, to simulate complex interactions, and to outwit similarly equipped rivals who were likewise scheming (Barrett et al., 2003).

Intraspecific competition with other hominins has the explanatory advantage of positive feedback: any gain by one of my competitors, a conspecific, will need to be matched, and advances in the trait will quickly spread through a population. The suite of changes in the brain that created a gregarious, cooperative, teachable species also created one subject to escalating dynamics of intraspecies competition: we are our own best friends and worst competitors, propelling brain evolution. In fact, Alexander has suggested that humans became "so ecologically dominant that they in effect became their own principal hostile force of nature," leading to a kind of "runaway social selection" that drove brain development to an extreme (1990, p. 4).

The early focus on "Machiavellian" traits, such as deception (see Byrne & Whiten, 1989), has broadened to take in other roles for "social" intelligence, such as coalition building, cooperation, and the need to balance gains from collective living with the dangers of evolutionary "free riders" (Price, Cosmides, & Tooby, 2002). But social intelligence theories could just as easily focus on communicative and cooperative adaptations, such as the greater neocortical innervation of facial muscles that might lead to more sophisticated emotional expression and communication (see Stedman, Kozyak, Nelson, Thesier et al., 2004), or to the increased likelihood of children surviving infancy with greater circles of cooperative caregiving (Lahdenpera, Lummaa, Helle, et al., 2004).

Flinn and colleagues (2005) argue that, once hominins achieved "ecological dominance"—a state in which extrinsic factors exerted less selective pressure on our ancestors than internal competition—brain growth could slough off a set of prior constraints, such as the dangers of an underdeveloped nervous system for infant mortality. This relaxed external pressure unleashed directional selection for brain development due to social competition, both within and between groups of humans, which were no longer balanced by countervailing selective pressures, sparking a kind of cognitive arms race in areas like communication, self-awareness, theory of mind, and the ability to form coalitions.

Brain Development in a Cultural Niche

Successful invasive species tend to have large brain sizes, likely because greater encephalization confers behavioral flexibility (Sol, Bacher, Reader, & Lefebvre, 2008). Humans seem to offer an extreme confirmation of this pattern, since members of our genus left Africa as the "colonizing ape" (Wells & Stock, 2007). The pattern has been inexorable: invade and transform, with predators and competitors hunted to extinction; forests turned to grasslands or remade to our liking; and watercourses transformed to create dams, weirs, and drains, making the planet over. But, at the same time, we accumulated immense amounts of information about novel environments and resources.

Boyd, Richerson, and Henrich (2011) argue that human adaptive advantage is not simply superior intelligence or the individual's brain, but rather the species-specific ability to create a "cultural niche" full of information, socially transmitted adaptive strategies, and the necessary social scaffolding to acquire environmentally appropriate toolkits. Or, as Merlin Donald (2001, p. xiii) puts it,

The key to understanding the human intellect is not so much the design of the individual brain as the synergy of many brains. We have evolved an adaptation for living in culture, and our exceptional powers as a species derive from the curious fact that we have broken out of one of the most critical limitations of traditional nervous systems—their loneliness, or solipsism. From our earliest birth as a species, humanity has relied upon creating "distributed" systems of thought and memory, in which intellectual work is shared across many nervous systems.

Not the brain alone, but brains in parallel and serial networks generate what Michael Tomasello (1999) has called a "ratchet effect," generating innovation but, more importantly, preserving and disseminating beneficial novelties and information.

The concept builds upon the suggestion by Tooby and DeVore (1987) that humans have occupied a "cognitive niche," a distinctive ecological niche based on the ability to acquire, manipulate, and transfer large amounts of information. Kim Sterelny (2011, p. 814) highlights that these kinds of information transfer require a gamut of channels, including both individual and social adaptation: "High fidelity, high bandwidth social learning depends on both individual adaptations and adapted environments." The niche creation model thematizes this factor, how one generation's adapted environment becomes the next's environment of adaptation: a child born into a cultural niche becomes a target for rich information transfers and is an active culture-assimilating agent, surrounded by human models, material culture, symbols to influence cognitive development, and the full range of human technology as well as intense pressures to learn quickly.

As Boyd and colleagues (2011) point out, no individual, however intelligent or resourceful, could devise the repertoire of techniques routinely deployed in even the "simplest" foraging societies. They vividly cite the myriad ingenuous techniques that circumpolar people employ; in contrast, European explorers who found themselves lost in the same Arctic environments, even though these explorers were supremely confident and accomplished outdoorsmen, often succumbed quickly unless they were able to ally with or learn from local peoples. In one particularly vivid example, the Franklin Expedition became trapped at the King William Island and, although they were perhaps the best provisioned voyage of exploration ever dispatched from Europe, eventually all succumbed to starvation or scurvy. Ironically, King Franklin Island is the heart of Netsilik territory, an area so rich in resources that the main harbor was called *Uqsuqtuuq*, meaning "lots of fat," by the indigenous people (Boyd et al., 2011, p. 3).

The point is that human "intelligence" is not simply determined by our disproportionately large brains, but rather by our collective ability to amass, transfer, improve upon, and deploy information, strategies, skills, and technology. Cultural niches, however, are both selective pressure and adaptive resource. Tools may aid us to survive, but only because we have the perceptual, executive, manual, and memory resources to successfully inhabit a cultural niche.

Emotional, Motivational, and Perceptual Changes

The focus on social and practical *intelligence*, however, can lead us to over-emphasize the strategic, executive, and planning dimensions of human cognitive life. Greater social and cultural complexity also likely required the emergence of both social emotions and forms of self regulation so that we could take in cultural knowledge and excel in social competition (Gross, 1999; see also Turner, 2000a). The human brain has not put aside emotion; on the contrary, our motivational structure and emotions have been complexly interwoven with our perceptual and cognitive abilities (see LeDoux, 1996). Social and cultural life is not just cognitively demanding because we must scheme and cooperate, but because these forms of interaction often require restraint, even self-manipulation or self-deception, but we are also drawn to social interaction for its intrinsic emotional rewards, not solely by a calculation of self-interest.

The descending innervation for our overly neocorticalized brains may allow us to override impulse and distraction, to maintain strategic ambivalence when a less reentrant brain might be unable or uninterested in checking reactive behavior. We have evolved a versatile "executive control" function that includes both planning and strategizing, as well as self-control (Ardila, 2008). Dietrich Stout (2005), for example, has argued that learning how to make stone tools depends, not simply on exquisite control of motor action and behavior, but also of affective states. Becoming distracted, agitated, bored, or enraged will derail the learning process; imitation and observational learning require temperament as well as capacity, and self-regulation would be essential (see also Ambrose, 2001; Stout, 2010). Moreover, as teachers, we must tolerate the proximity and attention of the novice learning.

Michael Tomasello and colleagues (2005) argue that human learning and cultural capacities are most distinct because of "shared intentionality": from the earliest age, we share attention, collaborate in joint action, understand at basic levels others' plans, even share emotions, providing a platform for intensely social forms of awareness. For example, Tomasello and Herrmann (2009) found that human 2-year-olds significantly outpaced apes in social cognition, but not in cognition about causation, space, and quantity. Even when nonhuman animals understand others' intentions, they do not fall in with those intentions, forming a "we" intentionality, so they cannot learn as much from each other or cooperate as seamlessly. Although the proposal is controversial (see the commentary following Tomasello et al., 2005), the account encourages us to recognize the hypercooperativeness of our species: we collaborate in child-rearing, with non-kin, and in situations that seem to offer little individual benefit (see Hill, Barton, & Hurtado, 2009). Our social inclinations lead to extraordinary patterns of empathy, conformity, and imitation, bootstrapped by hypersociality.

A hypercooperative social life provides a particular developmental niche for the immature, rapidly developing nervous system, including, among other key factors, humans' sophisticated forms of scaffolding for skills like joint attention, naming, language acquisition, and imitation. Our species has comparatively muted antagonistic social dynamics that might undermine learning or force children to too quickly become independent, and thus cut off extended periods of cognitive apprenticeship. Chimpanzees, for example, do not share food frequently, even with their own infants (Silk, Brosnan, Vonk, Henrich, et al., 2006). Human adults who consistently shared food and, in some cases, enjoyed interaction with infants, would help to establish a motivational structure that need not focus exclusive on selfish and short-term tactics. Evidence from "enculturated chimpanzees," which are animals raised by humans, suggests that given our developmental environment, they also evidence some of our more subtle human abilities, including capacities like shared attention and imitation, and use tools in ways distinctive from their wild brethren (see Tomasello, Savage-Rumbaugh, & Kruger, 1993).

Conclusion

New research on human evolution, genetics, and comparative neurology leads us to suggest that the widespread understanding of brain evolution, even in closely adjacent academic fields, may not have kept pace with the science. Powerful, but overly simple, models of evolution that assume evolutionary traits will necessarily result in human universals need to give way, not to erase evolutionary explanations, but to provide richer accounts that incorporate data emerging from genetics, paleoanthropology, comparative neuroscience, and anthropology, including research on human diversity (see Bolhuis, Brown, Richardson, & Laland, 2011; Brown, Dickins, Sear, & Laland, 2011). Rather than a strict adaptationist approach to brain function, where functionality is seen to be shaped by outside selective forces in a brain composed of discrete tools, current research on brain evolution offers a more holistic and interesting account: over time, repurposing of neural resources, developmental dynamics, and physiological processes from scaling to connectivity shape what brains can do and why. Social and cultural researchers need to recognize that evolutionary theory is not unified, and certainly has not agreed upon a universalizing, reductionist, genocentric "replicator" approach to Darwinist theory (Walsh, 2010). Recent developments in evolutionary theory can help socially oriented researchers increase the biological plausibility of their work. Evolutionary analogues suggest how varying experience and socioecological conditions, like selective environments and cultural niches, can generate emergent structure through developmental pressure, interaction patterns, and feedback from local environments.

Some practitioners of evolutionary psychology have famously argued for an "evolutionary lag" between our current environment and our species' neurological endowment, or, as Cosmides and Tooby (1997) more prosaically put it, "Our modern skulls house a Stone Age mind." Their argument assumes that brain evolution is gradual. In fact, a meta-analysis of genetic change research across a wide variety of species found that evolution could produce alterations of one standard deviation in a quantitative trait in only twenty-five generations, suggesting that human brains need not necessarily carry unchanged Paleolithic baggage in their "modern skulls" (see Kingsolver, Hoekstra, Hoekstra, Berrigan et al., 2001). A host of genetic studies have suggested that genes affecting brain development, neuron configuration, and metabolism have been under intense, possibly even accelerating pressure in the Holocene, our most recent geological epoch. Williamson and colleagues (2007), for example, estimate that as much as 10% of the human genome has undergone recent selective pressure (see also Hawks, Wang, Cochran, Harpending et al., 2007; Sabeti, Varilly, Fry, Lohmueller et al., 2007; Voight, Kudaravalli, Wen, & Pritchard, 2006; Wang, Kodama, Baldi, & Moyzis, 2006).

Ramped up social demands, unprecedented cultural and niche modification, massive increases in human population, ever-accelerating technological change, wholesale shifts in diet, new pathogens and increased disease load in crowded living conditions—it is not hard to see why our brains might be under selective pressure (see Armelagos & Harper, 2005; Western, 2001). An earlier understanding of brain-culture relations in evolution assumed that culture buffered us against these pressures, that tools and technology were "extra-somatic adaptation" that brought an end to human evolutionary change (e.g., White, 1959). Genetic data and coevolutionary theory, even recent scares about anthropogenic climate change and microbial "super-bugs," have made us less certain that selection has been called off, or even that we can draw a clear line between "natural" and "artificial" selective pressures. The evolution of the human brain suggests that this process is quite old; reinforcing cycles of brain change and behavioral complexity are likely as old as our species. The impact and time-depth of these reinforcing cycles imply that an information processing model of cognition understood to be layered on top of neural physiology does not adequately represent the brain evolution has wrought. Selection helps dictate the design of neural systems, but not in a simple cause-effect way. Humans have achieved a degree of "fitness" through a complex of neural dynamics that intersect with and are open to development, bodily function, social experience, and cultural contexts.

Neuroanthropological research will help to clarify ongoing processes in brain evolution by helping to illuminate the "evolvability" of the human brain, including how its patterns of open, dynamic brain development and socially facilitated forms of cognition lead to cultural diversity. To do this, we need to get away from seeing "culture" as merely information that is transmitted over evolutionary time and recognize that enculturation is, equally, the ways that our interaction with each other shapes our biological endowment, and has been doing so for a very long time. We do not just adapt through culture, but must also adapt *to* culture.

Note

1. For a discussion of the genetic and gene regulatory differences in human brain development, which extends well beyond the scope of this chapter, see excellent reviews and recent discussions by Danchin et al. (2011), McLean, Reno, Pollen, Bassan, et al. (2011), Nowick, Gernata, Almaas, & Stubbs (2009), Vallender (2011), Varki and Altheide (2005), Varki, Geschwind, & Eichler (2008).

References

Aiello, L. C., & Wheeler, P. (1995). The expensive-tissue hypothesis: The brain and the digestive system in human and primate evolution. *Current Anthropology, 36*(2), 199–221.

Alberch, P. (1982). Developmental constraints in evolutionary processes. In J. T. Bonner (Ed.), *Evolution and development* (pp. 313–332). Berlin: Springer.

Aldridge, K. (2011). Patterns of differences in brain morphology in humans as compared to extant apes. *Journal of Human Evolution, 60,* 94–105.

Alexander, R. D. (1990). How did humans evolve? Reflections on the uniquely unique species (Special publication 1). Ann Arbor, MI: University of Michigan Museum of Zoology.

Ambrose, S. (2001). Paleolithic technology and human evolution. *Science, 291,* 1748–1750.

Anderson, M. L. (2010). Neural reuse: A fundamental organizational principle of the brain. *Behavioral and Brain Sciences, 33,* 245–313.

Ardila, A. (2008). On the evolutionary origins of executive functions. *Brain and Cognition, 68,* 92–99.

Armelagos, G. J., & Harper, K. N. (2005). Genomics at the origins of agriculture, part one. *Evolutionary Anthropology, 14,* 68–77.

Barrett, L., Henzi, S. P., & Dunbar, R. I. M. (2003). Primate cognition: From what now to what if? *Trends in Cognitive Science, 7,* 494–497.

Barton, R. A. (2006). Primate brain evolution: Integrating comparative, neurophysiological, and ethological data. *Evolutionary Anthropology, 15,* 224–236.

Barton, R. A., & Dunbar, R. I. M. (1997). Evolution of the social brain. In A. Whiten & R. W. Byrne (Eds.), *Machiavellian intelligence II: Extensions and evaluations* (pp. 240–263). Cambridge: Cambridge University Press.

Barton, R. A., & Harvey, P. H. (2000). Mosaic evolution of brain structure in mammals. *Nature, 405,* 1055–1058.

Berkowitz, R. L., Coplan, J. D., Reddy, D. P., & Gorman, J. M. (2007). The human dimension: How the prefrontal cortex modulates the subcortical fear response. *Reviews in the Neurosciences, 18*(3–4), 191–208.

Bolhuis, J. J., Brown, G. R., Richardson, R. C., & Laland, K. N. (2011). Darwin in mind: New opportunities for evolutionary psychology. *PLoS Biology, 9*(7), e1001109.

Boyd, R., & Richerson, P. J. (1985). *Not by genes alone: How culture transformed human evolution.* Chicago: University of Chicago Press.

Boyd, R., Richerson, P. J., & Henrich, J. (2011). The cultural niche: Why social learning is essential for human adaptation. *Proceedings of the National Academy of Sciences of the United States of America, 108,* 10918–10925.

Brown, G. R., Dickins, T. E., Sear, R., & Laland, K. N. (2011). Evolutionary accounts of human behavioural diversity. *Philosophical Transactions of the Royal Society B: Biological Sciences, 366,* 313–324.

Byrne, R. W. (1997). Machiavellian intelligence. *Evolutionary Anthropology, 5,* 172–180.

Byrne, R. W., & Whiten, A. (Eds.). (1989). *Machiavellian intelligence: Social expertise and the evolution of intellect in monkeys, apes, and humans.* Oxford: Oxford University Press.

Carroll, S. B. (2003). Genetics and the making of Homo sapiens. *Nature, 422,* 849–857.

Carroll, S. B. (2005). *Endless forms most beautiful: The new science of evo-devo and the making of the animal kingdom.* W. W. Norton.

Cavalli-Sforza, L. L., & Feldman, M. W. (1981). *Cultural transmission and evolution: A quantitative approach.* Princeton, NJ: Princeton University Press.

Chisholm, A., & Tessier-Lavigne, M. (1999). Conservation and divergence of axon guidance mechanisms. *Current Opinion in Neurobiology, 9,* 603–615.

Chittka, L., & Niven, J. (2009). Are bigger brains better? *Current Biology, 19,* R995–R1008.

Clancy, B., Darlington, R. B., & Finlay, B. L. (2001). Translating developmental time across mammalian species. *Neuroscience, 105*(1), 7–17.

Corballis, M. C. (1991). *The lopsided ape: Evolution of the generative mind.* New York: Oxford University Press.

Cosans, C. (1994). Anatomy, metaphysics, and values: The ape brain debate reconsidered. *Biology and Philosophy, 9*(2), 129–165.

Cosmides, L., & Tooby, J. (1997). Evolutionary psychology: A primer. Santa Barbara, CA: University of California Santa Barbara Center for Evolutionary Psychology. Retrieved from http://www.psych.ucsb.edu/research/cep/primer.html

Cutler, R. G. (1976). Evolution of longevity in primates. *Journal of Human Evolution, 5*(2), 169–202.

Danchin, É., Charmantier, A., Champagne, F. A., Mesoudi, A., Pujol, B., & Blanchet, S. (2011). Beyond DNA: Integrating inclusive inheritance into an extended theory of evolution. *Nature Reviews Genetics, 12,* 475–486.

Darwin, C. (1890). *The descent of man and selection in relation to sex* (2nd ed.). Retrieved from www.openlibrary.org (Original work published 1874)

Deacon, T. W. (1990a). Fallacies of progression in theories of brain-size evolution. *International Journal of Primatology, 11,* 193–236.

Deacon, T. W. (1990b). Rethinking mammalian brain evolution. *American Zoologist, 30*(3), 629–705.

Deacon, T. W. (1997). What makes the human brain different? *Annual Review of Anthropology, 26,* 337–357.

Deacon, T. W. (2000). Evolutionary perspectives on language and brain plasticity. *Journal of Communication Disorders, 33,* 273–291.

Diamond, J. (1992). *The third chimpanzee: The evolution and future potential of the human animal.* New York: HarperCollins.

Dobzhansky, T. (1973). Nothing in biology makes sense except in the light of evolution. *American Biology Teacher, 35,* 125–129.

Donald, M. (2001). *A mind so rare: The evolution of human consciousness.* New York, NY: W. W. Norton.

Dunbar, R. I. M. (1998). The social brain hypothesis. *Evolutionary Anthropology: Issues, News and Reviews, 6,* 178–190.

Durham, W. (1991). *Coevolution: Genes, culture and human diversity.* Stanford, CA: Stanford University Press.

Edelman, G. M. (1987). *Neural Darwinism: The theory of neuronal group selection.* New York: Basic Books.

Edelman, G. M. (1993). Neural Darwinism: Selection and reentrant signaling in higher brain function. *Neuron, 10*(2), 115–125.

Edelman, G. M. (2003). Naturalizing consciousness: A theoretical framework. *Proceedings of the National Academy of Sciences of the United States of America, 100*(9), 5520–5524.

Falk, D. (1990). Brain evolution in Homo: The radiator theory. *Behavioral and Brain Sciences, 13,* 333–344.

Finlay, B. L. (2007). E Pluribus Unum: Too many unique human capacities and too many theories. In S. W. Gangestad & J. A. Simpson (Eds.), *The evolution of mind: Fundamental questions and controversies* (pp. 294–301). New York: Guilford Press.

Finlay, B. L., & Darlington, R. B. (1995). Linked regularities in the development and evolution of mammalian brains. *Science, 268,* 1578–1584.

Finlay, B. L., Darlington, R. B., & Nicastro, N. (2001). Developmental structure in brain evolution. *Behavioral and Brain Sciences, 24*, 263–308.

Flinn, M. V., Geary, D. C., & Ward, C. V. (2005). Ecological dominance, social competition, and coalitionary arms races: Why humans evolved extraordinary intelligence. *Evolution and Human Behavior, 26*, 10–46.

Gerhart, J., & Kirschner, M. (1997). *Cells, embryos and evolution: Toward a cellular and developmental understanding of phenotypic variation and evolutionary adaptability.* Malden, MA: Blackwell Science.

Goodman, M. (1963). Serological analysis of the systematics of recent hominoids. *Human Biology, 35*, 377–436.

Gould, S. J. (1975). Allometry in primates, with emphasis on scaling and the evolution of the brain. *Contributions to Primatology, 5*, 244–292.

Gould, S. J., & Lewontin, R. C. (1979). The spandrels of San Marco and the Panglossian paradigm: A critique of the adaptationist program. *Proceedings of the Royal Society of London B: Biological Sciences, 205*(1161), 581–598.

Griffiths, P. E., & Gray, R. D. (1994). Developmental systems and evolutionary explanations. *Journal of Philosophy, 91*, 277–304.

Gross, J. J. (1999). Emotion regulation: Past, present, future. *Cognition and Emotion, 13*, 551–573.

Hawks, J., Wang, E. T., Cochran, G. M., Harpending, H. C., & Moyzis, R. K. (2007). Recent acceleration of human adaptive evolution. *Proceedings of the National Academy of Sciences of the United States of America, 104*, 20753–20758.

Hill, K., Barton, M., & Hurtado, A. M. (2009). The emergence of human uniqueness: Characters underlying behavioral modernity. *Evolutionary Anthropology, 18*, 187–200.

Holloway, R. L. (1966). Cranial capacity, neural reorganization, and hominid evolution: A search for more suitable parameters. *American Anthropologist, 68*, 103–121.

Holloway, R. L. (1979). Brain size, allometry, and reorganization: Toward a synthesis. In M. E. Hahn, C. Jensen, & B. C. Dudek (Eds.), *Development and evolution of brain size: Behavioral implications* (pp. 59–88). New York: Academic.

Holloway, R. L. (1992). The failure of the gyrification index (GI) to account for volumetric reorganization in the evolution of the human brain. *Journal of Human Evolution, 22*, 163–170.

Holloway, R. L. (1995). Toward a synthetic theory of human brain evolution. In J.-P. Changeux & J. Chavaillon (Eds.), *Origins of the human brain* (pp. 42–54). Oxford: Clarendon Press.

Holloway, R. L. (2008). The human brain evolving: A personal retrospective. *Annual Review of Anthropology, 37*, 1–19.

Holloway, R. L., Broadfield, D. C., & Yuan, M. S. (2004). *The human fossil record* (Vol. 3, *Brain endocasts: The paleoneurological evidence).* New York: Wiley.

Humphrey, N. K. (1976). The social function of intellect. In P. P. G. Bateson & R. A. Hinde (Eds) *Growing Points in Ethology* (pp 303–317). Cambridge: Cambridge University Press.

Huxley, T. H. (1863). *Evidence as to man's place in nature.* New York: D. Appelton (reprinted in 2007).

Iwatsubo, T., Kuzuhara, S., Kanemitsu, A., Shimada, H., & Toyokura, Y. (1990). Corticofugal projections to the motor nuclei of the brainstem and spinal cord in humans. *Neurology, 40*(2), 309–312.

Jablonka, E. (2001). The systems of inheritance. In S. Oyama, P. E. Griffiths, & R. D. Gray (Eds.), *Cycles of contingency: Developmental systems and evolution* (pp. 99–116). Cambridge, MA: MIT Press.

Jablonka, E., & Lamb, M. J. (2004). *Evolution in four dimensions: Genetic, epigenetic, behavioral, and symbolic variation in the history of life.* Cambridge, MA: MIT Press.

Jacob, F. (1977). Evolution and tinkering. *Science, 196*(4295), 1161–1166.

Jerison, H. J. (1973). *Evolution of the brain and intelligence.* New York: Academic Press.

Jiao, Y., Medina, L., Veenman, C. L., Toledo, C., Puelles, L., & Reiner, A. (2000). Identification of the anterior nucleus of the ansa lenticularis in birds as the homolog of the mammalian subthalamic nucleus. *Journal of Neuroscience, 20*(18), 6998–7010.

Kaas, J. H. (2000). Why is brain size so important: Design problems and solutions as neocortex gets bigger or smaller. *Brain and Mind, 1*, 7–23.

Kaas, J. H. (2006). Evolution of the neocortex. *Current Biology, 16*(21), R910–R914.

Kendal, J., Tehrani, J. J., & Odling-Smee, J. (2011). Human niche construction in interdisciplinary focus. *Philosophical Transactions of the Royal Society B: Biological Sciences, 366*, 785–792.

Kingsolver, J. G., Hoekstra, H. E., Hoekstra, J. M., Berrigan, D., Vignieri, S. N., Hill, C. E., et al. (2001). The strength of phenotypic selection in natural populations. *American Naturalist, 157*(3), 245–261.

Kirschner, M. W., & Gerhart, J. C. (2005). *The plausibility of life: Resolving Darwin's dilemma.* New Haven, CT: Yale University Press.

Kuypers, H. G. (1958). Corticobular connexions to the pons and lower brain-stem in man: An anatomical study. *Brain, 81*(3), 364–388.

Lahdenpera, M., Lummaa, V., Helle, S., Tremblay, M., & Russell, A. F. (2004). Fitness benefits of prolonged post-reproductive lifespan in women. *Nature, 428*(6979), 178–181.

Laland, K. N., Odling-Smee, J., & Myles, S. (2010). How culture shaped the human genome: Bringing genetics and the human sciences together. *Nature Reviews Genetics, 11*, 137–148.

Law, M. I., & Constantine-Paton, M. (1981). Anatomy and physiology of experimentally induced striped tecta. *Journal of Neuroscience, 1*(7), 741–759.

LeDoux, J. E. (1996). *The emotional brain.* New York: Simon & Schuster.

Lewontin, R. (1983). Gene, organism, and environment. In D. S. Bendall (Ed.), *Evolution from molecules to men* (pp. 273–285). Cambridge, England: Cambridge University Press.

Lewontin, R. (2000). *The triple helix: Gene, organism and environment.* Cambridge, MA: Harvard University Press.

Li, S.-C. (2003). Biocultural orchestration of developmental plasticity across levels: The interplay of biology and culture in shaping the mind and behavior across the life span. *Psychological Bulletin, 129*(2), 171–194.

Love, A. C. (2007). Functional homology and homology of function: Biological concepts and philosophical consequences. *Biology and Philosophy, 22,* 691–708.

MacLarnon, A. M., & Hewitt, G. P. (1999). The evolution of human speech: The role of enhanced breathing control. *American Journal of Physical Anthropology, 109,* 341–363.

Marks, J. (2002). *What it means to be 98% chimpanzee.* Berkeley, CA: University of California Press.

Marks, J. (2005). Phylogenetic trees and evolutionary forests. *Evolutionary Anthropology, 14,* 49–53.

Maynard Smith, J. (2000). The concept of information in biology. *Philosophy of Science, 67,* 177–194.

Mayr, E. (1960). The emergence of evolutionary novelties. In S. Tax (Ed.), *Evolution after Darwin* (Vol. 1, pp. 349–380). Chicago: University of Chicago Press.

McLean, C. Y., Reno, P. L., Pollen, A. A., Bassan, A. I., Capellini, T. D., Guenther, C., et al. (2011). Human-specific loss of regulatory DNA and the evolution of human-specific traits. *Nature, 471,* 216–219.

Merabet, L. B., & Pascual-Leone, A. (2010). Neural reorganization following sensory loss: The opportunity of change. *Nature Reviews Neuroscience, 11*(1), 44–52.

Nowick, K., Gernata, T., Almaas, E., & Stubbs, L. (2009). Differences in human and chimpanzee gene expression patterns define an evolving network of transcription factors in brain. *Proceedings of the National Academy of Sciences of the United States of America, 106*(52), 22358–22363.

Nudo, R. J., & Masterton, R. B. (1990). Descending pathways to the spinal cord, IV: Some factors related to the amount of cortex devoted to the corticospinal tract. *Journal of Comparative Neurology, 296,* 584–597.

Odling-Smee, F. J. (1988). Niche-constructing phenotypes. In H. C. Plotkin (Ed.), *The role of behavior in evolution* (pp. 73–132). Cambridge, MA: MIT Press.

Odling-Smee, F. J. (2007). Niche inheritance: A possible basis for classifying multiple inheritance systems in evolution. *Biological Theory, 2,* 276–289.

Odling-Smee, F. J., Laland, K. N., & Feldman, M. W. (2003). *Niche construction: The neglected process in evolution.* Princeton, NJ: Princeton University Press.

Oppenheim, R. W. (1985). Naturally occurring cell death during neural development. *Trends in Neurosciences, 8,* 487–493.

Oxnard, C. E. (2004). Brain evolution: Mammals, primates, chimpanzees, and humans. *International Journal of Primatology, 25*(5), 1127–1158.

Owen, R. (1857). On the characters, principles of division, and primary groups of the class Mammalia. *Journal of the Proceedings of the Linnean Society, 2,* 1–37.

Oyama, S. (2000). *The ontogeny of information: Developmental systems and evolution* (2nd ed.). Durham, NC: Duke University Press.

Oyama, S., Griffiths, P. E., & Gray, R. D. (Eds.). (2000). *Cycles of contingency: Developmental systems and evolution.* Cambridge, MA: MIT Press.

Passingham, R. (1973). Anatomical differences between the neocortex of man and other primates. *Brain, Behavior and Evolution, 7,* 337–359.

Passingham, R. (2008). *What is special about the human brain?* Oxford, England: Oxford University Press.

Petrides, M., Cadoret, G., & Mackey, S. (2005). Orofacial somatomotor responses in the macaque monkey homologue of Broca's area. *Nature, 435,* 1235–1238.

Pilbeam, D., & Gould, S. J. (1974). Size and scaling in human evolution. *Science, 186*(4167), 892–901.

Preuss, T. M. (2001). The discovery of cerebral diversity: An unwelcome scientific revolution. In D. Falk & K. R. Gibson (Eds.), *Evolutionary anatomy of primate cerebral cortex* (pp. 138–164). Cambridge: Cambridge University Press.

Preuss, T. M. (2011). The human brain: Rewired and running hot. *Annals of the New York Academy of Sciences, 1225*(S1), E182–E191.

Preuss, T. M., Stepniewska, I., & Kass, J. H. (1996). Movement representation in the dorsal and ventral premotor areas of owl monkeys: A microstimulation study. *Journal of Comparative Neurology, 371*(4), 199–214.

Price, E. P., Cosmides, L., & Tooby, J. (2002). Punitive sentiment as an anti-free rider psychological device. *Evolution and Human Behavior, 23*(3), 203–231.

Puelles, L., & Rubenstein, J. L. (2003). Forebrain gene expression domains and the evolving prosomeric model. *Trends in Neurosciences, 26*(9), 469–476.

Rakic, P. (2009). Evolution of the neocortex: Perspective from developmental biology. *Nature Reviews Neuroscience, 10*(10), 724–735.

Richardson, K. (2000). *The making of human intelligence.* New York: Columbia University Press.

Richerson, P. J., Boyd, R., & Henrich, J. (2010). Gene-culture coevolution in the age of genomics. *Proceedings of the National Academy of Sciences of the United States of America, 107*(Suppl. 2), 8985–8992.

Rilling, J. K. (2006). Human and nonhuman primate brains: Are they allometrically scaled versions of the same design? *Evolutionary Anthropology, 15,* 65–77.

Rilling, J. K., & Insel, T. R. (1999). The primate neocortex in comparative perspective using magnetic resonance imaging. *Journal of Human Evolution, 37,* 191–223.

Ringo, J. L. (1991). Neuronal interconnection as a function of brain size. *Brain, Behavior and Evolution, 38*(1), 1–6.

Ringo, J. L., Doty, R. W., Demeter, S., & Simard, P. Y. (1994). Time is of the essence: A conjecture that hemispheric specialization arises from interhemispheric conduction delay. *Cerebral Cortex, 4*(4), 331–343.

Riska, B., & Atchley, W. R. (1985). Genetics of growth predicts patterns of brain size evolution. *Science, 229*(4714), 1302–1304.

Rosenberg, K., & Trevathan, W. (1995). Bipedalism and human birth: The obstetrical dilemma revisited. *Evolutionary Anthropology, 4*(5), 161–168.

Ruff, C. B., Trinkaus, E., & Holliday, T. W. (1997). Body mass and encephalization in Pleistocene Homo. *Nature, 387,* 173–176.

Sabeti, P. C., Varilly, P., Fry, B., Lohmueller, J., Hostetter, E., Cotsapas, C., et al. (2007). Genome-wide detection and characterization of positive selection in human populations. *Nature, 449,* 913–918.

Schoenemann, P. T. (2006). Evolution of the size and functional areas of the human brain. *Annual Review of Anthropology, 35,* 379–406.

Semendeferi, K., Lu, A., Schenker, N., & Damasio, H. (2002). Humans and great apes share a large frontal cortex. *Nature Neuroscience, 5,* 272–276.

Sherwood, C. C., Subiaul, F., & Zawidzki, T. W. (2008). A natural history of the human mind: Tracing evolutionary changes in brain and cognition. *Journal of Anatomy, 212*(4), 426–454.

Silk, J. B., Brosnan, S. F., Vonk, J., Henrich, J., Povinelli, D. J., Richardson, A. S., et al. (2006). Chimpanzees are indifferent to the welfare of unrelated group members. *Nature, 437*, 1357–1359.

Smil, V. (2002). *The Earth's biosphere: Evolution, dynamics, and change.* Cambridge, MA: MIT Press.

Sol, D., Bacher, S., Reader, S. M., & Lefebvre, L. (2008). Brain size predicts the success of mammal species introduced into novel environments. *American Naturalist, 172*, S63–S71.

Stedman, H. H., Kozyak, B. W., Nelson, A., Thesier, D. M., Su, L. T., Low, D. W., et al. (2004). Myosin gene mutation correlates with anatomical changes in the human lineage. *Nature, 428*, 415–418.

Stephan, H., Frahm, H., & Baron, G. (1981). New and revised data on volumes of brain structures in insectivores and primates. *Folia Primatologica, 35*(1), 1–29.

Sterelny, K. (2011). From hominins to humans: How sapiens became behaviourally modern. *Philosophical Transactions of the Royal Society B: Biological Sciences, 366*, 809–822.

Stiles, J., & Jernigan, T. L. (2010). The basics of brain development. *Neuropsychology Review, 20*, 327–348.

Stout, D. (2005). The social and cultural context of stone-knapping skill acquisition. In V. Roux & B. Bril (Eds.), *Stone knapping: The necessary conditions for a uniquely hominin behaviour* (pp. 331–340). Cambridge: McDonald Institute for Archaeological Research.

Stout, D. (2010). The evolution of cognitive control. *Topics in Cognitive Science, 2*, 614–630.

Striedter, G. S. (2005). *Principles of brain evolution.* Sunderland, MA: Sinauer.

Striedter, G. S. (2006). Précis of Principles of brain evolution. *Behavioral and Brain Sciences, 29*, 1–12.

Sur, M., Garraghty, P. E., & Roe, A. W. (1988). Experimentally induced visual projections into auditory thalamus and cortex. *Science, 242*(4884), 1437–1441.

Thaler, L., Arnott, S. R., & Goodale, M. A. (2011). Neural correlates of natural human echolocation in early and late blind echolocation experts. *PLoS ONE, 6*(5), e20162.

Tomasello, M. (1999). *The cultural origins of human cognition.* Cambridge, MA: Harvard University Press.

Tomasello, M., Carpenter, M., Call, J., Behne, T., & Moll, H. (2005). Understanding and sharing intentions: The origins of cultural cognition. *Behavioral and Brain Sciences, 28,* 675–691.

Tomasello, M., & Herrmann, E. (2009). Ape and human cognition: What's the difference? *Current Directions in Psychological Science, 19*(1), 3–8.

Tomasello, M., Savage-Rumbaugh, S., & Kruger, A. C. (1993). Imitative learning of actions on objects by children, chimpanzees, and enculturated chimpanzees. *Child Development, 64,* 1688–1705.

Tooby, J., & DeVore, I. (1987). The reconstruction of hominid behavioral evolution through strategic modeling. In W. G. Kinzey (Ed.), *Primate models of hominid behavior* (pp. 183–237). New York: SUNY Press.

Turner, J. H. (2000a). *On the origins of human emotions: A sociological inquiry into the evolution of human affect.* Stanford, CA: Stanford University Press.

Turner, J. S. (2000b). *The extended organism: The physiology of animal-built structures.* Cambridge, MA: Harvard University Press.

Vallender, E. J. (2011). Comparative genetic approaches to the evolution of human brain and behavior. *American Journal of Human Biology, 23,* 53–64.

Varki, A., & Altheide, T. K. (2005). Comparing the human and chimpanzee genomes: Searching for needles in a haystack. *Genome Research, 15,* 1746–1758.

Varki, A., Geschwind, D. H., & Eichler, E. E. (2008). Explaining human uniqueness: Genome interactions with environment, behaviour and culture. *Nature Reviews Genetics, 9,* 749–763.

Voight, B. F., Kudaravalli, S., Wen, X., & Pritchard, J. K. (2006). A map of recent positive selection in the human genome. *PLoS Biology, 4*(3), e72.

Von Melchner, L., Pallas, S. L., & Sur, M. (2000). Visual behavior mediated by retinal projections directed to the auditory pathway. *Nature, 404,* 871–876.

Wagner, A. (2005). *Robustness and evolvability in living systems.* Princeton, NJ: Princeton University Press.

Walsh, D. M. (2010). Two neo-Darwinisms. *History and Philosophy of the Life Sciences, 32,* 317–340.

Wang, E. T., Kodama, G., Baldi, P., & Moyzis, R. K. (2006). Global landscape of recent inferred Darwinian selection for Homo sapiens. *Proceedings of the National Academy of Sciences of the United States of America, 103*(1), 135–140.

Watts, D. J., & Strogatz, S. H. (1998). Collective dynamics of "small-world" networks. *Nature, 393,* 440–442.

Wells, J. C. K., & Stock, J. T. (2007). The biology of the colonizing ape. *Yearbook of Physical Anthropology, 50*, 191–222.

Western, D. (2001). Human-modified ecosystems and future evolution. *Proceedings of the National Academy of Sciences of the United States of America, 98*(10), 5458–5465.

White, L. (1959). The concept of culture. *American Anthropologist, 61*(2), 227–251.

Wildman, D. E., Uddin, M., Liu, G., Grossman, L. I., & Goodman, M. (2003). Implications of natural selection in shaping 99.4% nonsynonymous DNA identity between humans and chimpanzees: Enlarging genus Homo. *Proceedings of the National Academy of Sciences of the United States of America, 100*(12), 7181–7188.

Williamson, S. H., Hubisz, M. J., Clark, A. G., Payseur, B. A., Bustamante, C. D., & Nielsen, R. (2007). Localizing recent adaptive evolution in the human genome. *PLOS Genetics, 3*, 901–915.

Zhang, K., & Sejnowski, T. J. (2000). A universal scaling law between gray matter and white matter of cerebral cortex. *Proceedings of the National Academy of Sciences of the United States of America, 97*(10), 5621–5626.

II Case Studies on Human Capacities, Skills, and Variation

5 Memory and Medicine

M. Cameron Hay

"You may write them down for now, but you must study them. The words have to enter you. Once they have entered here," the spunky ancient healer gave my upper abdomen a firm pat, "then you don't have to worry when to use them. You will already know." She looked around in the dim space to see if anyone was trying to eavesdrop on this transfer of powerful knowledge. We listened for the deep breathing of her sleeping son, two grandchildren, a neighbor's child, and her son's younger friend, all lying where they had nodded off on the covered bamboo platform that she called home. Reassured, she continued to dictate the knowledge of healing. Healing in this rural Indonesian community depended on memory, on having words enter the body, so that when confronted with illness one will "already know" the words that initiate healing.

A few years later and a few thousand miles away, I was part of a research team interviewing urban physicians on their informational resources for medical decision making. The interviews were usually conducted in offices, with computers humming in the background and immaculately dressed physicians glancing at their watches to make sure our interview did not run over the allotted lunch hour. A number of physicians mentioned that the physical examination of a patient usually just confirmed the diagnosis they already had in mind. As one physician put it, "Honestly, after talking with most patients, I have a plan in my mind of what I want to do even before I take a look at them. And sometimes I'm wrong ... and I'll change that plan, but already I've formulated a plan based on what they've said."

This paper is about how medical information is socially and neurologically organized in ways that enable healers—whether in rural Indonesia or urban California—to immediately know a patient's diagnosis and what to do about it. In comparing the two medical traditions, I argue that each medical tradition emphasizes different kinds of memories and augments neurological memory processes in different ways, yet each tradition works to bring information immediately to mind when needed. Although the medical traditions of Sasak healers and US physicians are in many ways different, I propose a neuroanthropological model of knowledge that compares how sociocultural traditions and neurological processes co-create distinctive possibilities for remembering.

I argue three key points in this chapter. First, memory and medicine co-create each other constantly within local contexts. The neurobiology of memory structures how knowledge is retrieved and organized in the brain and, crucially, how it accumulates and becomes automatic through continual social processes of legitimation and repetition. The sociocultural traditions of medicine—the structures of medical learning within each tradition and the assumptions about learning that drive those structures—shape what kinds of knowledge are seen as vital, which in turn shapes how memories are organized, and ultimately how that medical tradition maintains itself across generations. The second point is that these co-creation processes are not limited to realms of medicine—anytime learning takes place, the brain and the local sociocultural world are interacting, creating, and prioritizing certain kinds of memory and thus certain ways of thinking. Medicine is but one example of how traditions of thought and practice are irreducibly biocultural. And third, in order to understand all of this, we need to reframe our own thinking, which has been too long reinforced in academic institutions by irrationally separating biological sciences from the social sciences and humanities. As Dominguez and his colleagues recently put it, culture is in the brain just as the brain is in culture (2009).

For some time now, a handful of leading scholars have been crossing the spurious disciplinary boundaries. People primarily associated with social science fields, such as Mel Konner, Carol Worthman, and Sarah Hrdy, have called for a deep appreciation—if not downright awe—of our neurobiological systems (see Konner, 1982; Hrdy, 2009; Worthman, 1992). Likewise neuroscientists and other biological scientists, such as Robert Sapolsky, Norman Cousins, Michael Irwin, and Antonio Damasio, have nurtured a deep appreciation, if not awe, of the astonishing ways in which individuals' social interactions, perceptions, emotions, and values impact them physiologically (see Sapolsky, 1994; Cousins, 1989; Irwin, 2008; Damasio, 2003). To build upon the paths of joint inquiry these scholars have outlined, we need to continue to develop two things: models that offer mechanisms for bridging disciplines and enabling biocultural understandings, and a restructuring of academic institutions to encourage transdisciplinary questions and collaborative research. Here, using medical traditions and memory structures, I offer one model, and hope that as more researchers insist on exploring transdisciplinary questions with biocultural models, a richer understanding of our enculturated, biological being in the world will be possible.

Traditions of Knowledge

Medicine—as it is organized, taught, and practiced within a given context—is what Fredrik Barth has called a tradition of knowledge (1987): a recognizable set of cultural ideologies about what knowledge is and how one knows and uses it. These ideologies in turn shape the social processes of educating others to know and use that knowledge. Through deep descriptions of traditions of knowledge, and drawing on multidisciplinary data, we can make comparisons explaining how healing knowledge differs across contexts and why (Hay, 2009).

Cross-cultural analyses have long emerged at the blurred edges of anthropology and psychology. Cultural psychologists and psychological anthropologists have gained enormous insight by making comparisons across cultural worlds on such topics as childrearing practices (e.g., Weisner, 1976; Ochs & Izquierdo, 2009) and illness experiences (e.g., Kleinman & Good, 1985; Hinton & Good, 2009; Hirsch et al., 2010). In each case, researchers focus on exploring similar problems of living—how do people raise responsible children? how do people respond to catastrophic cognitions?—bracketing analysis so that different contexts can be compared. In proposing a comparison between different medical traditions, the difficulty is to make comparisons in a way that approximates the experiences and constraints of living in a particular world without stereotyping.

In "The Guru and the Conjurer," Fredrik Barth (1990) demonstrates such an approach while comparing Muslim leaders in Bali and Ok initiators in Inner Papua New Guinea. Specifically, Barth found that in New Guinea "the value of knowledge is enhanced by veiling it and sharing it with as few as possible," whereas in Bali there is "no merit from even the deepest religious knowledge unless you *teach* it" (1990, p. 641). From these basic orientations, Barth shows how the processes through which cultural knowledge is transmitted are strikingly different. In New Guinea, the job of the Ok initiator is to conjure "a spellbinding performance" in which deeply secret knowledge is partially revealed through a series age-graded rituals that gradually transfer to boys the mysterious knowledge of senior men (1990, p. 643). In Bali, the job of the Muslim religious guru is to talk to people, explain religious texts to them, and regularly integrate new knowledge into his teaching so that students always have more to learn. Thus Islam—which is spread through open explication and thus easily communicable to large numbers of people at a time—is widely distributed throughout the world whereas the Ok religion is limited to those people who can personally go through the requisite rituals to experientially grasp the secretive knowledge. By careful comparison of these two traditions of knowledge and their consequences, Barth can explain how some religious traditions become widely distributed and others do not, without resorting to value judgments about the content of the traditions.

Adopting Barth's model, I compare medical traditions and explore the consequences each tradition likely has on the memory processes of healing specialists, enabling comparison through a rather unique lens. In this paper, I do not make any argument about efficacy or medical validity, and instead focus on the neurological strengths of each tradition.

Medical Traditions as Biocultural Traditions

Most research on the transmission of knowledge in anthropology has focused exclusively on the sociocultural dimensions of knowledge transmission, while assuming an underlying biological universality—for example, that the brains of Sasak twenty-somethings and Californian twenty-somethings are identically capable of learning, retaining, and recalling the same knowledge. However, if we take seriously the notion that a person's social, cultural, and physical environments have "an active role in driving cognitive processes," then mental experiences—one's patterns of learning, thinking, remembering, knowing—are also bioculturally co-emergent (Henningsen & Kirmayer, 2000, pp. 472–473; Worthman, 1992; Rubin, 2006). This means that neither biology nor context can be studied with blinders on. There is no acultural biological being, just as cultural worlds are constructed and shaped through the ways our bodies interact with the environment.

Medical practice emerges within a biocultural context that shapes both orientations about what constitutes valid medical knowledge and the way the brain of a healer learns and remembers that knowledge. Examining these processes in the context of insights into memory from neuroscience and psychology point to the potential neurological processes that are at work within particular traditions of healing. Thus I explore the possible biocultural variations embedded in the ways people think about and act upon a universal human event, namely, illness.

Developing Connections

I started questioning the possible connections between cultural assumptions within medical traditions and locally augmented neural networks gradually, as data from multiple research and educational projects comingled on my desk.

I conducted two years of fieldwork in a rural hamlet on the island of Lombok, Indonesia, from 1993 to 1995. My research focused on how local medical traditions enabled Sasaks to cope with illness and the constant specter of death. In a resulting book (Hay, 2001), I showed how the local medical traditions rely on memory—specifically secret, verbatim memories—to prevent and heal illness. After completing the manuscript, I became increasingly intrigued with the question of why, of all the possible ways of treating illness, a people would rely on secret, verbatim memories.

From 2003 to 2008 I was involved in various research, teaching, and learning projects with physicians and medical students in Southern California that included a year of seminars in psychoneuroimmunology (see Hay, 2010; Strathmann & Hay, 2008, 2009; Hay, Cadigan, Khanna, Strathmann et al., 2008; Hay, Strathmann, Lieber, Wick et al., 2008). Then, with a bit of background in neurology, I traced how the Sasak medical tradition, reliant on verbatim memories, would work neurologically (Hay, 2009) and became curious as to whether the very different American medical tradition would be reflected in a different neurological emphasis on memory. In this comparison chapter, I draw on the Sasak materials and on the initial exploratory interviews conducted in 2007 with 39 physicians and medical students, which were part of a multidisciplinary research project examining medical information usage and decision making in the United States (Hay, Weisner, Subramanian, Duan, et al., 2008).

Comparisons

In many ways, Sasak and US medical traditions are incomparable. The Sasak medical tradition emerged and is practiced in an impoverished world currently striving to recover from centuries of internal and external domination, exploitation, and extraction. Especially among the rural peoples, malnutrition is commonplace, material possessions are few, literacy is low, and life expectancy is about 50 years. Their medical tradition sustains their self-sufficiency; they do not wait for outsiders (or doctors, scientists, new research, or new funding) to come in and save them. Medical knowledge is in the form of *jampi*, inherently potent sequences of words; jampi are passed down vertically, memorized verbatim, and kept secret. For Sasaks, the real threat to their medical tradition is that people can forget or water down the strength of a jampi by giving it to too many people, and so they concentrate social and cognitive resources on remembering secrets verbatim.

In contrast, the American biomedical tradition is a written medical tradition with considerable financial, technological, and social resources. This medical tradition exists in a world where most people are literate, well-nourished, and have a life expectancy of about 78. Medical knowledge is assumed to be always improving, externally validated, and enhanced by experience. The primary threat within this medical tradition is that physicians will not be competent—unable to master and assimilate the emergent and relevant scientific knowledge to accurately diagnose and treat the patient at hand.

I argue that the neurological processes of Sasak healers and American physicians are different not because one group is "smarter" or more technologically advantaged than the other, but because neurological processes of medical information encoding, organization, and retrieval co-evolved within particular ideological, social, technological and institutional traditions to meet the local expectations and needs of ill people.

Overview of Memory and Its Neural Structures

Memory is currently understood as a collection of processes embedded within or, more accurately, catalyzed by specific brain structures (e.g., Budson & Price, 2005; Squire 2004). There are multiple memory systems including the four that seem to be critical to medical traditions: episodic, semantic, procedural, and working. *Episodic memory* involves conscious recollection of an experienced event or episode, such as the memory of examining one's first patient or the memory of a bit of knowledge learned once in a specific context. The term was coined by Endel Tulving (1972, 1983), who distinguished it from generalized knowledge, which he called *semantic memory* (for example, knowing that penicillin is an antibiotic, or knowing the symptoms associated with a disease). Both episodic and semantic memories are explicit and declarative, enabling content knowledge. Knowing how to do something—what Brenda Milner initially called motor learning and skill (e.g. Milner 1968)—is enabled by procedural memory. *Procedural memory* refers to learned motor behaviors or skills that become unconscious; for example, how to ride a bicycle or take out an appendix. Episodic, semantic, and procedural memories can be stored for minutes to years (Budson & Price, 2005, p. 693). I argue that whereas the Sasak medical tradition emphasizes episodic memory, the American medical tradition engages all three, with a particular emphasis on semantic and procedural memories. Finally, *working memory* involves allocating attentional and linguistic resources for seconds or minutes to address problems at hand (Budson & Price, 2005), and may be relevant in both traditions as part of healers' conscious efforts to address unfamiliar illnesses.

There are multiple brain structures involved in memory encoding, integration, and recall. The regions that are activated during acts of remembering suggest that there is considerable overlap in the areas where different types of memories are "stored." I find more intriguing the organizers and catalyzers of memory—the structures that enable memory usage for problem solving. First and most critical is the hippocampus, a small structure deep in the brain located within a larger structure called the medial temporal lobe. An intact and healthy hippocampus is pivotal to episodic memory encoding and later recall, information sequencing, and to the effortful problem-solving endeavors of working memory, which engages both semantic and episodic memories (Eichenbaum & Fortin, 2003; Brassen et al., 2006; Budson & Price, 2005). The hippocampus may even be involved in the initial encoding of memories later integrated into hippocampus-independent semantic memory clusters (Hardt, Einarsson, & Nader, 2010). The second crucial structure is the basolateral amygdala, also deep in the medial temporal lobe. The basolateral amygdala consists of a complex of nuclei that are stimulated by stress responses. The amygdala is crucial for mediating biological or emotional stress, focusing various brain structures to gather memories to resolve the stressor. The amygdala particularly stimulates the hippocampus for episodic and effortful semantic memory recall, and the basal ganglia for more subconscious semantic or procedural memory recall (McGaugh et al., 2002; Roozendaal, 2000, p. 232). The basal ganglia, a complex of nuclei within the cerebral cortex, is the last structure that catalyzes memory processes, specifically by enabling procedural memory to coordinate movement independent of the hippocampus (Budson & Price, 2005; Grahn, Parkinson, & Owen, 2009, p. 59). The hippocampus, the amygdala, and the basal ganglia thus constitute neural organizers for encoding and catalyzers for drawing upon episodic and semantic memories, emotionally valenced memories, and procedural and semantic memories, respectively.

By examining which types of memories are fostered within the different medical traditions, we can link these to the neural structures that are associatively augmented.

The Sasak Medical Tradition

Approximately 2.5 million people inhabit the island of Lombok, the vast majority of whom are ethnically Sasak. Lombok has among the worst health statistics in the Indonesian archipelago, with an infant and child mortality rate between 30 and 40 percent in the mid-1990s and currently hovering at three to five times the national rate (World Health Organization, 2007). In rural hamlets like the one where I conducted fieldwork, death often happened quickly, and a burial a week in the community of 800 was common. The island has a biomedical public health care system; however, rural Sasaks resort to such care facilities rarely and only consider clinics after treatments with jampi because biomedical care is not understood to be able to *heal* illness, but only to be able to facilitate a faster recovery (see Hay, 2001).

Jampi are potent, rotely memorized formulae that are central to medical practice. When someone is ill, people respond by seeking jampi. For Sasaks, jampi act on the ills that threaten the lives of those they cherish. The efficacy of words, which are believed to pragmatically act upon the world (Kang, 2006), is a fundamental premise in many Indonesian societies (e.g., Kuipers, 1990). For Sasaks, jampi are, to apply Webb Keane's description, "compulsively effective in themselves" (Keane, 1997, p. 59). Thus, to memorize a jampi is to memorize a procedure for acting on illness.

Jampi are also exclusively oral knowledge (aside from my note taking). The Sasak emphasis on oral, rote memorization is not uncommon in societies that also have a written tradition (Goody, 1998), but Sasak jampi are unusual in that there is no accompanying written text. If forgotten, the healing power of the jampi is potentially lost forever. This precariousness may in part explain the social emphasis on and possible local neurological strength in verbatim memorization (see also Bartlett, 1955, on "rote representation").

Jampi are learned during ritualized distribution and usually passed down within families. The giver and receiver of jampi isolate themselves from all others, sometimes by going to a field hut, sometimes by waiting until the dead of night. They then work sleeplessly, often for "a day and a night," to transmit the memorized lines. Each jampi is considered a discrete gift. Sasaks could always name who gave them a jampi, and many could describe the specific context (the time of day and the place) in which they received it. Because jampi are not relearned or introduced in other contexts, they are also associated with a single encoding episode, and as such best fit with psychologists' descriptions of episodic memories, or memories of a specific event.

Not every Sasak is adept at learning jampi, and the memory literature often reports deterioration in precise recall over time (Hamann & Canli, 2004; Hay, 2001); in the Sasak worldview, such forgetting would lead to an erosion in jampi potency and healing ability. Perhaps for this reason, Sasaks engage in certain practices to enhance precise memory—what psychologist Rubin calls "multiple constraints" (1995).

The first such constraint is the internally consistent content of all jampi. They begin with a standard evocation of Allah, followed by four to six lines uniquely describing the problem and its solution (e.g., the passages of bodily fluids are blocked, and the blockages are removed by this jampi). Jampi conclude with two formulaic lines commanding the illness to leave the body. Information that is organized in culturally known story grammars shows superior recall (Rubin, 1995, p. 29), and the meaningful structure of linking a problem with its solution likely facilitates associations in which remembering one element cues the recall of correlated elements (Ledoux & Gordon, 2006).

The second constraint on jampi is the limited number of words they include. Studies of verbatim, verbal memory show that it is possible for most people to commit at least 18 words to memory (McIver & Carroll, 2004). In the 41 jampi that I was given, there was an average of 27 words per jampi. If the eight words that are identical in all jampi are subtracted from the total, this would place the length within range of estimates for verbatim memory capacity. Because jampi are so important to memorize precisely, people likely practice jampi subvocally, until the recitation of a jampi becomes habitual or, in a sense, embodied in the rhythmic movement of lips.

A third constraint is that the encoding of jampi takes place in secret; the more people who know a jampi, the weaker it is said to become. Jampi are considered valuable heirlooms to be craved and coveted. Being chosen to receive the gift of jampi is a great honor, and receivers look forward to the gift with anticipation and often with ritual preparation, such as fasting. For receivers, this transmission period is their only opportunity to check their accuracy, thus further increasing the pressure to memorize jampi correctly. Indeed, one person described how her father had slapped her arm so that it stung each time she had incorrectly recited a jampi she was trying to memorize. Heightened emotion, particularly stress, facilitates verbal memory encoding (Abercrombie, Kalin, Thurow, Rosenkranz et al., 2003) and enhances memory accuracy, likely through the basolateral amygdala (Lupien & Lepage, 2001; Lupien, Maheu, Tu, Fiocco, et al., 2007). In short, moderate anxiety experienced while receiving a jampi should enhance memory accuracy (Dolan, 2002; Kensinger & Schacter, 2006).

Fourthly, while anyone over the age of 5 has been given at least a few jampi, healers are the primary repositories of jampi and may have what psychologists call "memory expertise" (Rubin, 1995, pp. 167–170). Because memories are more likely to be recollected if they are compatible with beliefs about the self (Bruner, 2002; Woike, Zavezzary, & Barsky, 2001), this expertise may be related to a self-concept in which healers see themselves as people who have strong memory abilities (in Sasak, *ndeqku tauq lupaq;* roughly, "I do not forget"). Healers may have particularly strong capacities for spontaneous recall. Individual differences in the ability to remember emotionally encoded jampi may account for why a relatively small proportion of Sasaks become renowned as healers (*dukuns* or *belians*).

Lastly, memorizing jampi is motivated by potential future need; it is prospective, in that memory requires intentional encoding of the item with when and how the item should be used (Brandimonte, Einstein, & McDaniel, 1996). Most research in memory encoding for prospective usage has shown continual monitoring of the environment for cues as to when to use the remembered item. These cues can be subtle and contextually embedded, requiring that the hippocampus recognize a cue and link it with the remembered item, which for Sasaks would be the associated memorized jampi.

To summarize, the Sasak medical tradition assumes that illness can only be healed by secret, inherently potent, and precisely memorized words. This orientation toward medical knowledge shapes a mode of transmission in which knowledge is memorized during discrete episodes, with multiple constraints to ensure memory precision.

The American Medical Tradition

There are approximately 972,000 licensed physicians in the United States, each of whom underwent training lasting seven to ten years: three to four years in classrooms and the rest in clinical apprenticeships (American Medical Association, 2011). After formal training, physicians are expected to continue refining their medical knowledge by attending conferences, reading journals, talking with colleagues, going online, and learning from their own patient cases. All of this information is weighed, valued, and merged to become what is widely referred to as *clinical experience* or, more simply, *experience*. A number of anthropologists have studied how physicians learn and practice medicine (Good & Good, 1993; Good, 1995; Konner, 1988; Hahn & Gaines, 1985). Here I focus on the assumptions about medical knowledge and knowledge distribution practices that affect the kinds of memories physicians use to solve clinical problems.

The first assumption is that competent practice is scientifically based. In order to become competent, trainees must master the scientific language, methods, techniques, and evidence that facilitate the best outcomes for patients. Although currently minimized in many medical schools, the mastery of the semantic knowledge of basic medical science—such as anatomy and physiology, algorithms or decision trees, disease symptoms and processes—remains crucial to passing school, licensure, and board examinations. Clinical observation involves learning episodically, through memorable events such as a students' first encounter with a patient with AIDS. Ironically, although more memorable, such episodic knowledge is generally less valued because it is considered anecdotal and thus, as one physician put it, "almost the lowest level of evidence that you could have." The goal of training is to use episodes of clinical observation and reasoning to master general principles and the scientific (valid, replicable, and generalizable) medical knowledge.

Semantic medical knowledge, particularly in the form of a memorizable guideline, is commonly seen as the solution to the "problem" of variation in physician diagnosis and treatment (Kostopoulou & Wildman, 2004). The assumption is that, if followed, guidelines would produce consistent patient outcomes (Woolf, Grol, Hutchinson, Eccles et al., 1999). Nonetheless, students of medicine in the United States show relatively poor accuracy for rote memorization (Woods, Brooks, & Norman, 2007). Indeed, relying on memory is increasingly considered suboptimal, with some even arguing that examinations should be open-book:

I would argue that *it is a lazy and dangerous habit for physicians to rely on memorized facts* rather than look them up to verify their accuracy. Closed-book examinations encourage memorization and *reinforce the bad habit of relying on the infallibility of memory* rather than on reference materials. (Keller, 2011, emphasis added).

While people within the Sasak medical tradition are concerned about forgetting, in the American medical tradition, relying on memory can be a "lazy and dangerous habit." Indeed, valued knowledge for physicians is "evidence-based" knowledge, confirmed by scientific research and published in books, journals, and online. Thus externalized, knowledge is not subject to irretrievable loss if forgotten.

The second assumption is that all scientific knowledge is potentially wrong and can be improved. This is one of the fundamental premises of science: "Scientific knowledge is a body of statements of varying degrees of certainty—some most unsure, some nearly sure, none absolutely certain" (Feynman, 1955/1999, p. 146). There are approximately 75 medical trial reports published each day in some 5,500 journals indexed on Medline (Bastian, Glasziou, & Chalmers, 2010). Medical students and physicians develop highly attuned critical thinking skills to help them sort through this blizzard of shifting medical information; indeed, one medical student quipped that learning how to critique journal articles was "what grad school is all about." Critical thinking about journal articles did not necessarily translate for medical students into thinking critically about medical information from other authoritative sources, because "It's almost like some of the homework's been done for you." In contrast, a senior physician commented that "I am the expert [they call to write the articles in something like Up-To-Date, but] I would take even that with a grain of salt because I realize that lots of times [the author] is just somebody who decided what they thought and wrote it down. But that doesn't mean it's necessarily right." These comments suggest that while critical thinking is taught in medical school, information associated with authority is less likely to be questioned, whereas as physicians gain their own authority, the amount of received information they tag with skepticism may actually increase.

The third and final assumption is that knowledge is not transferred as a finished product from one individual to the next, but instead must be integrated into one's prior knowledge. Medical educators expect medical students to develop their own "model of the disease process" by integrating memorized basic science and disease knowledge with personal understanding gained through clinical experience (Patel, Evans, & Kaufman, 1990, p. 130). As a family practitioner with twelve years of experience put it, "You can't just go by what the guidelines say—you have to sort of know how to interpret them and adapt them to your practice."

Knowledge integration occurs over time and is associated with competency. Physicians with accurate verbatim recall of disparate facts are actually less competent at diagnosis because "they tend to treat pieces of information as isolated entities rather than spontaneously picking up the theme or gist that connects the pieces" (Lloyd & Reyna, 2009, p. 1333). This kind of reasoning leads to five to ten times less accurate diagnoses than using the schemas or pattern-recognition strategies relied on by experienced physicians (Coderre, Mandin, Harasym, & Fick, 2003, p. 698). This suggests that acquiring clinical experience seems to involve a reorganization or integration of knowledge into schemas that enable pattern recognition. However, physicians are only better diagnosticians within the realm of their specialty (Ericsson, 2004, pp. S75–S76). In other words, if a physician had not previously treated a person with a particular disorder, that physician may neither have relevant schemas or discrete facts (perhaps previously pruned through disuse) to draw upon in making a diagnosis.

To the general agreement of her peers, one medical student said, "I think the best way to remember is through the actual experience in having a patient or your resident telling you to do something for that patient. ... It just sticks in your mind more than being lectured about it." An "actual experience" with a patient is an opportunity for connecting an episodic event to semantic memory. In expecting clinical practice to facilitate the reorganization and integration of medical knowledge, novices correctly assume that competency requires a personal understanding of integrated medical knowledge.

Reorganizing Memories for Effective Usage

So how do we understand this reorganization of knowledge that hovers on the blurred edges of memory encoding and retrieval? Based on evidence that cognitive stimuli can automatically engage motor action and that stimulation of motor action automatically triggers conceptual processing, the embodied cognition framework implies that cognitive (semantic) memories are in part (although not exhaustively) grounded in motor processes, or procedural memories (Mahon & Caramazza, 2009, pp. 41–42). Indeed, developing this kind of memory association across motor and cognitive processes would explain why performing the activities of clinical practice seems to facilitate students' better understanding of related medical concepts.

When procedural and semantic knowledge are further integrated with episodic memories, these constitute what psychological studies of medical decision making call schemas (also referred to as scripts or gist memories; e.g., Schmidt & Rikers, 2007; Lloyd & Reyna, 2009; Wang & Morris, 2010). Schemas are associated with clinical experience, expertise, and high diagnostic accuracy, and are understood to be formed cognitively through memory consolidation. Memory consolidation refers to the integration of information: that is, the retrieval or reactivation of memory that results in modification, recalibration, and integration of new with old knowledge (Hardt, Einarsson, & Nader, 2010, p. 150). The emerging evidence is that the hippocampus is crucial to this process of memory consolidation, because hippocampal activation is associated with memory plasticity (Henke, 2010) and with associating new information with previously known information in the neocortex. The associations are tentative "unless the new hippocampally processed information is interleaved within existing, activated cortical frameworks (schema)" (Wang & Morris, 2010, p. 53). In other words, with the learning of each new bit of cognitive information, the hippocampus seems to make tentative explorations for preexisting informational complexes within which the new information would fit. In the absence of such a framework, the hippocampus may hold on to that new information as an episodic memory (if it has emotional salience). Otherwise, beyond a window of about two days, the information will be "deposited" with few synaptic connections, rendering it a weak and essentially forgotten memory (Tse, Langston, Kakeyama, Bethus et al., 2007). The more often an informational complex or consolidated knowledge framework is reactivated by the hippocampus, the stronger central elements of that framework become, while weaker elements (those with fewer synaptic linkages) are gradual pruned (Hardt, Einarsson, & Nader, 2010, p. 160).

Selective forgetting enables webs of knowledge to form schemas that are likely to be independent of hippocampus stimulation (Hardt, Einarsson, & Nader, 2010, p. 148). Once schemas are formed, relevant new information can be assimilated rapidly, even in semantic form (Tse, Langston, Kakeyama, Bethus, et al., 2007, p. 82; Hardt, Einarsson & Nader, 2010). The use of schemas among American medical experts explains a number of research observations: why first-year students are more overwhelmed by new information than more experienced students, and why specialists may forget details of a patient case or an algorithm and yet still make more rapid and accurate diagnoses than novices who remember these details. Schema formation also may clarify how some knowledge—perhaps that which had been tagged with skepticism—may be peripheralized and more subject to selective forgetting. The integration and accumulation of information reconsolidated into schemas may yield the automatic or near-unconscious response of hippocampal-independent decision making, thereby explaining how expert physicians can diagnose and develop treatment plans for clinical problems within their specialty while drawing nearly exclusively on "experience."

More broadly, within a world in which knowledge must be integrated and subject to improvement over time, neurological processes that prune less important synapses amid reinforcement of central memory connections would create schemas of tightly integrated knowledge. Among Sasak healers, the schemas that are regularly reinforced by social emphasis are simply precisely memorized jampi, each associated with symptom cues. For the American medical tradition, experts have pruned the less important memories, integrated core knowledge, and retain the ability to tweak any component knowledge tagged with skepticism if new and contradictory knowledge is gained. Neither kind of schema is preordained by some force of "culture," but rather represents the interplay of cultural assumptions about knowledge with the perceived usefulness of particular nexuses of knowledge in the synaptic webs of memory.

  To summarize the argument thus far, the Sasak and American medical traditions have quite different assumptions about medical knowledge, which in turn shape the ways that knowledge is learned and thus encoded in memory. Sasak medical knowledge is primarily tightly constrained, episodic learning dependent on the hippocampus, medial temporal lobes, and prefrontal cortex. In contrast, American medical knowledge is encoded semantically, episodically, and procedurally and over time reconsolidated into schemas (probably within the neocortex) that can be accessed through the hippocampus as well as through the basal ganglia. Pragmatically, then, the key problem for healers in both traditions is the problem of knowing what knowledge to use when.

Familiar Illness

When I asked the ancient Sasak healer how I would know which jampi to use when, she reiterated that I had to let the jampi become a part of me. Then, she said, "You see the patient, recognize the illness name, and you know already which jampi is right." This kind of reliance on spontaneous retrieval depends on focused encoding between the remembered item, the cue, and the intended action (Einstein, McDaniel, Thomas, Mayfield, et al., 2005). When memorizing a jampi, the receiver is told what signs are indicative of the associated illness. The secret and emotionally laden context during which healers learn jampi, perhaps in combination with subvocal rehearsal, is likely to use the hippocampal associative system to tightly connect the jampi, the cues, and the intended action so that they can be accessed in a relatively rapid, automatic process (Moscovitch, 1994).

Because the subvocal recital of jampi is also the procedure for acting on illness, it is possible that the procedural memory structures may also engage in the recall process, but given that jampi are primarily words verbatim memorized, the available evidence suggests that recall necessarily engages the hippocampus (Einstein et al., 2005).

However, if the memory necessary for healing is not primarily episodic, as the American data implies, then a different retrieval process may be at work. In our study, in many cases, physicians reported that after listening to patients, they usually just knew what to do. We found that, regardless of years in clinical practice, if a patient presents with symptoms that the physician treats regularly, 96% of physicians will rely on experience to diagnose and treat the patient, with the remaining 4% relying on known algorithms or guidelines. Experts draw automatically on their evidence-based integrated experience to recognize disease patterns and know what course of action will lead to the best outcomes.

Even outside their areas of expertise, experienced physicians show the ability to take in information and more quickly reorient their clinical thinking than more junior counterparts. Experienced physicians may be open to evidence and can automatically shift schemas to fit the newly perceived need, as one doctor with 20 years of experience reported:

"I have a patient with high blood pressure who comes in and she didn't take her medicine. I'm talking to her about compliance, and then I notice a bandage on her leg. It turns out it's a stab wound, and she was stabbed by her husband. The reason she's not taking her medicine is because her medicine's at home, and her husband's at home, and she's not going home. So you know, it changes from control of blood pressure to compliance issues to a social service emergency."

By continuing to take in information and reorienting his response accordingly, the physician demonstrated the mental flexibility to move among multiple available schema (de Bruin, Camp, & van Merriènboer, 2010). In simple cases, this may be what enables expert physicians to make good, automatic decisions without "conscious thought" (Mamede, Schmidt, Rikers, Custers et al., 2010).

Experience in a particular domain likely emerges as a web of interconnected semantic, episodic, and procedural knowledge that has been reconsolidated into schemas. These schemas can be activated through two pathways. One path activates the hippocampus, taking in new information from the patient at hand and rapidly fitting that information into an existing schema (Hardt, Einarsson, & Nader, 2010). Alternatively, schemas may be activated through the basal ganglia, perhaps specifically the caudate. The caudate is connected structurally and functionally to both cognitive (the medial, ventral, and dorsolateral prefrontal cortex) as well as procedural (the motor cortex) brain structures. The cumulative evidence suggests that the caudate coordinates goal-directed behavior—such as solving the problem of the patient at hand—without requiring attentional resources (Grahn, Parkinson, & Owen 2009, p. 54).

While physicians may rely on automatically accessible schema that they refer to as "experience" to make diagnoses and treatment plans for familiar clinical problems, even veteran physicians regularly check medications and dosages, as one reported: "If I am considering some kind of medication, I will often go online and double check on the dosage ... [and] I will check to make sure that there's not going to be an interaction [with] other medications. So I do not trust myself to keep that in my head." This last comment suggests that physicians may not trust memory of precise facts, but because they are so good at pattern recognition, and with technologies enabling easy access to precise knowledge, physicians resort to memory and external knowledge to solve clinical problems.

Unfamiliar Illness

When a person's illness is unfamiliar, both Sasak and American healers resort to more effortful remembering and, for American healers, seek external knowledge resources. Neurologically, the switch from the automatic recall of familiar information to recognizing a need for effortful remembering happens within 300 to 500 milliseconds (Kompas, Olsson, Larsson, & Nyberg, 2009).

Among Sasak healers, effortful remembering is likely facilitated by low levels of nonclinical anxiety and its associated biological stress response resulting from evaluating a situation as a threat (Lazarus, 1999; McEwan, 2000, p. 173). In their community, most people were closely or distantly related to all the others. Acute illness among any one was cause for potential anxiety, even for healers. Sasak conversations and affiliative gathering around seriously ill people served regulatory functions. When few people gathered, anxiety levels were low and jampi treatments were few. The larger the crowd, the more focused conversations about illness, the higher the anxiety levels, and the more jampi treatments. Socially regulated, anxiety may do more than merely orient attention toward the problem to be solved. It may also facilitate effortful recall.

In episodic memory processes, the right hemisphere seems to be more important for recollection. Although not uncontroversial (see e.g., Owen, 2003), there is a good deal of evidence supporting right-hemisphere memory retrieval, particularly for episodic and effortful recall (Blanchet, Belleville, & Lavoie, 2003; Fletcher, Frith, & Rugg, 1997; Tulving, Kapur, Craik, Moscovitch, & Houle, 1994). In response to a potential stressor, norepinephrine activates the amygdala, and likely the thalamus, to appraise the stimulus (Berridge & Waterhouse, 2003; Tsigos & Chrousos, 2000). If appraisal confirms the threat, other structures, including the hippocampus, are activated to retrieve potentially relevant episodic memories connecting them to the current situation. Because episodic memories are permanently linked to hippocampal activation, the networks linking the amygdala and hippocampal activities in the right hemisphere are capitalized upon in a medical tradition when effortful recall is necessary and anxiety is likely.

Once remembered, there are virtually no costs to a Sasak jampi treatment. Administering one jampi does not prohibit others. The cost for the healer is only a few moments of time. The financial cost for the patient is minimal, often nonexistent for ordinary ills, and for unusual illness the cost is *at most* a small basket of food with a few coins (worth no more than one day's labor). While reputations of healers can increase or decline as the result of patient outcomes, even in a case in which gossip blamed a mother's death on a midwife's lack of knowledge, the midwife's reputation was not permanently damaged (Hay, 1999).

More is at stake in American medical treatment decisions, and yet anxiety is considered a distraction. Physicians speak of the importance of "not getting too involved" and "maintaining emotional distance" in order to preserve their clinical objectivity. Emotional detachment is related to the need to think critically about and refine semantic information, distinguishing it from the less valued anecdotal knowledge of episodic memories. Indeed, anxiety's usefulness for seeking memories may be correlated to how knowledge has been encoded in the brain.

Semantic retrieval can be aided by autobiographical components (e.g., episodic memories of specific patients) through the associative work of the hippocampus (Ryan, Cox, Hayes, & Nadel, 2008), yet people can recall semantic items despite hippocampal lesion (Eichenbaum & Fortin, 2003). Thus, whereas episodic recall reliably activates the right hippocampus and structures in the right hemisphere, semantic recall activates the left hemisphere and, depending on the task, may activate the hippocampus as well as other structures in the right hemisphere (Wiggs, Weisberg, & Martin, 1999, p. 115). This implies that when physicians are seeking memories for unfamiliar clinical problems, amygdala stimulation of the right hemisphere might distract cognitive attention toward episodic memories and away from recall efforts of semantic information.

Perhaps not surprisingly, in unfamiliar cases, physicians tend not to trust their memory at all and instead access external sources such as trusted online sites or calling on trusted colleagues to "kind of remind you of the disease that you haven't seen for ages." In interviews, physicians told of unfamiliar cases in which they considered any memories untrustworthy. Yet by gaining new semantic knowledge, they likely reconsolidate it with the episodic memory of the current case, potentially initiating for them a new schema.

Whereas jampi are never modified intentionally, American physicians frequently manipulate treatment protocols to benefit the patient at hand, and do so while weighing potential costs and benefits. In tailoring treatment, it is conceivable that the amygdala might conceivably activate the hippocampus to focus cognitive resources on integrating new information to solve the problem at hand. Says one senior pediatrician:

"Take albuterol. It's not really approved for children under two. But it's used all the time. I mean what else do you do? ... We do it all the time. You look at what the standard of care is and then decide if you're willing to go offline."

Although the physician remembers evidence-based semantic knowledge, he also realizes that the recommended treatment will not work for the patient, and—probably through the hippocampus—integrates multiple kinds of information to develop a creative solution. Once integrated, the information becomes schematic and possibly hippocampus-independent so that, as the pediatrician said, "we do it all the time."

In summary, the hippocampus seems to be the pivotal structure on which Sasak healing always depends (often with additional stimulation from the amygdala), and on which American healing may rely to reconsolidate memories in order to build schemas.

A Neuroanthropology of Memory and Medicine

This analysis of basic assumptions about medical knowledge and corresponding processes of learning suggests that each medical tradition augments neural memory structures differently. For Sasaks, all jampi must be learned with absolute precision, otherwise knowledge is lost forever. Therefore, the Sasak medical tradition is tightly constrained, with social mechanisms for memorizing and precisely remembering verbal memories, thus taking advantage of hippocampal associative systems that facilitate rote memory, encoding, and recall while augmenting these processes with the emotional valencing of the amygdala. In contrast, the American medical tradition is reliant on publicly available, written knowledge and has a broad distrust for the accuracy of memory. In this context, medical knowledge is acquired through integrating memorized semantic knowledge with experientially gained episodic and procedural memories. While integration is likely facilitated by the hippocampus, once knowledge is formed into schemas, physicians may access these schemas using hippocampus-independent pathways. The argument is thus simply that, even in such things as medical traditions, the brain is encultured, its memory processes fortified and structurally shaped in ways that foster the remembering of valued knowledge in a particular context.

The neuroanthropological model of knowledge that I outline—deeply describing traditions in ways that highlight the social and neurological craft of learning and using knowledge—enables grounded, fair comparisons across contexts. Examining memory structures implies that each medical tradition may have its own strengths. Unlike their American expert counterparts, the Sasak emphasis on verbatim memories may enable expert healers to be excellent diagnosticians immediately linking memorized symptoms with a particular name and jampi treatment, even if that particular healer had never before witnessed that sickness. In contrast, the American emphasis on knowledge reconsolidation and refinement may be particularly well suited to developing diagnoses and treatment strategies for new diseases, such as AIDs or avian flu (H5N1). Biomedical knowledge could be made available to Sasak healers in ways that acknowledge and take advantage of their particular neurological strengths in precise recall. In turn, American healers may be able to learn tools that would strengthen their ability to remember precise medical information in contexts where there is poor access to external informational resources (books, journals, or the Internet). If examined as experts specializing in particular memory processes, healers in both traditions discussed here would bring something of value to any mutual conversations.

My focus is on medical traditions, but any tradition of knowledge—whether of academics or artists—is, I suggest, biocultural. Each tradition of knowledge is guided by local assumptions about information, has particular methods for distributing information, and fosters specific kinds of learning and remembering processes, augmenting neural pathways accordingly. A neuroanthropological analysis of locally valued knowledge and associated recall pathways thus contributes to the expanding range of models for examining human biocultural existence.

To truly harness the potential of a neuroanthropology of knowledge, there needs to be disciplinary and institutional support for multidisciplinary, mixed-methods team research projects. We need prospective ethnographic, psychological and fMRI studies with apprentices from different knowledge traditions, such as those of Sasak and American healers. Neurologically, we could learn how different memory systems are increasingly augmented over time, and at what cost to neuroplasticity and the acquisition of new information. Psychologically, we could gain a far better understanding of the relationships between knowledge acquisition, memory organization, and recall. Anthropologically, we could assess just how strong knowledge assumptions and correlated distribution processes are in shaping how people in a given context think. This kind of research could lead to profound insights into the developmental windows of neuroplasticity as well as the impact augmented memory structures have on the opportunities of individuals and the societies in which they live.

At this stage we can only speculate on what we could learn by jumping with both feet into the research possibilities of neuroanthropology. I use the Sasak and American medical materials to illustrate the possible futures for a neuroanthropology of memory because these are the data I know best, but we need more than these two rough examples. A neuroanthropology of medicine and memory needs deep, multidisciplinary studies of memory in multiple traditions throughout the world. These will offer new insights into how memory structures the possibilities of expert knowledge traditions like medicine, and simultaneously how knowledge traditions structure the possibilities of memory.

Acknowledgments

The completion of this chapter is due to the unstinting encouragement and patience of Daniel Lende and Greg Downey. The data reported is due to the generous collaboration of Sasak and US colleagues and friends, and was based upon work supported in part by a Pilot Research Award from the National Multiple Sclerosis Society and grants from the National Science Foundation (SBR 93139 and SES 0137921) and Pfizer (#PG009184; R. L. Kravitz and N. Duan, PIs, T.S. Weisner, co-PI). I particularly thank Tom Weisner, Eli Lieber, Claudia Solari, Saskia Subramanian, Arman Haghighatgoo, and Janet Keller. Data analysis was conducted using Dedoose software. This chapter was written in residence at the Miami University John E. Dolibois Center in Luxembourg.

References

Abercrombie, H. C., Kalin, N. H., Thurow, M. E., Rosenkranz, M. A., & Davidson, R. J. (2003). Cortisol variation in humans affects memory for emotionally laden and neutral information. *Behavioral Neuroscience, 117,* 505–516.

American Medical Association. (2011). *Physician characteristics and distribution in the U.S.* Chicago: American Medical Association.

Barth, F. (1987). *Cosmologies in the making.* Cambridge: Cambridge University Press.

Barth, F. (1990). The guru and the conjurer: Transactions in knowledge and the shaping of culture in Southeast Asia and Melanesia. *Man, 25,* 460–553.

Bartlett, F. C. (1955). *Remembering: A study in experimental and social psychology.* Cambridge: Cambridge University Press.

Bastian, H., Glasziou, P., & Chalmers, I. (2010). Seventy-five trials and eleven systematic reviews a day: How will we ever keep up? *PLoS Medicine, 7*(9), e1000326.

Berridge, C. W., & Waterhouse, B. D. (2003). The locus coeruleus-noradrenergic system: Modulation of behavioral state and state-dependent cognitive processes. *Brain Research Reviews, 42*(1), 33–84.

Blanchet, S., Belleville, S., & Lavoie, M. E. (2003). Item-related versus task-related activity during encoding and retrieval in verbal and non-verbal episodic memory: An event-related potential study. *Cognitive Brain Research, 17,* 462–474.

Brandimonte, M., Einstein, G. O., & McDaniel, M. A. (1996). *Prospective memory: Theory and applications.* Mahwah, NJ: Lawrence Erlbaum.

Brassen, S., Weber-Fahr, W., Sommer, T., Lehmbeck, J. T., & Braus, D. F. (2006). Hippocampal-prefrontal encoding activation predicts whether words can be successfully recalled or only recognized. *Behavioural Brain Research, 171,* 271–278.

Bruner, J. (2002). *Making stories.* New York: Farrar, Straus, and Giroux.

Budson, A. E., & Price, B. H. (2005). Memory dysfunction. *New England Journal of Medicine, 352,* 692–699.

Coderre, S., Mandin, H., Harasym, P. H., & Fick, G. H. (2003). Diagnostic reasoning strategies and diagnostic success. *Medical Education, 37,* 695–703.

Cousins, N. (1989). Belief becomes biology. *Advances, 6*(3), 20–29.

Damasio, A. (2003). *Looking for Spinoza: Joy, sorrow, and the feeling brain.* New York: Harcourt.

de Bruin, A. B. H., Camp, G., & van Merriënboer, J. J. G. (2010). Available but irrelevant: When and why information from memory hinders diagnostic reasoning. *Medical Education, 44,* 948–950.

Dolan, R. J. (2002). Emotion, cognition, and behavior. *Science, 298*(5596), 1191–1194.

Dominguez, J. F., Lewis, E. D., Turner, R., & Egan, G. (2009). The brain in culture and culture in the brain: A review of core issues in neuroscience. In J. Y. Chiao (Ed.), Cultural neuroscience: *Cultural influences on brain function* (Vol. 178, pp. 43–64). New York: Elsevier.

Eichenbaum, H., & Fortin, N. (2003). Episodic memory and the hippocampus: It's about time. *Current Directions in Psychological Science, 12*(2), 53–57.

Einstein, G. O., McDaniel, M. A., Thomas, R., Mayfield, S., Shank, H., Morrisette, N., et al. (2005). Multiple processes in prospective memory retrieval: Factors determining monitoring versus spontaneous retrieval. *Journal of Experimental Psychology, 134*(3), 327–342.

Ericsson, K. A. (2004). Deliberate practice and the acquisition and maintenance of expert performance in medicine and related domains. *Academic Medicine 79*(10, Supplement):S70–S81.

Feynman, R. (1999). The value of science. In *The pleasure of finding things out* (pp. 141–149). New York: Basic Books. (Original work published 1955)

Fletcher, P. C., Frith, C. D., & Rugg, M. D. (1997). The functional neuroanatomy of episodic memory. *Trends in Neurosciences, 20*(5), 213–218.

Good, B., & Good, M.-J. D. (1993). "Learning medicine": The constructing of medical knowledge at Harvard Medical School. In S. Lindenbaum & M. Lock (Eds.), *Knowledge, power, and practice* (pp. 81–107). Berkeley, CA: University of California.

Good, M.-J. D. (1995). *American medicine: The quest for competence.* Berkeley, CA: University of California Press.

Goody, J. (1998). Memory in oral traditions. In P. Fara & K. Patterson (Eds.), *Memory* (pp. 73–94). Cambridge: Cambridge University Press.

Grahn, J. A., Parkinson, J. A., & Owen, A. M. (2009). The role of the basal ganglia in learning and memory: Neuropsychological studies. *Behavioural Brain Research, 199*, 53–60.

Hahn, R. A., & Gaines, A. D. (Eds.). (1985). *Physicians of Western medicine: Anthropological approaches to theory and practice.* Dordrecht, The Netherlands: D. Reidel Publishing.

Hamann, S., & Canli, T. (2004). Individual differences in emotion processing. *Current Opinion in Neurobiology, 14*, 233–238.

Hardt, O., Einarsson, E. O., & Nader, K. (2010). A bridge over troubled water: Reconsolidation as a link between cognitive and neuroscientific memory research traditions. *Annual Review of Psychology, 61*, 141–167.

Hay, M. C. (1999). Dying mothers: Maternal mortality in rural Indonesia. *Medical Anthropology, 18*(3), 243–279.

Hay, M. C. (2001). *Remembering to live: Illness at the intersection of anxiety and knowledge in rural Indonesia.* Ann Arbor, MI: University of Michigan Press.

Hay, M. C. (2009). Anxiety, remembering, and agency: Biocultural insights for understanding Sasaks' responses to illness. *Ethos, 37*(1), 1–31.

Hay, M. C. (2010). Suffering in a productive world: Chronic illness, visibility, and the space beyond agency. *American Ethnologist, 37*(2), 259–274.

Hay, M. C., Cadigan, J. R., Khanna, D., Strathmann, C. M., Altman, R., McMahon, M., et al. (2008). Prepared patients: The Internet and new rheumatology patients. *Arthritis Care and Research, 59*(4), 575–582.

Hay, M. C., Strathmann, C., Lieber, E., Wick, K., & Giesser, B. (2008). Why patients go online: Multiple sclerosis, the Internet, and physician-patient communication. *Neurologist, 14*(6), 374–381.

Hay, M. C., Weisner, T. S., Subramanian, S., Duan, N., Niedzinski, E. J., & Kravitz, R. L. (2008). Harnessing experience: Exploring the gap between evidence-based medicine and clinical practice. *Journal of Evaluation in Clinical Practice, 14,* 707–713.

Henke, K. (2010). A model for memory systems based on processing modes rather than consciousness. *Nature Neuroscience, 11,* 523–532.

Henningsen, P., & Kirmayer, L. J. (2000). Mind beyond the net: Implications of cognitive neuroscience for cultural psychiatry. *Transcultural Psychiatry, 37*(4), 467–494.

Hinton, D. E., & Good, B. (Eds.). (2009). *Culture and panic disorder.* Stanford, CA: Stanford University Press.

Hirsch, J. S., Wardlow, H., Smith, D. J., Phinney, H. M., Parikh, S. & Nathanson, C. A. (2010). *The secret: Love, marriage and HIV.* Nashville, TN: Vanderbilt University Press.

Hrdy, S. B. (2009). *Mothers and others.* Cambridge, MA: Harvard University Press.

Irwin, M. R. (2008). Human psychoneuroimmunology: 20 years of discovery. *Brain, Behavior, and Immunity, 22,* 129–139.

Kang, Y. (2006). "Staged" rituals and "veiled" spells: Multiple language ideologies and transformations in Petalangan verbal magic. *Journal of Linguistic Anthropology, 16*(1), 1–22.

Keane, W. (1997). Religious language. *Annual Review of Anthropology, 26*(1), 47–71.

Keller, D. L. (2011). Letter to the Editor: Allowing Resources During the Certification Examination. *Journal of the American Medical Association, 305*(1), 40–41.

Kensinger, E. A., & Schacter, D. L. (2006). Amygdala activity is associated with the successful encoding of item, but not source, information for positive and negative stimuli. *Journal of Neuroscience 26*(9), 2564–2570.

Kleinman, A., & Good, B. (Eds.). (1985). *Culture and depression.* Berkeley, CA: University of California.

Kompas, K., Olsson, C. J., Larsson, A., & Nyberg, L. (2009). Dynamic switching between semantic and episodic memory systems. *Neuropsychologia, 47,* 2252–2260.

Konner, M. (1982). *The tangled wing.* New York: Harper and Row.

Konner, M. (1988). *Becoming a doctor: A journey of initiation in medical school.* New York: Penguin.

Kostopoulou, O., & Wildman, M. (2004). Sources of variability in uncertain medical decisions in the ICU: A process tracing study. *Quality & Safety in Health Care, 13,* 272–280.

Kuipers, J. (1990). *Power in performance.* Philadelphia, PA: University of Pennsylvania.

Lazarus, R. S. (1999). *Stress and emotion: A new synthesis.* New York: Springer.

Ledoux, K., & Gordon, P. C. (2006). Interruption-similarity effects during discourse processing. *Memory, 14*(7), 789–803.

Lloyd, F. J., & Reyna, V. F. (2009). Clinical gist and medical education. *Journal of the American Medical Association, 302*(12), 1332–1333.

Lupien, S., & Lepage, M. (2001). Stress, memory, and the hippocampus. *Behavioural Brain Research, 127,* 137–158.

Lupien, S. J., Maheu, F., Tu, M., Fiocco, A., & Schramek, T. E. (2007). The effects of stress and stress hormones on human cognition: Implications for the field of brain and cognition. *Brain and Cognition, 65,* 209–237.

Mahon, B. Z., & Caramazza, A. (2009). Concepts and categories: A cognitive neuropsychological perspective. *Annual Review of Psychology, 60,* 27–51.

Mamede, S., Schmidt, H.G., Rikers, R. M., Custers, E. J., Splinter, T. A., & van Saase, J. L. (2010). Conscious thought beats deliberation without attention in diagnostic decision-making: At least when you are an expert. *Psychological Research, 74,* 586–592.

McEwan, B. S. (2000). Neurobiology of stress: From serendipity to clinical relevance brain research. *Brain Research, 886,* 172–189.

McGaugh, J. L., McIntyre, C. K., & Power, A. E. (2002). Amygdala modulation of memory consolidation: Interaction with other brain systems. *Neurobiology of Learning and Memory, 78,* 539–552.

McIver, R. K., & Carroll, M. (2004). Distinguishing characteristics of orally transmitted material when compared to material transmitted by literary means. *Applied Cognitive Psychology, 18,* 1251–1269.

Milner, B., Corkin, S., & Teuber, H. L. (1968). Further analysis of the hippocampal amnesia syndrome: 14-year follow-up study of H.M. *Neuropsychologia, 6,* 215–234.

Moscovitch, M. (1994). Cognitive resources and dual-task interference effects at retrieval in normal people: The role of the frontal lobes and medial temporal cortex. *Neuropsychology, 8*(4), 524–534.

Ochs, E., & Izquierdo, C. (2009). Responsibility in childhood: Three developmental trajectories. *Ethos, 37*(4), 391–413.

Owen, A. M. (2003). HERA today, gone tomorrow? *Trends in Cognitive Sciences, 7*(9), 383–384.

Patel, V. L., Evans, D. A., & Kaufman, D. R. (1990). Reasoning strategies and the use of biomedical knowledge by medical students. *Medical Education, 24,* 129–136.

Roozendaal, B. (2000). Glucocorticoids and the regulation of memory consolidation. *Psychoneuroendocrinology, 25,* 213–238.

Rubin, D. C. (1995). *Memory in oral traditions: The cognitive psychology of epic, ballads, and counting-out rhymes.* New York: Oxford University Press.

Rubin, D. C. (2006). The basic-systems model of episodic memory. *Perspectives on Psychological Science, 1*(4), 277–311.

Ryan, L., Cox, C., Hayes, S. M., & Nadel, L. (2008). Hippocampal activation during episodic and semantic memory retrieval. *Neuropsychologia, 46,* 2109–2121.

Sapolsky, R. M. (1994). *Why zebras don't get ulcers.* New York: W. H. Freeman.

Schmidt, H. G., & Rikers, R. M. J. P. (2007). How expertise develops in medicine: Knowledge encapsulation and illness script formation. *Medical Education, 41,* 1133–1139.

Squire, L. R. (2004). Memory systems of the brain: A brief history and current perspective. *Neurobiology of Learning and Memory, 82,* 171–177.

Strathmann, C., & Hay, M. C. (2008). "I'm paying your salary here!": Social inequality, consumerism, and the politics of space in medical clinics. *Human Organization, 67*(1), 49–60.

Strathmann, C., & Hay, M. C. (2009). Working the waiting room: Managing fear, hope, and rage at the clinic gate. *Medical Anthropology, 28*(3), 212–234.

Tse, D., Langston, R. F., Kakeyama, M., Bethus, I., Spooner, P. A., Wood, E. R., et al. (2007). Schemas and memory consolidation. *Science, 316,* 76–82.

Tsigos, C., & Chrousos, G. P. (2000). Hypothalamic-pituitary-adrenal axis, neuroendocrine factors, and stress. *Journal of Psychosomatic Research, 53,* 865–871.

Tulving, E. (1972). Episodic and semantic memory. In E. Tulving & W. Donaldson (Eds.), *Organization of memory.* New York: Academic Press.

Tulving, E. (1983). *Elements of episodic memory.* Oxford: Clarendon Press.

Tulving, E., Kapur, S., Craik, F. I., Moscovitch, M., & Houle, S. (1994). Hemispheric encoding/retrieval asymmetry in episodic memory: Positron emission tomography findings. *Proceedings of the National Academy of Sciences of the United States of America, 91,* 2016–2020.

Wang, S.-H., & Morris, R. G. M. (2010). Hippocampal-neocortical interactions in memory formation, consolidation, and reconsolidation. *Annual Review of Psychology, 61,* 69–79.

Weisner, T. S. (1976). Urban-rural differences in African children's performance on cognitive and memory tasks. *Ethos, 4*(2), 223–250.

Wiggs, C. L., Weisberg, J., & Martin, A. (1999). Neural correlates of semantic and episodic memory retrieval. *Neuropsychologia, 37,* 103–118.

Woike, B., Zavezzary, E., & Barsky, J. (2001). The influence of implicit motives on memory processes. *Journal of Personality and Social Psychology, 81*(5), 935–945.

Woods, N. N., Brooks, L. R., & Norman, G. R. (2007). The role of biomedical knowledge in diagnosis of difficult clinical cases. *Advances in Health Science Education, 12*:417–426.

Woolf, S. H., Grol, R., Hutchinson, A., Eccles, M., Grimshaw, J. (1999). Potential benefits, limitations, and harms of clinical guidelines. *British Medical Journal, 318,* 527–530.

World Health Organization. (2007). Country Health System Profile: Indonesia. World Health Organization.

Worthman, C. M. (1992). Cupid and Psyche: Investigative syncretism in biological and psychosocial anthropology. In T. Schwartz, G. White, & C. Lutz (Eds.), *New directions in psychological anthropology* (pp. 150–178). Cambridge: Cambridge University Press.

6 Balancing between Cultures: Equilibrium in Capoeira

Greg Downey

One of the first handstands I attempted while training in the Afro-Brazilian art of capoeira earned me a headbutt to the belly. The blow sent me crashing to the ground in a heap in the basement of the student activities building where I had arranged to get my first capoeira lessons. Prior to my field research, through a long chain of friends and acquaintances, I had managed to find a Brazilian who had toured with a traveling dance troupe and knew capoeira. Promising to handle all the logistical details, I cajoled her into teaching some classes, and she had invited a veteran practitioner visiting from the West Coast to our little, informal classes in the basement of an old building at the University of Chicago.

As I picked myself up from my collapsed handstand and turned, questioning, to the instructor, he pointed at his eye. The gesture is common among Brazilian capoeira practitioners and generally means either "watch closely" or, as in this case, "watch me"—meaning, that is, "don't take your eyes off your opponent." I had done an inverted movement, cartwheeling, but incorrectly looked at the ground. Looking down is natural to the novice, almost inevitable; without looking down, the inexperienced player has little sense of when the hands will reach the floor, and looking down makes balancing easier for reasons that only became obvious to me once I was deep into my research. By pointing to his eye, the instructor wordlessly offered advice that would be repeated myriad times over in the next few years: a capoeirista needs to keep an adversary in sight at all times or risk attack. The demand to watch the other player in capoeira, instead of giving in to the inclination to look down, forces practitioners to develop a distinctive sensory system to maintain inverted postures, a peculiar deep neurological enculturation of equilibrium.

Learning any sort of handstand demands sensory refinement as well as physical conditioning: recalibrating the practitioner's sense of balance, becoming familiar with the body's movement dynamics while upside-down, and developing a repertoire of corrective reactions to maintain the inverted posture. To stand on one's hands while watching an adversary, however, is especially challenging, because watching another person diminishes the contributions that focal vision can make to balance, eliminating compensatory strategies used in more static handstands, such as in Olympic gymnastics (for the proper position, see figure 6.1). Moreover, novices are told a *bananeira* should be dynamic, unpredictable, and respond to an adversary's actions, including using the legs to defend oneself. The resulting technique can be asymmetrical, sometimes tucked or with legs splayed,

Figure 6.1
The capoeira *bananeira*. Boca do Rio does a bananeira while playing against Cizinho (back to the photo). Notice the tucked and asymmetrical head position, clear vision of adversary, and bent legs (in order to be able to defend the torso). Practitioners of Capoeira Angola call a bananeira with bent legs a "closed" posture, whereas straightened legs leave the body "open" and invite attack. Kicks or headbutts to the legs are considered undesirable and inconclusive, whereas a headbutt to an open belly, especially when the target is balanced precariously and is likely to fall, is a considered decisive "entry" (see Downey, 2005, 2011). Photograph by Greg Downey, 1994.

and almost always in motion, in marked contrast to the ideal static symmetry and linear extension of a gymnastic handstand. In other words, cultural, aesthetic, and practical dimensions of doing a bananeira require a distinctive configuration of sensory perception for maintaining the inverted posture and allow dynamic corrective techniques penalized in gymnastics, demonstrating malleability in the human equilibrium system as well as cultural variation in skill acquisition.

This chapter explores the enculturation of the human sensory systems, offering inverted balance in the bananeira as an example of the ways that skill formation can affect nonconscious or only semiconscious sensory-motor systems. This case study demonstrates that a thorough neuroanthropological analysis of the ways in which cultures induce patterns of neural variation will necessarily consider carefully behavior patterns, training techniques, and subtle cultural kinesthetic norms, all of which affect perception and motor learning (see also chapter 7 in this volume). The reason is simple: we are always, inexorably, training our nervous system, enculturing our senses, and recalibrating the nonconscious or only semiconscious perceptual and self-monitoring processes that undergird our basic sense of normality and our capacity to do anything. These perceptual, proprioceptive, and motor processes run in the background of awareness, typically only coming to light when they malfunction or are forced to attempt an unusual activity. But the fact that they run in the background of awareness makes these processes no less susceptible to cultural influence; in fact, the pervasive influence they exercise on our experience may make them an ideal model for embodied cultural influences.

In many ways, although this discussion uses neurological evidence and a vocabulary borrowed from biology that may be unfamiliar to cultural anthropologists, some of its conclusions are already accepted in the field, especially in accounts of embodiment, body techniques, and sensory variation. For example, readers may find echoes of work on sensory expertise (see Geurts, 2002; Grasseni, 2004, 2007), ethnographies of sports training (for example, Wacquant, 2004), or the embodiment of culture (see Csordas, 1990, 1994). As Tim Ingold (1998, p. 26) has pointed out, however, most attempts to grapple with embodiment in anthropology have focused on the body "as a site or medium for the inscription of social values," overlooking the organic dimensions of enculturation such as neurological adaptation. In contrast, Ingold argues for the inclusion of biology, a recognition in cultural anthropology "that throughout life, the body undergoes processes of growth and decay, and that as it does so, particular skills, habits, capacities and strengths, as well as debilities and weaknesses, are enfolded into its very constitution—in its neurology, musculature, even its anatomy" (Ingold, 1998, p. 26). New research techniques in the brain sciences make Ingold's proposal both increasingly practical and more likely to be acknowledged, as fields such as cultural neuroscience highlight variation in brain performance (see, for example, Chiao, Li, & Harada, 2008; Han & Northoff, 2008; Kitayama & Cohen, 2007). In this sense, neuroanthropology seeks to fulfill the integrative promise offered in studies of skill, the senses, and embodiment by combining cultural, neurological, psychological, developmental, and even phenomenological approaches to cultural variation.

The equilibrium system is an especially rich subject for a study of neural enculturation because its input is multisensory, its capacity malleable, and its functioning only semiconscious, at best—subject to instruction but also, like much of proprioception, existing as a nonconscious background to much human experience. Research on the degeneration of balance in some medical conditions, together with studies of elite performers' equilibrium, provide a wealth of information on the vestibular system's response to altered conditions, in regard to both rapid adjustment and long-term adaptation.

Even in the simplest activities, we are a precarious species, balanced high atop two small feet without even a tail to aid in keeping us upright. And yet members of our species manage to balance in extraordinary conditions—on tightropes, skateboards, scaffolding, moving animals, rolling logs, and even boats tossed at sea. Standing upright is like balancing an inherently unstable inverted pendulum: close study of human balance shows that we constantly shift as the body itself produces and counteracts sway (Morasso, Baratto, Capra, & Spada, 1999). Asking how the equilibrium system operates and might be refined by bodily disciplines leads us to explore forms of activity-based neuroplasticity as a dimension of cultural variation. Activity-based neural variation complements the more traditional focus of both anthropology and psychology on cultural variation in conscious processes, such as categorization, calculation, and reasoning (see, for example, Kitayama & Cohen, 2007; Nisbett & Masuda, 2003; Norenzayan, Choi, & Peng, 2007).

Senses of Balance

Psychologists and anatomists typically label the vestibular system, located in the inner ear, as the organ of balance. The vestibular system includes semicircular canals (three to each ear) and the otoliths (pairs of small bones, the saccule and utricle, on each side). The otoliths sense linear acceleration, such as gravity and movement, and the semicircular canals detect angular acceleration in three planes, combining to give us a perception of rotation and which direction is up, among other things. The workings of the vestibular system are largely nonconscious, although they contribute information to a host of other perceptual and motor processes, as becomes evident when the system catastrophically fails (such as in patients with severe inner ear damage).

Equilibrium, however, is more accurately termed a "sensory system," a concept outlined by ecological psychologist James Gibson (1966, 1979), rather than a single sense. In day-to-day activities, people maintain upright posture by using a number of sensory channels and a repertoire of largely unconscious rapid postural adjustments in relation to both the outside world and the body's position (Horak & Macpherson, 1996; Massion, 1994; Peterka & Loughlin, 2004). Because balance is always a relational and dynamic problem, demanding simultaneous apprehension of the body's position, the environment, and the unfolding of movement, the vestibular organs, strictly speaking, are only part of the equilibrium system.

Stationary bipedal positioning is maintained by sensations from the inner ear organs but also from vision, proprioception at the ankles and joints, and pressure perception on the soles of the feet, as shown on the left side of figure 6.2 (Mergner, Maurer, & Peterka, 2003). Even normal subjects will sway more if they are asked to close their eyes (Edwards, 1946). When we stumble forward, we know we are falling because the otoliths sense the head's acceleration and shift in its axis relative to gravity, the ear canals register the head's forward pitch, our legs feel out of position under us, our joints no longer align, the pressure shifts forward on the soles of our feet, our frontal visual field suddenly starts to fill with the ground beneath us, and our peripheral vision detects the blur of "optic flow" (see also Amblard & Carblanc, 1980).

Horak and Macpherson (1996) argue that redundancy in the system is absolutely necessary to resolve the ambiguities that occur when different sensory channels offer conflicting information about posture and balance. Sensory information is checked against the body's intentions, so as not to treat a step or a diving slide into home plate like a face-first stumble. Unexpected movement triggers stereotyped adaptive reflexes, such as the vestibulospinal reflex (which occurs when we yank our head back and put out our arms when falling forward), through very short excitatory links to the motor neurons. This process is shown on the right side of figure 6.2, after the "error detection" symbol in the middle of the figure. In other

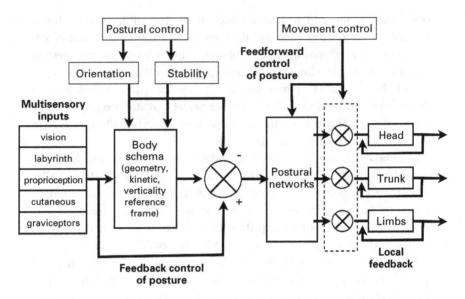

Figure 6.2

"Central organization of postural control" diagram reproduced from Massion (1994, p. 878, figure 1). This schematic highlights some of the complexity of postural control, including sensory conditioning and feedback. According to Massion, postural control consists of two separate stages: perception of the orientation of individual body segments, and whole body stability (perceiving equilibrium in part by recognizing how bodily segments relate to each other). Massion argues that both of these postural perceptual systems make use of the internal representation built up over time through multisensory inputs—the body schema—which includes an apprehension of the body's own geometry and mechanics, as well as awareness of the external reference frame, especially gravity and verticality. The external and proprioceptive senses also provide ongoing feedback and "error detecting" to maintain balance. When an error is detected, the body adjusts through postural networks. In addition, the postural networks allow a "feedforward" anticipation of adjustments necessary to compensate for voluntary movement. The postural networks themselves must also respond to feedback from the various parts of the body that perceive and adjust for shifts in equilibrium (Massion, 1994, p. 878.).

words, the "sense" of balance is actually a synthesis of diverse afferences and often-unconscious compensatory behavior (Balasubramaniam & Wing, 2002; and figure 6.2). If we had to concentrate on balancing, we probably wouldn't be able to walk, and we certainly wouldn't be able to surf, catch ourselves when we slip, or do gymnastics on a balance beam; those who suffer severe malfunction of the vestibular system report profoundly upsetting, disorienting sensations, even when stationary.

The extended vestibular system passes information on to other bodily systems so that they can compensate for the body's movement and positioning (see Smart, Mobley, Otten, & Smith et al., 2004). The vestibulo-ocular reflex, for example, allows the eyes to adjust to counteract movements of the head to maintain a stable visual field (Precht, 1979). Yates and Miller (1998) review evidence that the vestibular system also helps to regulate autonomic and respiratory systems (see also Ray & Hume, 1998). On an experiential level, Maurice Merleau-Ponty (1945/1962) argued that the bodily sense of self was the phenomenological "horizon" of all sensation, the sensory ground of reference and context for perceiving the outside world (see also Massion, 1994).

Confronted with challenges like running, moving in the dark, or standing on a shifting surface, the equilibrium system must "re-weight" the value of various inputs, discounting misleading proprioceptive, vestibular, graviceptive, or visual information (see Jeka, Oie, & Kiemel, 2000; Mahboobin, Loughlin, Redfern, Anderson et al., 2008; Oie, Kiemel, & Jeka, 2002). For example, as we walk, our otoliths sense acceleration, but proprioception that our legs are striding leads the equilibrium system to discount the vestibular suggestion that we might be falling forward. The equilibrium system can respond quickly to even novel sensations, such as weightlessness, by re-evaluating different information streams and substituting for those that become compromised (see Clément, Gurfinkel, Lestienne, Lipshits et al., 1984). In fact, the use of the inner-ear vestibular system appears to be a response to experience; very young children who have just learned to stand bipedally depend more heavily on vision for balance than older children (Lee & Aronson, 1974; see also Wann, Mon-Williams, & Rushton, 1998).

The Brain in Balance

The plasticity of the human equilibrium system, including its ability to find multiple ways to solve a balance-related problem and to interpret even unprecedented artificial input, allows this brain system to be deeply encultured. This demonstrable trainability contrasts with the views of some cognitive theorists who argue that the equilibrium system—or at least portions of it—is more rigid in its functioning. Philosopher Jerry Fodor has suggested in passing that "there is a new contender for 'best example of a module': the apparently domain-specific, encapsulated mechanism that many vertebrates, people included, use to recover from spatial disorientation" (2000, p. 118, fn. 9).

A "module," according to Fodor's landmark works on the subject, is a specialized ("domain-specific"), independent ("encapsulated"), quick, fixed functional system in the mind, which is inaccessible to conscious thought or information from outside its specific domain (see especially Fodor, 1983, 2000). Fodor builds upon Chomsky (especially 1980, 1988), but also draws heavily on evidence from optical illusions and theory of mind to suggest that portions of the mind are specialized to solve specific problems and largely immune to consciousness, which might undermine or slow smooth functioning.

Evolutionary psychologists, especially following on Leda Cosmides, John Tooby, and Stephen Pinker, have sometimes claimed that the brain is "*massively* modular," composed of a myriad innate, domain-specific computational mechanisms shaped by evolutionary pressures (see Barkow, Cosmides, & Tooby, 1992; also Hirschfeld & Gelman, 1994; Pinker, 1997; Pylyshyn, 1999; cf. Samuels, 1998). Similarly, cognitive anthropologist Dan Sperber has been a proponent of evolved, task-specific modularity in the human brain—arguing for the likelihood of a "snake detector, a face recognition device, a language acquisition device," for example (see Sperber & Hirschfeld, 2004, p. 41).

The evidence from empirical studies, however, does not suggest that the equilibrium system is a particularly good example of merely automatic, fast, unvarying reflexes. As Massion (1994, p. 883) concludes in a review of vestibular research, "While the old description of these reflexes is still valid and their analysis is still a useful means of experimentation and neurological evaluation, the emphasis now is on the flexibility of postural control and its adaptability to different contexts." Evidence from elite athletes, dancers, and more extreme examples of vestibular prosthesis to compensate for disability (see below) suggest that the equilibrium system is too open and networked to be considered strictly modular. The equilibrium system inherently makes use of diverse senses, compares and reweights those inputs as necessary on the fly, reconfigures itself to solve new challenges given time and training, develops new strategies both consciously and nonconsciously, and can even learn to use completely novel sensations, if those sensations reliably offer dependable information about balance. In other words, rather than a fixed mental module with proprietary, highly channeled information, the equilibrium system is partially open and flexible, tending strongly toward certain configurations but capable of being trained into alterative arrangements, or of temporarily assuming different dynamics as contexts demand.

Recognizing the network-like nature of the vestibular system allows us to better understand the mixture of malleability and continuity that we find in the brain and to see how cultural regimes, patterns of experience, explicit coaching, and unconscious conditioning might affect the system any one person assembles. Fodor correctly suggests that equilibrium is largely immune to short-term conscious intervention; in the long run, however, structured training regimes can affect refinement, or even reconfiguration, of the equilibrium system. Long-term neurological and perceptual adaptation to the tasks we set ourselves is a form of enculturation, even if we are unaware the change is happening.

For example, extensive training can create a strong connection in gymnasts between visual inputs and the network of reflex actions that maintain equilibrium; the equilibrium system may be "modularized," in that novices cannot simply will themselves to use vision to balance, but the nervous system is open in that it can reinforce existing connections and adapt over time to make use of sensation from the inverted inner ear and focal vision. Coaches and teachers may be unaware of how the nervous system accomplishes this reweighting, but they know from experience how to support the necessary adaptation through drill and behavior manipulation (see Downey, 2008a).

In addition, consciously experimenting with the edge of disequilibrium, such as in capoeira training, may loosen overly sensitive reflexes maladaptive to maintaining inverted posture, suppressing prior reflexes rather than simply building upon innate neurological settings. For example, a novice may have once tucked up and rolled to the floor as soon as he or she experienced dorsal disequilibrium in a handstand—falling toward the back. Enough capoeira training, however, will lead the same person to transition smoothly to a headstand (see Downey, 2011). The new skill can only develop in the space for action created by preventing automatic reflex.

Neurologically, the equilibrium system may learn in a number of different places. In a review of research on sensory learning, Geoffrey Ghose suggests that "the physiological basis of perceptual learning is considerably more complex than originally anticipated" (2004, p. 513). In particular, it involves changes to areas other than the primary sensory regions of the cortex and includes the ability to disregard distracting sensory "noise" and focus more intensely on relevant stimuli. As Dahmen and King (2007) write:

It has been known for a long time that perceptual abilities in a variety of tasks can be improved by training. The time course and specificity of learning point to plasticity throughout life at different levels of cortical processing and the challenge today lies in identifying the changes in neuronal response properties that are causally related to the perceptual improvements.

Clearly, perceptual abilities shift during capoeira apprenticeship, especially as virtuoso practitioners devote thousands of hours to training over their careers, although ethnographic research does not produce the same sorts of data as laboratory-based studies. The very fact that the Olympic handstand and bananeira require extensive training suggests that the neural changes involved are not simple or located in only one brain area, but rather are sequential and likely affect multiple brain systems (see figure 6.3, for example, for likely contributors in that sequence; see also Downey, 2010). For the equilibrium system, training likely affects neural connections into the vestibular nuclei and among these nuclei, as well as outgoing stimulation to motor areas; Vuillerme and colleagues (2001, p. 75) tested gymnasts and found marked differences with control subjects in the flexibility and responsiveness of the "central integrative mechanism," rather than the sensitivity of equilibrium-related senses. But training may also include suppressing reflexes from the brain stem and a host of other linked areas (see Green, Hirata, Galiana, & Highstein, 2004). The vestibular nuclei, themselves, are composed of a range of different cell types, many with different degrees and forms of plasticity (Gittis & du Lac, 2006, p. 388).

Training Equilibrium

In the long run, athletic training can effect subtle changes to the equilibrium system as practitioners adapt to better handle demanding activities. Training can instill sensory strategies, make re-weighting faster and more discerning, heighten sensitivity to crucial input, bias perception of the environment toward key variables, allow use of novel sensory information, and create new patterns of motor compensation. Psychological studies demonstrate that practitioners develop specialized strategies or heightened sensitivity in order to maintain balance in a range of activities, such as ballet (Osterhammel, Terkildsen, & Zilstorff, 1968), gymnastics (Bringoux, Marin, Nougier, Barraud et al., 2000; Vuillerme, Teasdale, & Nougier, 2001), martial arts (Jacobson, Chen, Cashel, & Guerrero, 1997; Perrin, Deviterne, Hugel, & Perrot, 2002), and space travel (Clément, Deguine, Parant, Costes-Salon et al., 2001).

The experience of engaging in these activities allows experts to better contend with the sensory distortions generated by extremely vigorous movement. Tanguy and colleagues (2008), for example, found that figure

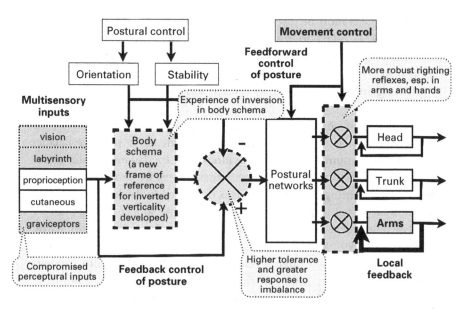

Figure 6.3
Like gymnasts learning the handstand, capoeira practitioners must learn to compensate for degrading information from the vestibular system (the labyrinth). Due to the necessity of paying visual attention to an opponent, however, capoeiristas cannot substitute visual fixation strategies for the missing or distorted vestibular sensations because focal vision must be devoted to tracking an adversary. Overall balance in the bananeira depends much more heavily on dynamic movement-control procedures after an error in balance is detected; and capoeiristas evidence a high tolerance for variation, asymmetry, and temporary imbalance, but respond with vigorous righting techniques with the legs, arms, trunk, and other body parts in their postural networks. That is, the bananeira depends heavily on the ability to respond to local feedback using the arms and hands with stepping, bending, switching to a headstand posture, and other sway-compensation techniques.

skaters were less susceptible to motion sickness from spinning, even when spun at unfamiliar angles. One of Osterhammel and colleagues' subjects, a ballerina, demonstrated her ability to overcome disorientation by exiting a spinning chair and immediately walking a straight line, heel-to-toe, with her eyes closed; another male dancer completed the same test for Osterhammel's team successfully after nine "strong drinks" over two hours (1968, p. 223–224). Clément and colleagues (2001) found that cosmonauts' terrestrial training regimen blunted the nausea-inducing effects of conflicting sensations from ear canals and otoliths, because the usual sense of gravity's acceleration on the otoliths is absent in weightless conditions. Collins (1974) refers to this adaptation in space flight as "vestibular conditioning." In sum, increased sensory skill in equilibrium helps experts to perceive more accurately in challenging circumstances, to sort out conflicting information, and to resist disorienting aspects of elite-level performance.

Training also affects the active motor patterns that compensate for imbalance; that is, the rapid corporeal adjustments to preserve balance once disequilibrium is perceived. Gymnasts on the balance beam, penalized for obvious movements to steady themselves, for example, maintain ankle-based righting strategies, whereas untrained individuals prefer more obvious hip-based responses to stay on a narrow surface. The gymnasts' strategy helps them avoid being marked down for a stumble (Marin, Bardy, & Bootsma, 1999). Elsewhere, I've written about finding my own vestibulospinal reflex relaxed after several years of falling over into capoeira techniques (Downey, 2005). In that case, inhibition of an "automatic" response was required for new action patterns to emerge, such as defensive "escapes" (*saidas*) in which players fall onto their hands to tumble away from a leg sweep or tripping attack.

Training techniques can specifically affect how elite performers cope with extraordinary situations. For instance, in many activities requiring spinning at high speeds, such as dance, gymnastics, and diving, practitioners adopt "spotting" strategies, holding the head steady and visually fixing on a point for short periods during rotation (see Davlin, Sands, & Shultz, 2004). "Spotting" substitutes focal visual orientation for vestibular information because the rotation of the head generates centrifugal force that confounds the inner ear system (see Osterhammel et al., 1968, p. 222).

The example of spotting techniques for balance, however, allows us to see the role of cultural influences in configuring a heightened sense of balance. Spotting is not the automatic, nonconscious result of task constraints; teachers must explicitly instruct and then systematically drill novices in order for the students to learn the technique. Although spotting becomes automatized with increased skill, in the initial learning stages, the visual strategy is both conscious and social, attempted after explicit instruction from a coach. In other spinning activities, such as in Dervish dancing, break dancing, and some traditional capoeira kicks, practitioners spin without spotting, necessitating different equilibrium-maintaining skills that do not use focal vision. Some of the sociocultural influences on the capoeira practitioner's changing equilibrium system are shown in figure 6.4, indicating how different external and internal parts of the system interact.

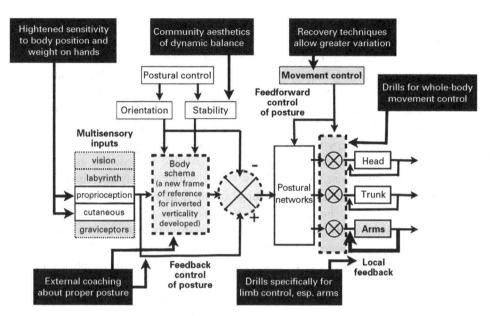

Figure 6.4

Over the long term, the emergence of the bananeira as a skilled form of training includes social influences on the individual's equilibrium system, such as coaching, community aesthetic preferences, and training drills that help to instill specific patterns of motor response. This figure shows some of the ways that social and cultural influences affect the development of the postural control complex. For example, external coaching affects the body schema by giving feedback on appropriate posture, offering criticism, and providing experiences that shape the novice's experiential expectations for proper positioning, muscle tension, and sensory feedback. In a sense, coaching becomes a feedback channel like proprioception. Community aesthetic preferences for dynamic balance encourage vigorous recovery techniques, and thus allow greater tolerance of instabilities than in gymnastics. In addition, training drills specifically acquaint students with whole-body movements for maintaining equilibrium, scaffolding experiences of recovery techniques and helping novices to develop the limb control, strength, and perceptual acuity required to perform variants of the bananeira. For example, during the author's apprenticeship, his hands developed the sensitivity to read equilibrium shifts from variation in pressure on palm and finger tips, responding to these shifts with increased wrist tension or small movements of the hand.

Perhaps the most radical demonstration of plasticity in the equilibrium system, however, is found in the work of the late neuroscientist Paul Bach-y-Rita, who developed prosthetic devices for people who had lost their vestibular sense (see Bach-y-Rita, 1972; Doidge, 2007, pp. 1–26; see also Bach-y-Rita, Collins, Saunders, White, et al., 1969). Bach-y-Rita's patients often suffered severe damage to the vestibular system due to a negative reaction to the antibiotic streptomycin; the condition left them feeling that they were perpetually falling even when lying down or physically braced. Bach-y-Rita and colleagues had remarkable success: the use of a prosthetic could even retrain the rest of the equilibrium system, in some cases, by reintegrating postural control of the head so that the prosthesis subsequently could be discarded. But Bach-y-Rita's prosthetic vestibular system was especially important for demonstrating trainability in equilibrium, because the system used a head-mounted accelerometer linked to an electrotactile interface placed under the tongue (or, in an early model, on the back; see Tyler, Danilov, & Bach-y-Rita, 2003). The vestibular nuclei in the brain learned to read the sensation of soft electrical shocks on the bottom of the tongue and interpret the novel sensation as a form of "sensory substitution." As Tyler and collaborators write: "For the brain to correctly interpret information from a sensory substitution device, it is not necessary that the information be presented in the same form as the natural sensory system" (Tyler et al., 2003, p. 160).

Sense of touch on the tongue, in this extreme example, can be integrated into the sensory system of equilibrium; as long as the sensation reliably offers patterned information about body positioning, the vestibular nucleus in the brain will find and integrate that information. This extraordinary example of compensatory adaptation demonstrates the degree to which the equilibrium system is synthetic, making use of multimodal information and capable of being both trained to greater sensitivity and linked to a range of behavioral patterns.

Inverted Balance

Balancing while being inverted on the hands poses a distinctive set of challenges to novices. The arms, shoulders, hips, and torso muscles have to bear the physical stress, and the upper body fatigues more rapidly than the legs under the weight, especially as the arm joints cannot be comfortably locked like the hips and knees. An inverted posture is inherently less stable than standing on the feet, with a smaller support surface (hands compared to soles), a higher center of gravity, and an effectively taller body to balance, because the extended arms add to its inverted length (Slobounov & Newell, 1996). In addition, the neural system must become tuned; not only are the vestibular organs rotated when the inner ear is inverted, but the head is no longer on the top of the body. The head's proximity to the ground means that becoming unbalanced does not produce the pronounced acceleration in the otoliths that would occur in normal upright posture.

As a result, learning a handstand typically involves developing perceptual strategies to substitute visual information for radically altered, dampened, or even counterproductive vestibular information. To generate reliable information for equilibrium, most gymnasts focus their eyes on a visual anchor point about five centimeters in front of the wrists and equidistant between them. Clément, Pozzo, and Berthoz (1988, p. 569) suggest that "a gaze 'anchoring' is helpful for the maintenance of fine equilibrium. ... the projection of gaze on a particular point of the floor may correspond to a 'desired' projection of the center of gravity around which equilibrium is controlled." In addition, visual anchoring may allow the gymnast to more quickly perceive a shift in the body by freezing the visual field against other movements (see Gautier, Thouvaecq, & Chollet, 2007, p. 1277). Gautier and colleagues found under experimental conditions that vision, both focal and peripheral, accounted for approximately 47% of balance in a handstand (2007, p. 1275).

The visual anchoring strategy means that gymnasts tend to crane their necks backward to see the ground, holding the position rigidly. In contrast, capoeira practitioners must not look at the floor during any technique, even if visual anchoring facilitates balance, as instructors are well aware. They insist students avoid anchoring as a technical shortcut because the sensory tactic impedes eventually achieving correct technique.[1] The instructors' reluctance to allow students to look at the ground may be well-founded neurologically; visually based balance skills are especially likely to degenerate if vision is removed from the sensory flux.[2] Perrin and colleagues (2002, p. 191) found that elite dancers, who rely especially on vision for balance, performed worse even than control subjects when forced to maintain posture with their eyes closed, possibly because dancers become especially adept at discounting misleading vestibular information and substituting visual spotting.

Instead of relying on vision, novice capoeira practitioners are told to hold their heads in line with their spines so that they can watch another player, even while cartwheeling or doing acrobatic *floreios* ("embellishments"). On several occasions during training, instructors told me to pull my head as far forward as mechanically possible—one instructor made students touch chin to chest. The head-forward position, watching an opponent, prevents using visual anchoring to supplement vestibular information, inverts the inner ear more severely than in a head-back gymnastic handstand, and, in addition, puts the spine in a position in which extreme exertion is biomechanically difficult (the reason that weightlifters do not press weights overhead with their chins tucked downward).

These differences in posture between the *bananeira* and a gymnastics handstand demand distinctive motor-perceptual methods of achieving balance, making transfer of inverted balance between the two techniques difficult, as suggested in figure 6.3. This was brought home especially vividly during my field research when a visiting Swiss capoeirista, a long-time practitioner of circus arts, complained bitterly that the bananeira's head position prevented her from transferring expertise in circus handstands. I had assumed the circus would be an ideal form of cross-training, but, like the dancers discussed by Perrin and colleagues (2002), visual balancing strategies from the circus proved too specialized to help with capoeira techniques.[3]

With the head mobile to track an adversary, both vestibular and visual information are severely degraded for balancing purposes in the bananeira. Although expert practitioners may benefit from increased cutaneous sensitivity (touch), improved proprioception, or long-term adaptation in the vestibular system, I suspect that their robust inverted balance is a result, not just of more acute perception, but also of a much-expanded repertoire of righting behaviors (see the right side of figure 6.3). That is, since the sensory system of equilibrium includes action patterns as well as information-gathering channels, equilibrium can adapt and grow more robust not only through perception but also through reaction, especially the development of very quick and refined hand-stepping reflexes.

While a handstand in gymnastics is ideally tightly controlled, the bananeira is dynamic, mobile, and maintains no specific posture; on the contrary, novices are told that changing postures, curling the body, or flailing the legs would help defend the body. The most virtuoso practitioners, especially those pursuing the traditionalist style, Capoeira Angola, develop extraordinary comfort in upside-down positions and can walk about, shift postures, and play upside-down. This distinctive community aesthetic encourages each group to explore a repertoire of righting behaviors, linking perception of disequilibrium to disparate automatized reactions over time.

Gymnasts employ fine postural adjustment strategies for maintaining inverted equilibrium, as Kerwin and Trewartha (2001) outlined. Because proper gymnastic form requires them to hold a "rigid, straight position" with extended elbows and plantar-flexed ankles, most righting strategies involve torque adjustments in the wrists and shoulders that twist or tense these joints (Kerwin & Trewartha, 2001, p. 1186). The more expert the gymnast, Kerwin and Trewartha observed, the more wrist torque becomes dominant when maintaining equilibrium in a handstand; less experienced gymnasts, and less successful handstands by the same gymnast, used the hip joint more extensively (p. 1188). This strategy is judged "successful," however, not simply for the biomechanics of equilibrium, but as a result of the aesthetic requirement to remain still in the handstand imposed by judging. Marin and colleagues (1999) refer to these technical considerations as "intrinsic constraints," demonstrating that expert gymnasts better maintain equilibrium strategies favored by judges even when severely physically perturbed.[4] As they write: "Many years of practice leads to modifications in the functional organization of the postural system to correspond with judges' expectations" (Marin et al., 1999, p. 621). Unlike capoeira practitioners, as a result of their training gymnasts do not explore and thus develop more rigorous, dynamic righting strategies, because these would cause point deductions in competition. The aesthetic preference shapes the way the perceptual system learns to manage inverted posture. The point is that gymnasts do not simply choose to remain stationary on their hands to maintain balance; because they actively avoid doing so, they do not develop great facility at walking around or moving their hip and knee joints to maintain equilibrium.

In contrast, capoeira practitioners feel no compunction about moving and avoid maintaining the rigidity and stillness that gymnasts seek. Many of the capoeiristas with whom I worked trained hard to walk on their hands, and specifically sought to increase their range of motion while inverted. They practiced kicking their legs down, splitting, twisting, tucking, and otherwise finding novel ways to maintain balance upside-down. Training exercises artificially forced novices to practice inverted *movements*, rather than hold *postures*. For example, drills I witnessed required students to turn in a circle while on their hands, walk in place, or lift each hand high while stepping, even touching the chest. Studies of gymnasts (e.g., Gautier, Thouvaecq, & Chollet, 2007) find they avoid bending their elbows to shift their centers of gravity up or down as a righting strategy; in contrasts, capoeira training frequently demands that a person bend his or her arms, even lower the whole body directly from a handstand to a headstand (Downey, 2011). In one group I trained with extensively, we drilled at leaping over a plastic chair, coming down on our hands, descending into a headstand, and then tipping over into a back-bridge. This exercise built a wide repertoire of whole body movements and adjustments at individual joints that a player could draw upon.

The same training also loosened the perception-reaction link, inhibiting sudden, disruptive reflexes to prepare for impact if one was slightly off-balance; we became less anxious if we tipped slightly because our bodies were familiar with how to compensate. This slackening of overly quick counterproductive reaction provides the capoeirista with greater latitude to experience degrees of disequilibrium while inverted than a gymnast might tolerate; to be able to take a long "step" with the hands while inverted requires generating an initial unbalanced momentum in the direction the player wishes to step. In addition, capoeiristas may not need the heightened sensitivity to disequilibrium provided by visual anchoring because they can respond more slowly than a gymnast—allowing the body to become off-balance—since they are not concerned to maintain a static position. Free to take a large step with the hand, adept at exiting a handstand by falling in any direction, and comfortable with a much larger degree of sway in their handstand, capoeiristas rely more heavily on the movement-control portion of the postural equilibrium system to remain upright (refer to figure 6.3).

In sum, the difference in inverted posture between the two disciplines results in distinctive skill sets and perceptual-motor strategies; these techniques each provide their own set of opportunities to develop the equilibrium system. The two forms, handstand and bananeira, demonstrate the way that a malleable constellation of sensory input can be employed by the equilibrium system, and that physical training can lead practitioners to assemble unusual perceptual-motor systems, including righting behaviors. The community's aesthetic preference for rigid and symmetrical or dynamic and variable inverted posture creates distinct opportunities for sensory-motor development, just as it channels skill acquisition (see Downey, 2008b). Aesthetic preferences, first encouraged socially but later internalized as conscious self-coaching before eventually becoming automatized propensities, act through concrete practices (such as the drills instructors design) to affect neurological development of the system at crucial points (see figure 6.4). In this way, community understandings of "right practice," authority structures among practitioners, patterns of train-ing, well-worn coaching maxims, finicky standards of sports-regulating bodies (where such organizations exist), videotaped performances of role models, and a host of other external influences become internalized by the nervous system, muscles, and even skeleton. Both inverted techniques, handstand and bananeira, are also cultural in that they elaborate upon a particular potential in human anatomy, exploring an aesthetically preferred type of movement and thereby reinforcing it. Other activities, from tightrope-walking, judo, and Okinawan karate to tango and running the ball in American football, all have distinctive strategies for maintaining balance that respond to the specific demands of the activity, but are also shaped by indigenous expert analysis, molded by patterns of training, and influenced by "intrinsic" cultural constraints. These social influences may be objectively "wrong," offering advice that is misleading or ambiguous, as Strauss and Quinn (1997, p.77–78) point out, but the advice is no less influential, even if its influence delays or deflects proficiency. The fact that expertise in one setting does not automatically transfer to other settings highlights the specialization of the expert equilibrium system.

Understanding Training

To discuss the neurological consequences of behavior patterns is not to discount the importance of consciousness or explicit cultural symbols, but rather to point out the practical links and discontinuities between cultural ideologies and the day-to-day training regimens that shape neurological development. Discussions of embodiment in anthropology have tended to describe ideas as becoming "embodied" or culture as "inscribed" on the flesh, without considering too carefully how that might occur physiologically. What a handstand or bananeira "means" and what it does to the nervous system are not identical, but both are "cultural" in the sense that they are dimensions of induced, patterned variation in human beings. Descriptions of cultural embodiment in cultural anthropology have tended to be simultaneously determinist, thus assuming that culture has its way with the body, and fleshless, with little to offer those who study human physiology except the occasional off-handed criticism or dismissal of science as a form of discourse. The close study of training regimens across cultures, and the forms of induced variation that arise from the vagaries of practice, offer a more solid footing for biocultural collaboration.

My exploration of equilibrium is an attempt to respond to Strauss and Quinn's suggestion that anthropologists study how "psychological structures that give rise to powerful internal thoughts, feelings, and tendencies to act a certain way" are built up over time (1997, pp. 32–33). The example of the equilibrium system, however, locates these psychological structures much more clearly in encultured neurophysiology and suggests that the feelings and tendencies these structures generate may be very much nonconscious, motor, and corporeal; just because they are not symbolic does not mean they are not cultural. Strauss and Quinn (1997, p. 33) are emphatic in declaring that "social discourses do not directly construct psychological realities," but the dominant understanding of cultural embodiment upon which they draw (e.g., Bourdieu, 1977), from the perspective of the equilibrium system, looks suspiciously like conscious thought and knowledge, rather than neurological adaptation and sensory learning.

The nervous system is always training, becoming better adapted to our patterns of living, recalibrating to the situations we place ourselves in, and transforming our biographies into our biology. Neural enculturation is constant. The interpretive turn in anthropology, including the treatment of culture as a system of symbols or webs of meaning, has tended to suppress anthropologists' attention to behavior, habit, built environment, and training as forces that shape our neurological endowments. In addition, the historical trajectory of discoveries in human biology, such as emerging research in genetics and neurology, has only recently acknowledged widely the variability, malleability, and diversity in human capacities. Whereas biologists may have once been reluctant to consider how culture and biology interact, especially how cultural variety might affect human physiological variation, the environment is changing on both sides of the intellectual divide.

As neuroanthropology develops, cultural sites of training will be ideal settings in which to study enculturation, because at these sites individuals voluntarily shape their nervous systems to achieve goals that are meaningful to them and their societies. Arduous training regimens and elite skills are naturally occurring experimental opportunities to observe how the nervous system can be shaped by what we do to ourselves. Although we may marvel at the achievements of virtuoso performers, we should also wonder at the capacity for enculturation that they demonstrate lies within us all.

Notes

1. In addition, in the head-flexed position of a gymnast, a quick descent to the ground can lead a novice to land directly on the face; capoeiristas drill extensively to shift downward to the top of the head, using the headstand as a low-vulnerability defense or transitional movement to exit the handstand (see Downey, 2011).

2. The admonition to avoid visual focus may also be a reflection of the post-expertise condition. Vuillerme and Nougier (2004) found that the demand for atten-

tion to equilibrium decreases once a gymnast has gained skill. Capoeiristas may depend upon attention eventually being freed up, delaying the acquisition of inverted equilibrium but preparing students for the opportunity to use attention once they gain expertise.

3. Other dimensions of the skill seemingly did transfer, since the circus performer's excellent strength, body control, and especially her shoulder flexibility proved useful in eventually achieving excellent technique in the bananeira.

4. Using the term "intrinsic" to describe the constraint of judging highlights the difficulty of determining whether cultural or aesthetic influences are "inside" or "outside" the actor. Marin and colleagues (1999) refer to internalized judging standards as "intrinsic," but they could just as accurately be called "extrinsic," as they are imposed by a sport's governing body.

References

Amblard, B., & Carblanc, A. (1980). The role of foveal vision and peripheral visual information in maintenance of postural equilibrium in man. *Perceptual and Motor Skills, 51*, 903–912.

Bach-y-Rita, P. (1972). *Brain mechanisms and sensory substitution.* New York: Academic Press.

Bach-y-Rita, P., Collins, C. C., Saunders, F. A., White, B., & Scadden, L. (1969). Vision substitution by tactile image projection. *Nature, 221*(5184), 963–964.

Balasubramaniam, R., & Wing, A. M. (2002). The dynamics of standing balance. *Trends in Cognitive Sciences, 6*(12), 532–536.

Barkow, J. H., Cosmides, L., & Tooby, J. (Eds.). (1992). *The adapted mind: Evolutionary psychology and the generation of culture.* New York: Oxford University Press.

Bourdieu, P. (1977). *Outline of a theory of practice* (R. Nice, Trans.). New York: Cambridge University Press. (Original work published 1972)

Bringoux, L., Marin, L., Nougier, V., Barraud, P. A., & Raphel, C. (2000). Effects of gymnastics expertise on the perception of body orientation in the pitch dimension. *Journal of Vestibular Research, 10*, 251–258.

Chiao, J. Y., Li, Z., & Harada, T. (2008). Cultural neuroscience of consciousness: From visual perception to self-awareness. *Journal of Consciousness Studies, 15*(10–11), 58–69.

Chomsky, N. (1980). On cognitive structures and their development: A reply to Piaget. In M. Piattelli-Palmarini (Ed.), *Language and learning: The debate between Jean Piaget and Noam Chomsky* (pp. 35–54). Cambridge, MA: Harvard University Press.

Chomsky, N. (1988). *Language and problems of knowledge*. Cambridge, MA: MIT Press.

Clément, G., Deguine, O., Parant, M., Costes-Salon, M.-C., Vasseur-Claussen, P., & Pavy-LeTraon, A. (2001). Effects of cosmonaut vestibular training on vestibular function prior to spaceflight. *European Journal of Applied Physiology, 85*(6), 539–545.

Clément, G., Gurfinkel, V. S., Lestienne, F., Lipshits, M. I., & Popov, K. E. (1984). Adaptation of postural control to weightlessness. *Experimental Brain Research, 57*, 61–72.

Clément, G., Pozzo, T., & Berthoz, A. (1988). Contribution of eye positioning to control of the upside-down standing posture. *Experimental Brain Research, 73*(3), 569–576.

Collins, W. E. (1974). Adaptation to vestibular disorientation, XII. Habituation of vestibular responses: An overview. Washington, DC: Office of Aviation Medicine, Federal Aviation Authority. Retrieved from http://www.faa.gov/library/reports/medical/oamtechreports/1970s/media/AM74-03.pdf

Csordas, T. J. (1990). Embodiment as a paradigm for anthropology. *Ethos, 18*(1), 5–47.

Csordas, T. J. (Ed.). (1994). *Embodiment and experience: The existential ground of culture and self*. Cambridge: Cambridge University Press.

Dahmen, J. C., & King, A. J. (2007). Learning to hear: Plasticity of auditory cortical processing. *Current Opinion in Neurobiology, 17*(4), 456–464.

Davlin, C. D., Sands, W. A., & Shultz, B. B. (2004). Do gymnasts "spot" during a back tuck somersault? *International Sports Journal, 8*(2), 72–79.

Doidge, N. (2007). *The brain that changes itself: Stories of personal triumph from the frontiers of brain science*. New York: Penguin.

Downey, G. (2005). *Learning capoeira: Lessons in cunning from an Afro-Brazilian art*. New York: Oxford University Press.

Downey, G. (2008a). Coaches as phenomenologists: Para-ethnographic work in sports. In I. Maxwell (Ed.), *Being there: Before, during and after*. Proceedings of the 2006 Conference for the Australasian Association for Drama, Theatre and Performance Studies. Retrieved from http://ses.library.usyd.edu.au/handle/2123/2470

Downey, G. (2008b). Scaffolding imitation in capoeira: Physical education and enculturation in an Afro-Brazilian art. *American Anthropologist, 110*(2), 204–213.

Downey, G. (2010). "Practice without theory": A neuroanthropological perspective on embodied learning. *Journal of the Royal Anthropological Institute, 16*(S1), S22–S40.

Downey, G. (2011). Learning the "banana-tree": Self-modification through movement. In T. Ingold (Ed.), *Redrawing anthropology: Materials, movements, lines* (pp. 77–90). London: Ashgate.

Edwards, A. S. (1946). Body sway and vision. *Journal of Experimental Psychology, 36,* 526–535.

Fodor, J. (1983). *Modularity of mind: An essay on faculty psychology.* Cambridge, MA: MIT Press.

Fodor, J. (2000). *The mind doesn't work that way: The scope and limits of computational psychology.* Cambridge, MA: MIT Press.

Gautier, G., Thouvaecq, R., & Chollet, D. (2007). Visual and postural control of an arbitrary posture: The handstand. *Journal of Sports Sciences, 25*(11), 1271–1278.

Geurts, K. L. (2002). *Culture and the senses: Bodily ways of knowing in an African community.* Berkeley, CA: University of California Press.

Ghose, G. M. (2004). Learning in mammalian sensory cortex. *Current Opinion in Neurobiology, 14,* 513–518.

Gibson, J. J. (1966). *The senses considered as perceptual systems.* Boston: Houghton Mifflin.

Gibson, J. J. (1979). *The ecological approach to visual perception.* Boston, MA: Houghton Mifflin.

Gittis, A. H., & du Lac, S. (2006). Intrinsic and synaptic plasticity in the vestibular system. *Current Opinion in Neurobiology, 16,* 385–390.

Grasseni, C. (2004). Skilled vision: An apprenticeship in breeding aesthetics. *Social Anthropology, 12*(1), 1–15.

Grasseni, C. (Ed.). (2007). *Skilled visions: Between apprenticeship and standards.* Oxford: Berghahn Books.

Green, A. M., Hirata, Y., Galiana, H. L., & Highstein, S. M. (2004). Localizing sites for plasticity in the vestibular system. In S. M. Highstein, R. R. Fay, & A. N. Popper (Eds.), *The vestibular system* (pp. 423–495). New York: Springer.

Han, S., & Northoff, G. (2008). Culture-sensitive neural substrates of human cognition: A transcultural neuroimaging approach. *Nature Reviews Neuroscience, 9,* 646–654.

Hirschfeld, L. A., & Gelman, S. A. (Eds.). (1994). *Mapping the mind: Domain specificity in cognition and culture.* Cambridge: Cambridge University Press.

Horak, F. B., & Macpherson, J. M. (1996). Postural orientation and equilibrium. In L. Rowell & J. Shepherd (Eds.), *Exercise: Regulation and integration of multiple systems* (Handbook of physiology, Rev. ed., pp. 255–292). Oxford: Oxford University Press.

Ingold, T. (1998). From complementarity to obviation: On dissolving the boundaries between social and biological anthropology, archaeology and psychology. *Zeitschrift fur Ethnologie, 123*(1), 21–52.

Jacobson, B. H., Chen, H. C., Cashel, C., & Guerrero, L. (1997). The effect of T'ai Chi Chuan training on balance, kinesthetic sense, and strength. *Perceptual and Motor Skills, 84*, 27–33.

Jeka, J., Oie, K. S., & Kiemel, T. (2000). Multisensory information for human postural control: Integrating touch and vision. *Experimental Brain Research, 134*(1), 107–125.

Kerwin, D. G., & Trewartha, G. (2001). Strategies for maintaining a handstand in the anterior-posterior direction. *Medicine and Science in Sports and Exercise, 33*, 1182–1188.

Kitayama, S., & Cohen, D. (Eds.). (2007). *Handbook of cultural psychology*. New York: Guilford Press.

Lee, D. N., & Aronson, E. (1974). Visual proprioceptive control of standing in human infants. *Perception & Psychophysics, 15*(3), 529–532.

Mahboobin, A., Loughlin, P. J., Redfern, M. S., Anderson, S. O., Atkeson, C. G., & Hodgins, J. K. (2008). Sensory adaptation in human balance control: Lessons for biomimetic robotic bipeds. *Neural Networks, 21*, 621–627.

Marin, L., Bardy, B. G., & Bootsma, R. J. (1999). Level of gymnastic skill as an intrinsic constraint on postural coordination. *Journal of Sports Sciences, 17*(8), 615–626.

Massion, J. (1994). Postural control system. *Current Opinion in Neurobiology, 4*(6), 877–887.

Mergner, T., Maurer, C., & Peterka, R. J. (2003). A multisensory posture control model of human upright stance. *Progress in Brain Research, 142*, 189–201.

Merleau-Ponty, M. (1962). *Phenomenology of perception* (C. Smith, Trans.). London: Routledge. (Original work published 1945)

Morasso, P. G., Baratto, L. R., Capra, R., & Spada, G. (1999). Internal models in the control of posture. *Neural Networks, 12*(7–8), 1173–1180.

Nisbett, R. E., & Masuda, T. (2003). Culture and point of view. *Proceedings of the National Academy of Sciences of the United States of America, 100*, 11163–11170.

Norenzayan, A., Choi, I., & Peng, K. (2007). Perception and cognition. In S. Kitayama & D. Cohen (Eds.), *Handbook of cultural psychology* (pp. 569–594). New York: Guilford Press.

Oie, K. S., Kiemel, T., & Jeka, J. J. (2002). Multisensory fusion: Simultaneous re-weighting of vision and touch for the control of human posture. *Cognitive Brain Research, 12,* 164–176.

Osterhammel, P., Terkildsen, K., & Zilstorff, K. (1968). Vestibular habituation in ballet dancers. *Acta Oto-Laryngologica, 66* (1–6), 221–228.

Perrin, P., Deviterne, D., Hugel, F., & Perrot, C. (2002). Judo, better than dance, develops sensorimotor adaptabilities involved in balance control. *Gait & Posture, 15,* 187–194.

Peterka, R. J., & Loughlin, P. J. (2004). Dynamic regulation of sensorimotor integration in human postural control. *Journal of Neurophysiology, 91,* 410–423.

Pinker, S. (1997). *How the mind works.* New York: Norton.

Precht, W. (1979). Vestibular mechanisms. *Annual Review of Neuroscience, 2,* 265–289.

Pylyshyn, Z. (1999). Is vision continuous with cognition?: The case for cognitive impenetrability of visual perception. *Behavioral and Brain Sciences, 22,* 341–423.

Ray, C. A., & Hume, K. M. (1998). Neck afferents and muscle sympathetic activity in humans: Implications for the vestibulosympathetic reflex. *Applied Physiology, 84,* 450–453.

Samuels, R. (1998). Evolutionary psychology and the massive modularity thesis. *British Journal for the Philosophy of Science,* 49(4), 575–602.

Slobounov, S. M., & Newell, K. M. (1996). Postural dynamics in upright and inverted stances. *Journal of Applied Biomechanics, 12*(2), 185–196.

Smart, L. J., Jr., Mobley, B. S., Otten, E. W., Smith, D. L., & Amin, M. R. (2004). Not just standing there: The use of postural coordination to aid visual tasks. *Human Movement Science, 22,* 769–780.

Sperber, D., & Hirschfeld, L. A. (2004). The cognitive foundation of cultural stability and diversity. *Trends in Cognitive Sciences, 8*(1), 40–46.

Strauss, C., & Quinn, N. (1997). *A cognitive theory of cultural meaning.* Cambridge: Cambridge University Press.

Tanguy, S., Quarck, G., Etard, O., Gauthier, A., & Denise, P. (2008). Vestibulo-ocular reflex and motion sickness in figure skaters. *European Journal of Applied Physiology, 104,* 1031–1037.

Tyler, M., Danilov, Y., & Bach-y-Rita, P. (2003). Closing an open-loop control system: Vestibular substitution through the tongue. *Journal of Integrative Neuroscience, 2*(2), 159–164.

Vuillerme, N., & Nougier, V. (2004). Attentional demand for regulating postural sway: The effect of expertise in gymnastics. *Brain Research Bulletin, 63,* 161–165.

Vuillerme, N., Teasdale, N., & Nougier, V. (2001). The effect of expertise in gymnastics on proprioceptive sensory integration in human subjects. *Neuroscience Letters, 311,* 73–76.

Wacquant, L. (2004). *Body and soul: Notebooks of an apprentice boxer.* New York: Oxford University Press.

Wann, J. P., Mon-Williams, M., & Rushton, K. (1998). Postural control and co-ordination disorders: The swinging room revisited. *Human Movement Science, 17,* 491–513.

Yates, B. J., & Miller, A. D. (1998). Physiological evidence that the vestibular system participates in autonomic and respiratory control. *Journal of Vestibular Research, 8,* 17–25.

7 From Habits of Doing to Habits of Feeling: Skill Acquisition in Taijutsu Practice

Katja Pettinen

Cultivating Habitual Movement Based on the Feet: Knee, Hip, Spine

Taijutsu, or the "art of the body," is a non-competitive martial art that cultivates somatic principles from nine different traditional Japanese martial lineages under one holistic approach. On the surface, Taijutsu includes a great range of techniques and movements, including training with weapons, such as the sword, short and long staff, and rope, as well as the use of numerous unarmed fighting methods, such as body throws, joint locks, and submissions. Despite this surface-level variation, the art relies on select fundamental approaches to using one's body in time and space. One such systematic form of moving the body is a synchronization of the knee, hip, and spine turning together, which can generate great centripetal energy, for example, in terms of one's own movement in body throws, as well as centrifugal energy, in terms of the opponent's movement. This synchronization takes place in practically all techniques of the art, and it forms a systematic way of placing one's center of gravity in relationship to another body's movement.

Within many karate styles, a commonplace approach to punching utilizes a similar principle, wherein the strength of a punch does not come from one's arm alone, but as an extension of a hip turn. In karate, however, this movement is delivered with a lag in the overall timing of the hip turn and the consequent impact of the fist. Within Taijutsu practice, this movement is foundational to throws, punching, and a number of types of weapons work and, consequently, utilizes a different kind of timing, wherein the knee, hip, spine, and the arm aim to work together seamlessly without any lag between them at the moment of a strike or a throw.

In a training session in the southeastern United States that included brand-new students, an instructor highlighted this distinctiveness of the Taijutsu approach. He set me holding a large kicking pad that hid about half of my body behind it. The pad was five inches of foam wrapped in a tight plastic shell, therefore easing much of the impact from any kick. The instructor set himself sideways and to the left of me while I held up the pad. I braced myself against the pad and tucked my face down in the middle. Using a kick typical of many karate styles and tae kwon do—the knee is picked up in the air and the kick hits the target horizontally— the instructor delivered a fast and hard kick to the pad. The sound of his foot hitting the pad was loud, and always made me wince even behind the thick foam, but I easily stayed on my feet since my balance had not been challenged.

"Now that would definitely hurt, if it hit someone. Can you imagine it? Something like that hitting your face, or even ribs. Not very good. But now, check this out." The instructor set me holding up the pad again and now faced me directly from in front. He simply placed his open palm against the middle of the pad, a couple of inches from its surface, and lowered his hips, while turning the spine and bending his knee, thus synchronizing all these movements into one unified system.

This single, relatively slow but carefully timed movement immediately sent me backward several steps as the instructor's palm hit the pad. There was no way I could maintain my balance against the momentum coming through his body. "This moves her because the whole weight of my body is behind it. There's some 250 pounds coming at you, even from a short distance, and it's gonna move you. That's the difference. With the kick there's just speed." After the demonstration, the students spread out into pairs for further training. I paired up with Lukas, a 28-year-old who had been studying the art for a little over three years. I asked him if the demonstration made sense to him. "Yeah, sure it makes sense. But that doesn't mean that I can do it," he stated, laughing.

During five years of studying Taijutsu in United States and Canada, I spent approximately five hours each week training in the numerous techniques and movements involved in the art with different students like Lukas. Most of us came across a number of challenges trying to make sense of our own learning of the art and of the ways in which different instructors teach it. Some of these challenges are distinctly biophysical, in the sense that it is difficult, and initially strenuous, to learn to use one's body in ways that are then systematically maintained; for example, the synchronization of the knee, hip, and spine turn across a range of different contexts. This synchronization in advanced practitioners comes through whether they are executing a body throw, a punch, or a sword cut, or using a long weapon such as the six-foot staff.

Instead of finding ways to utilize these foundational movement perspectives, most of us try to locate a route to successful learning by repeating techniques so that we become equipped to execute them with maximum efficiency. Eventually, this would lead to a level of "automation," "habitual response," or even "second nature." Even though demonstrations such as the one above do make sense on some level, their simplicity is deceptive. This is the case in part because such foundational movement principles are difficult to align with the concepts of "motor skill" or "muscle memory" that are frequently evoked in North America in relation to somatic skill learning. While all somatic traditions, such as dance, theater, and martial arts, cultivate the shared phylogenetic potential of the human body, each tradition also relies on a number of culturally specific models of learning.

A distinct way in which Taijutsu practice is incompatible with models of learning centered on repetition or habitual automation can be illuminated through what is called the *sakki* test. This test determines when a student is licensed to teach the art and is only given in Japan in the presence of the *soke*, Masaaki Hatsumi, the main teacher and the originator of the organization. The test stands out as a distinct indicator of the kind of skill-cultivation that Taijutsu centers on, and challenges the idea of skill learning as repetition, efficiency, or context-free habit formation. Specifically, by divorcing skill from visual cues, the test emphasizes sensing over doing.

Guiding the Senses Away from Vision

Among the numerous organizations, generically referred to as *Bujinkan*, that all teach the art of Taijutsu, the process of testing for advancement in rank varies greatly. Traditionally, practitioners in Japan utilized only two main ranks: white and black belts. Outside Japan, however, such an approach is especially difficult to maintain because students wish to have more ways of measuring their sense of advancement. In the North American cultural context, where I studied the art, practitioners desire a range of indicators indexing their skill level. As a result, a more nuanced rank system is in place, while still remaining particular to the exported style. In this system, there are white, green, and black belts. The green belts are further divided into nine *kyo*, or ranks, while the black belts range from the first to fifteenth *dan*, or level, the higher number indicating a more advanced rank, while in the green belts the order is reversed and count down in rank.

In the overall context of North American martial arts, this approach to ranking is atypical in terms of having a relatively small variety of visual signifiers of one's "progress." This atypicality is linked to the fairly recent emergence of Taijutsu as a distinct style of martial art available for practitioners: only since the 1970s has this style been available outside Japan. In the 1970s, a few individuals from the West traveled to Japan and began to export the style to North America, Europe, and Israel. During the 1980s and 1990s, Hatsumi Masaaki traveled extensively across North America and Europe, giving a number of seminars and demonstrations, which resulted in great popularity for the art outside Japan. Even today, while Hatsumi teaches the art only in Japan, the majority of students in his classes are from abroad, most visiting the *dojo* (a place of training) for a few weeks, some living in Japan year-round as foreigners.

Another feature of the relatively recent popularization and the distinct method of conducting belt testing or advancement in the Bujinkan is the sakki test. The test occurs when a student reaches the fifth dan black belt, and consists of only one action: avoiding a single non-metal sword swing, a cut straight down that is performed behind the testee while he or she is kneeling down in a *seiza* posture (sitting with the legs folded underneath the body).

There is a high degree of systematicity to the context in which the test is given, which is a key feature in the psychosocial parameters that mediate this sensorial engagement. On the most central level, there is a particular kind of cohesion to the art and to the organization because it has been developed by one single teacher (born in 1931), still active in the art. Without this structural and organizational cohesion, a somewhat esoteric test such as the sakki would be difficult to maintain, particularly given the highly international make-up of Bujinkan.

At the end of his training sessions in Japan, the soke Hatsumi calls out for any test-takers, and if someone is judged by other teachers to be ready, he or she walks up to the center of the dojo. Hatsumi, or another advanced teacher who is chosen as the test giver—the one who performs the cut—positions himself behind the test taker, who is already kneeling down in seiza. The test giver might tap the test taker's shoulders and the head with the padded sword. He (I am not aware of any females as test givers) then raises the sword above his head, and focuses his mind for the intensity of the cut while closing his eyes. The test giver swiftly cuts down to the level of the test taker's head with focused intent. The focused intent, or sakki, refers to the test giver wanting to deliver a painful hit upon the testee, with the intent of actually hitting him or her.

A brief moment lapses while the sword travels down from above the cutter's head to approximately the height of his knees, the level at which the test taker's head is positioned. In this time the testee must move—ever so slightly in actual distance but at the very instant of his or her perception becoming engaged—in order to avoid the hit. One of the advanced North American practitioners explained to me, based on his more than twenty trips to Japan to study the art, that the test brings focus to a kind of sensorial awareness that balances out any repetitive aspects of skill learning. He stated: "Those who study and train with feeling over the years pass the test, no problem. The cut comes down, and they just roll out. But if you can't feel what's going on when you train your basics, you just get clunked in the head."

For most outsiders, the test might sound almost mystical, because it involves avoiding a strike without any reliance on vision. In fact, most test takers find it easier to concentrate if they also close their eyes while kneeling in seiza. However, quite apart from any mysticism, the test foregrounds an ability to concentrate and remain relaxed, while anticipating a simple physical movement—dropping one's head down in an angle toward the left or right, and continuing this initial movement into a forward roll—within a narrow window of time based on slight physical stimuli. There are two possibilities about the nature of this stimulus: either the test taker is able to hear the initiation of the cut based on the movement of the cutter's arms and the upper body, or, based on a more emic explanation of the test, the test taker can sense the intensity of the cut, the sakki, that an advanced practitioner is able to project, for example, as an "adumbrative feature" (Hall, 1964; also see Davies, 2005, p. 4, for a representation of the phenomena in Western literature).

Despite the details about the nature of the stimulus that a test taker perceives, what is clear is that practitioners are systematically able to respond to it—hundreds of North Americans alone have taken the test during recent decades. As a whole, the sakki test is part of the Taijutsu cultivation of the senses more broadly; the test demonstrates a specific kind of skill, but also serves as a key index of the epistemology of this somatic practice. As a key "test" for Bujinkan practitioners, the sakki does not require any action that repetition alone could achieve. Instead, it presents a form of sensory acuity that directly challenges cumulative notions of learning by highlighting feeling over doing.

From a neurological standpoint, the sakki test is reflective of a "cultural modulation of perception" (Howes, 2006, p. 385). David Howes has argued that when considering somatic practices, we must recognize that "cultural practices and technologies ... generate different sense ratios" (2006, p. 385–386). Howes' comment, as well as his work more broadly, lends itself to an idea that the sakki test is embedded in a set of cultural practices that shape the body and the senses in distinct ways. From a sensorial perspective, it does not seem coincidental that the Japanese samurai fighting arts, from which Taijutsu draws many of its technical components, incorporated such sensorial-aesthetic activities as the tea ceremony and floral arranging along with the physical skills of fighting battles. In any such aesthetic tradition, "representation ... not only describes but conditions perception" (Pallasmaa, 2005, p. 16).

Furthermore, the modulation of the sense ratios within practices such as the sakki or other forms of somatic positioning cannot be simply physiological. The actions done with and through the body are also embedded within a set of epistemologically organized orientations toward the mind, body, brain, and the relationships among these (Geurts, 2002 p. 3; Wacquant, 2004, p. 89). In other words, forms of materiality are interrelated and interdependent with forms of classification and meaning (Deacon, 1997; Merrell, 2010).

Practitioners of Taijutsu in North America do not generally make observations about the biocultural features shaping somatic skill acquisition. While some comment upon the somatic aspects of habits, such as squatting, kneeling, or sitting in seiza, their comments tend to focus on a strictly physical level. Most any practitioner who has visited Japan recognizes correlations between the ways in which the body is used in everyday activity and in martial techniques, such as throwing or striking, where a lower stance is required while maintaining upright posture with one's back. Comments such as, "it's like sitting on a Japanese toilet," are in some cases made to highlight that, from a bodily perspective—that is, the basic strength of the leg muscles and ligaments—daily life practices are connected to martial arts methods (also see Downey, 2005).

On a broader conceptual level, however, practitioners tend to not articulate ways in which cultural conceptions about the body itself might be part of the interplay through which martial arts practices, such as Taijutsu, are structured. In part, this silence reflects Western understandings of bodily practices, which assume a number of Cartesian dichotomies: mind-body, emotion-rationality, and material-symbolic (Geurts, 2002). In such an epistemic framework, the body is first and foremost a sheer mass of materiality that can be brought to perform upon command as a result of the systematic repetition of movements. Moreover, in such epistemologies, the body is predominantly regarded as an entity, which contrasts, for example, with viewing and experiencing the body as an open-ended process (Weiming, 2007; Zarrilli, 2004, p. 655).

The sakki test, however, is positioned within a different epistemological framework, and with contrasting notions of the mind-body relationship and its operations. Furthermore, any such framework bears an intimate connection to somatic practices that are present in the broader cultural and historical context of the art. While it is critical to note that the sakki test is in part a practice that draws from non-Cartesian models of somatic learning—for example, on the level of discourse and representation about the body, the brain, and the mind—Taijutsu is also an expansion of culturally specific life practices. It is the systematic presence of such practices that establishes an orientation to moving in time and space that can be named as the "grammar of the feet." Instead of centering on the execution of arm and hand movements that the mind-body has achieved requisite "muscle memory" for, the definition of skill in Taijutsu centers on the intelligence of the feet in creating a kinesthetic space wherein the manipulation and control of the opponent's body comes forth mainly as a result of his or her own movements.

Toward Habituated Movements: The Grammar of the Feet

A crucial aspect of the overall epistemology guiding the somatic practice of Taijutsu is centered on what I call the grammar of the feet, borrowing this expression from a Japanese theater theorist Tadashi Suzuki (1986). Suzuki has suggested that Western and Japanese theater arts approach the expressive aspects of the human body differently. In Suzuki's reading, the Japanese theater arts cultivate especially nuanced movements of the feet and the legs, while the Western theater arts display a more distinctly top-centered approach to the body, foregrounding predominantly the face, arms, and upper torso.

Tim Ingold (2004) offers bioculturally grounded observations of similar overall differences in how the body and its movement are cultivated in Japan and the West. Ingold cites work by Japanese anthropologist Junzo Kawada and suggests that, due to a complex interaction of ecological landscape, phenotypic distribution (average population heights), economics, and related forms of social organization, the Japanese take a distinct approach to the human body that centers on the importance of the feet. Through Kawada's work, Ingold suggests that in the European context, walking dynamically focuses on the hips in contrast to the knees. As a result, the legs and posture are more distinctly upright in Western everyday movement, while Japanese walking centers more on the knees and maintains a proportionately slightly lower center of gravity.

Ingold observes that this form of walking is connected to the regional history of Japan, wherein "walking from the knees" is an effective response for a terrain with many elevation differences,

since with the lowered centre of gravity the risk of tripping and falling is much reduced. It is also ergonomically consistent with the technique, once widely used in Japan, of carrying heavy loads suspended from a long supple pole resting athwart the shoulder. Kawada is able to relate the postural differences in walking, respectively from the hips and from the knees, not only to alternative methods and devices for load-carrying, but also to traditional dance styles, artisanal techniques and practices of child rearing. ... All in all, Japanese posture and gesture seem to be strongly and positively oriented towards the ground, in striking contrast to European efforts to rise above it. (Ingold, 2004, p. 325)

For North Americans, the more Japanese walking style, also evident in Taijutsu practice, is at times marked with the English term "shuffle." Some instructors emphasize that more important in the context of learning the art than physical training, such as running, is to learn to systematically walk in this distinctive manner, bending one's knees and using shorter steps. One additional consequence of this way of walking, besides a lowered center of gravity, is the manner in which the foot makes contact with the ground. With hip-centered walking, the ball of the foot will make more immediate contact with the ground, as the heel is driven in directly on each step and the weight immediately shifted from the heel to the ball of the foot. In contrast, with knee-centered walking, one's center of gravity is positioned lower and also slightly further back. As a result, the walking style brings out a slightly more nuanced weight shift, because the heel-to-ball motion of the foot-weight transfer is more mediated by the side of the foot.

This walking style can be overemphasized in order to make an even greater separation for the way in which one's overall balance does not commit to going forward until one's foot has securely made contact with the ground: by keeping the weight on the back leg for a moment longer, while stepping out with the front foot, and only then shifting the weight from the heel to the ball of this foot. The main distinction therefore comes from the way in which the foot that steps out is ahead of the movement of the upper body, in contrast to the more forward committed steps that accompany the Western style.

Furthermore, based on these observations about contrasting walking styles, Ingold suggests that within the Western context, a systematic neglect of the feet and the lower body is part of a broader paradigm through which the body, as well as a number of scientific theories about human evolution and the role of bipedalism, are conceptualized. Part of this paradigm in the Western context is an overemphasis on the brain as a semi-independent center for the action of the human body (Coward & Gamble, 2008; Mithen & Parsons, 2008). As a whole, a body of work on the embodied and situated aspects of cognitive processes has emerged in response to such concepts of a mind-body dichotomy and its concomitant assumptions (Callagher, 2005; Donald, 2001; Lakoff & Johnson, 1999; Ziemke, Zlatev, & Frank, 2007).

The overemphasis on the upper body and the consequent disregard of the role of the feet is also visible in the context of Taijutsu practice. While Taijutsu arguably centers on learning a grammar of the feet, many practitioners find it challenging to recognize the importance of this foundational way of moving in the art. The classic Western martial arts, such as boxing and fencing, bring systematic representational and functional focus upon the hands and the arms, a theme that popular culture expands on. Learning to recognize that using one's feet is a key habitual orientation toward the manipulation of time and space in relation to the opponent's body is a central aspect in the cultivation of Taijutsu skill. However, due to a number of cultural patterns highlighting the importance of the upper body, forms of punching being some of the most iconic, making sense of such orientation poses some particular challenges for North American students of Taijutsu.

I suggest that one of the key challenges in this case is the culturally salient assumption that centers skill acquisition on "muscle memory" or "motor skill," which easily evoke repetition as an inherently enabling method for somatic learning. In sports and movement science, this conception of skill acquisition refers to "the classic notion of automaticity of skilled motor behavior" (Hatfield, Haufler, & Spalding, 2006, p. 221; Fitts & Posner, 1967). Such models of learning as an internalized set of automatic habitual responses carry a great amount of cognitive weight within martial arts contexts, in part because these assumptions are not "wrong": they do offer one form of empirically based insight into human movement performance. However, it is the kind of emphasis these assumptions easily place upon the body, the mind, and the brain that Taijutsu practice re-positions through a differing epistemology.

How Do You Know What to Do?

In many Taijutsu dojos, or training spaces, explanations about the importance of repetition are at times brought up precisely because they are relevant to learning. However, this referencing is done in a specific relationship to the overall movement practices of this non-competitive martial art form.

One Taijutsu training session that I participated in took place in the eastern United States, outside on the grass in a large wooded city park. The American instructor, who held the rank of *shihan*, a term indicating a highly advanced teacher, had recently been to Japan for several weeks and was conducting a weekend seminar in which he reviewed teachings from Japan. I traveled to the seminar with a few members of the group I had been training with for the previous year. Only some members of the training group, which comprised 25 people, were able to attend, and the five of us were able to fit into one car, driving out three hours to the nearby city. As a whole, the seminar drew in forty-five practitioners from the surrounding areas, students ranging from individuals who had only recently begun studying the art to those who had been training for over a decade.

At one point in the afternoon, shihan Stewart broke up the tempo of our previous training. In the morning hours of the seminar, he had demonstrated mainly the eight techniques most fundamental in the art (*kihon happo*), correcting the manner in which many teachers had been teaching them, as well as showing a number of different variations and applications of the techniques for different bodily contexts, for example, being attacked while sitting on a chair or on the ground, in contrast to being up on one's feet. In each such teaching demonstration, the instructors invite one of the students to be an *uke*, or a dedicated attacker in a joint exercise. The instructor tells the uke the overall manner in which to perform an attack (i.e., a kick vs. a punch or grab), then himself illustrates how to respond through Taijutsu movements.

In the afternoon demonstration, the shihan uniformly utilized one single punch as an attack in several different instances, but always executed a different way of responding. While the attacks in some ways always looked the same—one direct punch toward the face, even executed by the same uke—the techniques utilized in response varied greatly. At the end of the demonstration, shihan Stewart asked whether anyone had questions. One of the newer students responded by asking: "How do you know what to do?"

The question points toward the fact that, fairly soon in the process of skill acquisition, each student of Taijutsu practice realizes on some level that even the execution of a fairly simple technique in response to an uke's attack will not always "work." The execution of one's own movement does not always bring about a smooth and thorough control of the uke's body and consequently his or her opportunity for further strikes, or other forms of attack upon the *tori*, the person designated the attacker in a training sequence. Taijutsu is an art that places great emphasis on what the practitioners consider "real-life applications," for example, means of self-defense. As such, much of the teaching and training in the art focuses upon a careful consideration of any "holes" in one's technique: any moments in which the tori might be vulnerable to a counterattack, such as avoiding a first punch but leaving his or her face in the immediate reach of the uke's other hand.

Some of the reasons for an unsuccessful execution of a movement by a tori have to do with the fact that individuals have differently shaped bodies, postures, and habits of movement, as well as differing levels of tolerance for pain and discomfort. As a result, each uke will respond to anything one does as a tori somewhat differently. On the other hand, even with the same individual, such as an uke who attacks with a punch toward the face three separate times, there will be variation in the velocity of the punch, the exact height of the arm, the positioning of the feet, and, consequently, the exact trajectory of the uke's balance points in the course of an attack. Even the substrate on which punching (or kicking) takes place brings forth differing affordances (Gibson, 1979) in relation to the balance dynamic: if training takes place outside, a slippery wet grass forms a different kind of balance structure and positioning in contrast to a wooden floor, a sidewalk, gravel, or ice.

As a result, any advanced instructor can sense subtle differences in the manner in which each punch is executed and the variation generated by different bodies or multiple repetitions of the "same" punch by one person. In response to the presence of this infinite variation, though infinite only if fostered as such, for example, by focusing on the overall context of the movement rather than on the movement itself, an advanced practitioner immediately adapts the timing, distance, and angle of his or her response through a subtle positioning of the feet. The common Western adage "you can't step into the same river twice" captures the situation well, as it suggests that every attack exists in its own temporal dynamic. The great difficulty, however, remains in one's ability to recognize, make sense of, and consequently potentially learn to perceive through one's own sensorial-cognitive system how such subtle shifts make a difference in the complex interaction between two kinesthetic systems. As Gregory Bateson noted, information can be conceptualized as a "difference that makes a difference" (1972, p. 453), this insight highlighting the importance of the perceptive system in being able to bring forth novel affordances based on particularities of the context, for example through an expansion of visual perception (Downey, 2007).

The somewhat contradictory aspect about the importance of repetition in relation to skill acquisition is raised when the shihan responded to the student's question, "How do you know what to do?" His response included a phrase about the importance of doing the movements over and over again: "Because you've seen it a thousand times, you know when it's there. You see this guy punch you in a certain way, and you just know that what I'm gonna do will work. And of course, you have the motor skills to do it."

The shihan's response raises two points: the first is that some form of repetition does play a part in skill learning in Taijutsu practice. In order to be familiar with certain movements, those movements have to be performed often enough, as well as in relation to an attack by an uke. When I later asked the shihan whether he thought that the basic techniques can be effectively mastered after enough repetition, he quickly responded, "No, it's not about repetition. You have to learn the feeling of it."

The second point that his answer highlights, particularly in relation to the added point about the "feeling of it," is that an advanced level of skill has less to do with execution than with perception. The notion of knowing "when something is there" underlies the importance of a holistic reading of the kinesthetic situation, including the kinds of affordances that are available based on the given substrate, objects, and affective dimensions—for example, perceiving if the opponent is irritated, aggressive, trusting, or fearful.

The key bodily training of Taijutsu involves continuous balancing between what one is learning, in the sense of motor skills acquired for the execution of techniques, with the ability to disengage from any reified habitual aspects of such learning. In other words, a more advanced level of skill emerges in the context of a refined perceptive system that allows the practitioner to foster a more effective kinesthetic manipulation with his or her movements by doing less (Friday, 1997), by being at the right distance and angle at the right moment. While for initial learning motor skills might be helpful, they seem to have a developmental trajectory that advanced practitioners are able to disengage with in order to reach the fluidity of movement that the art emphasizes.

Such dynamism and fluidity of learning patterns has also been documented from a neurologically centered perspective. The primary motor cortex, known as M1, is an area of the brain distinctly involved with learned movement patterns. Neurologists Sanes and Donoghue note that the M1 "is not simply a static motor control structure. It also contains dynamic substrate that participates in motor learning and possibly in cognitive events as well" (2000, p. 393). Furthermore, Sanes and Donoghue suggest that the activation of neurons in M1 occurs in relation to complex neuronal networks already at the level of simple voluntary movements, such as the movement of one's fingers. The MI is also involved with patterns dealing with the perception of distance, timing, and angle in relation to one's own movement (Sanes & Donoghue, 2000, p. 406), elements of Taijutsu practice that advanced practitioners refer to as being the most crucial aspects of the art. In such dynamic situations, the primary motor cortex shares direct neural connections to cognitive operations, and the separation between a motor system and a sensory system is not meaningful (Järvilehto, 2006).

From a brain-centered perspective, Taijutsu practice holistically cultivates these distributed neurological networks, in contrast to any reification or efficiency of particular patterns of movement. Of course, any learning process, including that of Taijutsu, involves the whole person, not merely the neuronal networks that are "necessary but not sufficient" for skillfulness (Varela, Thompson, & Rosch, 1991). Further, as has been observed with many skill areas ranging from acting to writing, the question emerges whether skill is something that can be taught.

Throwing Technique Away

Doug Wilson is a current North American Taijutsu practitioner living and training in Japan. He is one of the main English translators of Masaaki Hatsumi's teachings and writings. Wilson explains how Hatsumi systematically emphasizes that the central aspects of the art cannot be "taught" in any traditional sense of the word, emphasizing an emergent nature of skill in Taijutsu practice. Rather than teaching or transmitting skills as entities of some kind, the instruction of Taijutsu consists of an invitation to the feelings associated with movements:

At almost every training session with Soke we hear him say, "I am not teaching form. What I'm showing cannot be taught, it must be discovered for yourself in your own training." Then how can we learn? What reference do we have to learning what Soke is teaching if he is not teaching? Before it was stated as the feeling, "I'm teaching the feeling." How can we teach the feeling? The feeling must be felt, not taught. Therefore, Soke gives the feeling and then asks to express what they felt, so that others might feel the same. This is not teaching, but facilitating experience. (Wilson, 2010)

Another way that Hatsumi has sought to express the skilled aspects of Taijutsu is to say that, in order to gain the feeling for the art, practitioners have to be able to "throw their technique away." Through this expression, Hatsumi emphasizes that the central aspects of Taijutsu revolve around the ability to perceive, to be able to "see" when "something is there," in terms of opportunities for kinesthetic manipulation based on the affordances of the overall context, and in terms of the proprioceptive understanding of one's own body.

Of course, the English metaphor of "seeing" is part of the conceptual challenge of being able to capture Taijutsu through a Western cognitive epistemology that places great emphasis on the role of vision over other senses (Howes, 2003), vision therein becoming a "cognitive style" (Grasseni, 2007, p. 3). The sakki test stands as a direct symbolic and sensory disruption for such a vision-centered paradigm. Furthermore, the test, like the overall emphasis of Taijutsu training, highlights sensorial integration in processes of perception and awareness rather than habituated bodily movements to execute techniques based on internalized motor skills. In Taijutsu practice, the goal of training is not to become imprinted with many skill sets as efficiently as possible. Rather, the goal of training is at times expressed as "becoming zero."

The idea of "becoming zero," or "throwing the technique away," has connections with current brain-imaging studies. Previous studies (Hill & Schneider, 2006; Solso, 2001) on expertise, for example in the context of art skills such as drawing, suggest that from the perspective offered by brain-imaging techniques (fMRI), the difference between expert and novice brains is not measured through the presence of patterns, such as increased neural scaffolding. Rather, the neurological difference between an expert and a novice brain may appear, paradoxically, as a reduction in brain activity.

While motor-skill learning in the context of Taijutsu practice or drawing is essential in order to create the potential for higher-level skill to emerge, advanced practitioners need to undo any attachment to the comfort that the repetition of techniques can bring. Instead of being comprised of specific techniques or forms of movement, skill is more accurately captured as increased "sensitivity to perceptual affordances" (Roepstorff, 2007, p. 204; see also Ingold, 2009, p. 203). It is further crucial to highlight the locality of skill as an emergent property of a context itself; for example, the kinesthetic and perceptive interaction between two bodies in a particular environment. As an emergent property, skill is not solely locatable within the boundaries of the practitioner's physical body, in a manner analogous to how cognition more broadly is not locatable in one's "head" (Clark & Chalmers, 1998). Thus, in advanced Taijutsu practice, technique is something that needs to be "thrown away," because if one is committed to any given course of action, the degrees of freedom for the emergence of skill cannot be present. In other words, by becoming zero, the practitioner is able perceive more.

Conclusion: Sensation as a Means of Movement Intelligence

One of the key North American assumptions about somatic skill foregrounds repetition, and hence *performance* itself, as a key dimension of teaching and learning. This assumption gains further implications when combined with notions such as "muscle memory" or "motor skill." This is the case because such a standardized approach to skill cultivation places emphasis away from the processes of sensation, which I suggest actually are the key components of advanced somatic skill. When performance becomes systematized into particular predetermined forms, whether in the case of educational or somatic learning, certain forms of movement in and of themselves become targets of learning. In contrast, Taijutsu practice, and learning in many contexts more broadly defined, is an open-ended process which cannot directly be "taught," at least in the sense of conventional and performance-based approaches and certainly cannot be broken down into discrete units that can be cultivated through repetition or "drills."

In the course of this chapter, I provide an ethnographically situated analysis of the movement patterns that characterize advanced skill in Taijutsu practice, demonstrating the manner in which each advanced movement pattern is also a form of sensation, and the ways in which the sakki test stands out as a key symbol of such skill cultivation. When advanced Taijutsu practitioners explain that they "know how to do" something well, they do not emphasize the experience of having previously executed particular forms of movement in great quantities, thereby having achieved some level of "automation." Instead, through their actions and their verbal accounts, advanced practitioners demonstrate having studied the various *contexts* of movement in great depth. In Taijutsu, the overall movement context involves an interaction between at least two bodies in time and space, and, consequently, predetermined movement patterns or any isolate "motor skills" have very little capacity to explain the nature of somatic skill in such complex open-ended interaction.

Neurologically, it is worth noting that the fusion of moving and sensing that characterizes advanced skill is also a cognitive event: the sensorial and perceptive engagement with the context parameters of movement execution defines advanced skill as an emergent property of the kinesthetic interaction, for example, away from one's own "head" or even the body. Through this somatic case study I suggest that any motor skills, in order to be *skills* at all, can only operate as extensions of more broadly distributed patterns of sensation. When performance is intertwined with perception, movement becomes a way of interpreting and understanding the kinesthetic context, thus operating as one of many forms of intelligence that characterize the nature of living systems.

References

Bateson, G. (1972). *Steps to an ecology of mind*. New York: Ballantine Books.

Callagher, S. (2005). *How the body shapes the mind*. Oxford: Oxford University Press.

Clark, A., & Chalmers, D. (1998). The extended mind. *Analysis, 58*(1), 7–19.

Coward, F., & Gamble, C. (2008). Big brains, small worlds: Material culture and the evolution of the mind. *Philosophical Transactions of the Royal Society B, 363,* 1969–1979.

Davies, R. (2005). *The fifth business*. New York, NY: Penguin. (Original work published 1970)

Deacon, T. (1997). *The symbolic species: Co-evolution of language and the brain*. New York: W. W. Norton.

Donald, M. (2001). *A mind so rare: The evolution of human consciousness*. New York: W. W. Norton.

Downey, G. (2007). Seeing with a "sideways glance": Visuomotor "knowing" and the plasticity of perception. In M. Harris (Ed.), *Ways of knowing: Anthropological approaches to crafting experience and knowledge* (pp. 202–241). New York: Berghahn Books.

Downey, G. (2005). *Learning capoeira: Lessons in cunning from an Afro-American Brazilian art*. Oxford: Oxford University Press.

Fitts, P. M., & Posner, M. I. (1967). *Human performance*. Belmont, CA: Brooks/Cole.

Friday, K. (1997). *Legacies of the sword: The Kashima-Shinryu and samurai martial culture.* Honolulu: University of Hawaii Press.

Geurts, K. L. (2002). *Culture and the senses: Bodily ways of knowing in an African community.* Berkeley, CA: University of California Press.

Gibson, J. J. (1979). *The ecological approach to visual perception.* Boston: Houghton-Mifflin.

Grasseni, C. (2007). Introduction. In C. Grasseni (Ed.), *Skilled visions: Between apprenticeship and standards* (pp. 1–22). New York: Berghahn Books.

Hall, E. T. (1964). Adumbration as a feature of intercultural communication. *American Anthropologist, 66*(6), 154–163.

Hatfield, B. D., Haufler, A. J., & Spalding, T. W. (2006). A cognitive neuroscience perspective on sport performance. In E. O. Acevedo & P. Ekkekakis (Eds.), *Psychobiology of physical activity* (pp. 221–240). Champaign, IL: Human Kinetics.

Hill, M. N., & Schneider, W. (2006). Brain changes in the development of expertise: Neuroanatomical and neurophysiological evidence about skill-based adaptations. In K. Anders, N. Charness, P. J. Feltovich, R. R. Hoffman, (Eds.), *The Cambridge handbook of expertise and expert performance* (pp. 655–684). Cambridge: Cambridge University Press.

Howes, D. (2006). Cross-talk between the senses. *Senses and Society, 1*(3), 381–390.

Howes, D. (2003). *The empire of senses: Engaging the senses in culture and social theory.* Ann Arbor, MI: University of Michigan Press.

Ingold, T. (2009). Stories against classification: Transport, wayfaring and the integration of knowledge. In S. Bamford & J. Leach (Eds.), *Kinship and beyond: The genealogical model revisited* (pp. 193–213). Oxford: Berghahn Books.

Ingold, T. (2004). Culture on the ground: The world perceived through the feet. *Journal of Material Culture, 9*(3), 315–340.

Järvilehto, T. (2006). What is motor learning? In K. Thomson, T. Jaakkola, & J. Liukkonen (Eds.), *Promotion of motor skills in sports and physical education* (pp. 9–18). Jyväskylä, Finland: University of Jyväskylä, Department of Sport Sciences.

Lakoff, G., & Johnson, M. (1999). *Philosophy in flesh: The embodied mind and its challenge to Western thought.* New York: Basic Books.

Merrell, F. (2010). *Entangling forms: Within semiosic processes.* Berlin, Germany: De Gruyter Moton.

Mithen, S., & Parsons, L. (2008). The brain as a cultural artefact. *Cambridge Archaeological Journal, 18*(3), 415–422.

Pallasmaa, J. (2005). *The eyes of the skin: Architecture and the senses.* Sussex, England: Wiley.

Roepstorff, A. (2007). Navigating the brainscape: When knowing becomes seeing. In C. Grasseni (Ed.), *Skilled visions: Between apprenticeship and standards* (pp. 191–206). New York: Berghahn Books.

Sanes, J., & Donoghue, J. P. (2000). Plasticity and primary motor cortex. *Annual Review of Neuroscience, 23,* 393–415.

Solso, R. L. (2001). Brain activities in a skilled versus a novice artist: An fMRI study. *Leonardo, 34*(1), 31–34.

Suzuki, T. (1986). *The way of acting: The theatre writings of Tadashi Suzuki* (J. T. Rimer, Trans.). New York: Theatre Communications Group.

Varela, F. J., Thompson, E., & Rosch, E. (1991). *The embodied mind: Cognitive science and human experience.* Cambridge, MA: MIT press.

Wacquant, L. (2004). *Body and soul: Notebooks of an apprentice boxer.* New York: Oxford University Press.

Weiming, T. (2007). Pain and humanity in the Confucian learning of the heart-and-mind. In S. Coakley & K. K. Shelemay (Eds.), *Pain and its transformations: The interface of biology and culture* (pp. 221–241). Cambridge, MA: Harvard University Press.

Wilson, D. (2010, April 27). Striking Compassion [Blog post]. Retrieved from http://henka.wordpress.com. Accessed June 10, 2011.

Zarrilli, P. (2004). Toward a phenomenological model of the actor's embodied modes of experience. *Theatre Journal, 56,* 653–666.

Ziemke, T., Zlatev, J., & Frank, R. (Eds.). (2007). *Body, language and mind: Embodiment* (Vol. 1). New York: Mouton de Gruyter.

8 Holistic Humor: Coping with Breast Cancer

Kathryn Bouskill

A few months after Marjorie's regimen of chemotherapy ended, she stood before her mirror and noticed hair growing on her scalp. Although it would still be months before Marjorie looked like herself—as she did before discovering she had breast cancer—she entered the breast cancer support center and proclaimed with a grin, "I've got *fuzz*! Peach fuzz on my head!" The absurdity of celebrating mere fuzz on her head made Marjorie and the other survivors in the support center break out in laughter. The following day, her fellow survivors greeted Marjorie with a cake displaying a cartoonized version of her head with "peach fuzz" on top. Marjorie told me, "It was hilarious!" Chemotherapy treatment had been awful and the threat of breast cancer was still looming, but in this moment the women celebrated and turned the sadness of breast cancer into laughter.

Breast cancer is a physically and emotionally painful disease, and there is nothing ostensibly humorous about the hardships of breast cancer. How, then, could these women devote their time and energy to making each other laugh about things like hair loss? The presence of humor among these breast cancer survivors, who had endured such a loss of control over their lives, led to the present ethnographic exploration of how and why humor as a coping mechanism could be so compelling.

Told through the stories of breast cancer survivors in the context of an American breast care clinic and support center, humor is presented as a means of coping that intertwines social interaction, interpretive meaning-making, and internal mental states. The neurological processes and socio-cultural contexts described here confound artificial distinctions between body and mind and reveal humor to be a meaningful social and biological mechanism. While this chapter explores the constituent elements of the use of humor, the goal is not to reduce humor solely to a culturally informed script or an instinctive neurological response. Rather, I argue that humor is a cognitive, internal reworking of a greater sociocultural reality, and thus something neuroanthropological to its core. It is the bidirectional interaction of cognitive and sociocultural processes that makes humor a meaningful coping mechanism among these breast cancer survivors.

The unique form of humor present within the context of breast cancer requires a social connection and mutual understanding among the survivors. In particular, humor relies upon acceptance of enduring the hardships of the disease and its illness experience, which was facilitated by the openness with which breast cancer survivors were encouraged to speak in the support center. Humor, however, is not a measure of optimism among these survivors, but rather shows how these women negotiate the challenges of breast cancer and their new roles as survivors. Although the underlying fears evoked by breast cancer did not disappear through humor, survivors took as much control as possible by laughing about that which was not life-threatening, such as hair and breast reconstruction.

The women who comprised this ethnography differed in age, race, and social status, yet humor emerged as a self-described and overarching theme as a means of coping with the stress of having the disease. The vivacious use of humor among the women serves its own important role in the experience of having breast cancer. Thus, the purpose of this study was not to test the therapeutic properties of laughter or to assert that humor can alter a cancer prognosis. Furthermore, this chapter does not provide a typology of humor or a complete review of the neurological functions of humor. Rather, humor is considered as a holistic coping mechanism, particularly in its ability to reinforce social bonds, relieve personal stress, and signify a collective sense of strength against the negative effects of breast cancer.

Neuroanthropology: Joining Humor and Coping

Breast cancer is a crisis and a major disruption in a person's life (Culver, Arena, Antoni, & Carver, 2002; Culver, Arena, Wimberly, Antonia, et al., 2004). A breast cancer diagnosis is a critical point of transition between life before cancer and life with cancer. In fact, cancer patients in the United States are labeled cancer survivors both socially and medically for their entire lives.[1] A woman's life changes substantially from the moment a lump is discovered in a breast self-exam or an annual mammogram. In rapid succession, a woman enters into the realm of the sick and is treated through the standard protocols of lumpectomy or mastectomy, and likely chemotherapy and radiation. In short, the time following a cancer diagnosis is a whirlwind of acute emotional and physical distress (Golden-Kreutz & Andersen, 2004).

Cancer survivors are faced not only with a life-threatening disease, but also with a taxing and emotional illness experience (McMullin & Weiner, 2009). For example, breast cancer affects key signifiers of female identity. Cancer survivors participate in a variety of nonconventional methods and complementary and alternative medicine to treat their disease and enhance their quality of life (Söllner, Maislinger, DeVries, Steixner, et al., 2000). Among the nonconventional methods of coping, clinicians have anecdotally recognized the use of humor among cancer survivors (Christie, 2005; Spross & Burke, 1995).

Coping is defined as the behavioral and cognitive means of managing a stressor that is perceived as exceeding one's resources or blocking one's path toward a desired goal (Culver et al., 2002; Lazarus & Folkman, 1984). Coping is the result of an individual's primary cognitive appraisal, or ability to determine what is at stake in encountering a life-event, followed by a secondary appraisal, where an individual determines if anything can be done to change, prevent, or control the situation (Folkman, Lazarus, Dunkel-Schetter, DeLongis, et al. 1986; Lazarus and Folkman, 1984). Positive cognitive appraisal occurs when an individual perceives a situation or life-event as changeable or able to be brought under control (Folkman, Lazarus, Dunkel-Schetter, DeLongis, et al. 1986). Although coping with an illness is generally presented as an individual goal, the ability and effectiveness of coping are never unbounded from sociocultural contexts, including interpersonal relationships, spiritual ideals, socially defined ethnicity, access to institutional support, and the degree to which an illness or condition is stigmatized (Cohen & McKay, 1984; Holahan, Moos, Holahan, & Brennan, 1997; Stahly, 1989). Thus, like humor, coping is a cognitive process that is firmly rooted in sociocultural settings.

Bridging the role of humor and positive emotions within sociocultural contexts is warranted, considering the universal presence of humor and the supposition that humans have a genetic predisposition to understand humor and to respond with laughter (Gervais & Wilson, 2005). Humor and laughter have received considerable attention since Darwin (1872/1955) asserted the evolutionary roots of the phenomena as a means of socially expressing happiness. Humor as a positive emotion has been relatively overlooked in an evolutionary-adaptionist context because natural selection is thought to only affect emotions that are selective for certain threats or opportunities that enhance fitness (Nesse, 1990). However, this claim says little about why humans spend a vast amount of energy on producing positive emotions in response to stress (Fredrickson, 1998), or why humor is so ubiquitous in social communication.

The psychological and neurological evidence for humor as a promoter of positive emotions and a buffer against the negative effects of stress is growing (Abel, 2002; Kuiper & Martin, 1998; Moran, Wig, Adams, & Janata, 2004; Lefcourt, Davidson, Shepherd, Phillips et al., 1995). However, this research remains fragmented, largely due to disciplinary boundaries and methodological constraints that impede efforts to recreate natural, spontaneous humor in a scientific setting. Abel (2002), among others, has described humor as a coping strategy within which there is a cognitive-affective shift and a restructuring of a stressful situation to make it less threatening. Through this cognitive-affective shift, humor creates a distance from stress and suffering (Freud, 1959; Kuiper, McKenzie, & Belanger, 1995).

Neurological research has also attended to the neural correlates of humor and laughter; however, the sociocultural complexity of laughter stimuli has made some neuroscientists reluctant to unravel the neurological mechanics of the everyday phenomenon (Wild, Rodden, Grodd, & Ruch, 2003). The ability to perceive and respond to humor is complex and relies upon focus, attention, memory, emotional evaluation, and understanding abstract communication (Wild, Rodden, Grodd, & Ruch, 2003), all of which occur within sociocultural interactions and contexts. Thus, while the task of understanding how the neurological correlates of humor and laughter correspond to stress relief is formidable, trying to do so without an incorporation of both the mental and social aspects would be largely incomplete.

In the sociocultural domain of humor, anthropologists have long used ethnography to interpret the importance of humor and joking relationships. Key anthropological insights include humor as a controlled form of social deviance that allows for the discussion of otherwise taboo subjects (Palmer, 1994), a functional means of diffusing tension between social groups (Radcliffe-Brown, 1940), and a form of "play" intended to challenge and resynthesize meaning through social communication (Bateson, 1972/2000). More recently, Goldstein (2003) showed the power of humor to act as a cohesive force in defining social relationships and as a means of resistance. In industrial societies, social relationships that involve the use of humor demonstrate a consciousness of solidarity and a shared social identity (Apte, 1985). The power of this solidarity also has the opposite function of producing social boundaries among those who understand the humor and those who do not (Levine, 1969). These boundaries can come from linguistic differences, social classes, and differing social identities. While traditional anthropological insights describe the functional aspects of humor within a society and culture, less attention is given to the experiential embodiment of humor.

Despite the vast research on humor, a central limitation persists in the tendency to reduce the phenomenon solely to its cognitive or sociocultural factors (Palmer, 1994). If the brain is the machinery in the perception of humor, then sociocultural contexts are the necessary spark. Considering humor without understanding the effects of both domains misses the full breadth of why it is such a powerful part of being human. Neuroanthropology cannot grasp the totality of humor, but it does facilitate a conceptual framework with which to understand how humor as a simultaneously sociocultural and neurological phenomenon is a compelling coping mechanism. This analysis proceeds by considering how sociocultural constructions of breast cancer in the United States set up the contexts and resources within which breast cancer survivors cope.

Breast Cancer in the United States: Politics and Pink Ribbons

Understanding the ways in which women cope with breast cancer requires a description of the broader social and cultural attitudes toward the disease. Therefore, discussing the perceptions of breast cancer in the United States elucidates how the illness experience of survivors is shaped. This section briefly describes the transformation of attitudes toward breast cancer in the United States, which provides a framework for demonstrating how sociocultural settings can mediate emotional responses. In line with the ethos of neuroanthropology, this aspect of the analysis incorporates the political and sociocultural factors that affect internal neurological function.

Approximately 200,000 American women are diagnosed each year with breast cancer (Jemal, Siegel, Xu, & Ward, 2010). While breast cancer continues to affect the lives of many, perceptions of breast cancer in the United States have undergone significant changes within the last decades, affecting those diagnosed with the disease at the political, social, and personal levels. For instance, breast cancer has become an important social movement within the United States, and the pink ribbon, while acknowledged to be problematic, is now an iconic image of breast cancer support (King, 2006; Sulik, 2011).

With all the publicity surrounding breast cancer, it is easy to forget that breast cancer survivors in the United States and elsewhere have suffered a long history of stigmatization and shame. Breast cancer was long thought to be the result of "insufficient passion," and patients were essentially viewed as devoid of feeling and self (Sontag, 1979, p. 24). The way the breasts and the body are consumed by cancer was a metaphor for weakness of character and a failure of self-expression. Additionally, speaking publicly about sexualized body parts (e.g., breasts) was taboo. By the mid-1970s, breast cancer rapidly reached the public eye when feminists called for a repeal of the stigma that forced women to suffer the illness alone and inhibited prevention and treatment efforts. The media also aided this new perception as prominent public figures such as Betty Ford and Happy Rockefeller spoke openly about their breast cancer and called for women to receive regular mammograms and perform self-breast exams (Sherwin, 2006). In addition, women's health became a popular bipartisan issue and breast cancer provided a noncontroversial means of addressing political support for female voters (Dodson, 2005).

Over time, political groups, grassroots campaigns, and many corporations have responded to increased awareness of breast cancer and thrust it squarely into the public eye (Kasper & Ferguson 2002; Sherwin, 2006). The United States has also become a center of breast cancer advocacy, a place where women can speak publicly and take control of the personal impact of the disease (Kaufert 1998). Thus, breast cancer survivors who once faced a silent, stigmatizing illness experience now have voices strong enough to influence businesses, government groups, and scientific communities to rally behind the cause for breast cancer awareness (Braun, 2003; Kasper & Ferguson, 2002.

Occurring in tandem with the socio-politicization of breast cancer is the rise of new metaphors, particularly the valorization and feminization of the disease. For instance, common parlance includes images of war, as in "battling" or "fighting breast cancer," and the disease is often seen as an enemy plaguing an otherwise healthy body. Additional metaphors include the overly abundant pink ribbons that ascribe stereotypical notions of femininity to the disease. Cognitively, these metaphors reflect an unconscious conceptual framework, indicating that they may be so commonplace that they are used unintentionally rather than deliberately (Gibbs & Franks, 2002). This shows the power of cultural influences on how people describe and embody the experience of breast cancer. The popularization of these symbolic metaphors has created an emerging sociocultural identity of the breast cancer survivor and has made breast cancer awareness an aspect of the life of an American woman. Cultural models of breast cancer, whether a survivor is for or against them, elicit meanings and ideas that shape how individuals respond to breast cancer, in turn affecting both the public perception of the disease and the ability to cope with the disease (Coreil, Wilke, & Pintado, 2004; Matthews, 2000).

Breast cancer survivors in the United States no longer need to fear a breast cancer diagnosis will be stigmatized as shameful, and survivors have taken greater control over their treatment decisions, coping strategies, and their experiences with breast cancer (Lerner, 2006). With this control has come the call for attention to the serious aspects of breast cancer and the awareness that it is a time of "uncertainty, obscurity, and ambiguity" (Skott, 2002, p. 230). That is, all of the valorization and feminization of breast cancer can obscure the actual hardships breast cancer survivors suffer (King, 2006; Sulik, 2011). The issue is that the presence of pink ribbons on postal stamps, yogurt lids, or bumper stickers does not actually provide opportunities to truly cope with the immense stress and pain that breast cancer brings about, nor does it always allow women to openly express these difficulties. Still, this public support has been a major factor in reducing the stigma of breast cancer, and it enables survivors to take more control over their illness experiences, alleviating feelings of helplessness and hopelessness in the face of breast cancer.

As a result of this public support, "breast cancer survivor" has become a commonplace social label. This assertion must be made in order to frame the use of humor socially, because it cannot be assumed that connections and solidarity among women survivors would occur without this particular sociocultural identity. By understanding breast cancer's transition from a once-stigmatized condition to a major political and social movement, this analysis can then return to the community level to understand how these broader sociocultural ideas of breast cancer have an effect on the emotional experiences of breast cancer.

"We Laughed for Hours!"

Interest in this research began when the director of the support center, who is a nurse and a breast cancer survivor herself, held a support group meeting for local survivors. Anticipating an outpouring of frustration over the hardships of breast cancer, what she found instead was "over three hours of laughter." The incongruity of humor instead of sadness spurred a community-based participatory ethnography to understand this particular coping style among thirty local breast cancer survivors.

The support center and its associated biomedical clinic are located within a traditional, urban public hospital in the midwestern United States.

The breast care clinic treats patients from a range of ethnicities, socioeconomic statuses, ages, and religious backgrounds. While the clinic is for scheduled appointments with medical personnel, survivors use the support center located next door to have both planned and spontaneous meetings. In addition to providing generalized support and information for living with breast cancer, the center organizes community outreach programs to provide mammograms for minority and underserved women in the region. The survivors in this study also described the support center as a place to learn about the disease itself, commenting that the medical jargon used by doctors made it difficult to understand the biological effects of breast cancer and its treatments.

The support center was the main location of the ethnography, which included both semistructured interviews and participant observation. Additional participation-observation sessions took place at events hosted for breast cancer survivors in the community. Semistructured interviews centered on the experiential aspects of breast cancer, mainly the associated stress, types of medical and psychological support available, the personal significance of the disease, and the degree to which interactions with fellow survivors impacted the experience of having breast cancer. Every study participant identified humor as one form of coping with breast cancer and had similar means of using humor as a coping mechanism. Rather than attempt to define or categorize this type of humor at the outset, a grounded approach was used to focus on what the survivors deemed humorous, their reactions to humor, and the contexts within which they used it.

Transitioning to "Cancer World"

The survivors often referred to their new experience of having breast cancer as "cancer world," which signifies both a literal and metaphorical transformation in their lives. "Cancer world," in addition to the social bonds of survivorship, also includes the drastic physical transformations and the emotional burden incurred by the loss of signifiers of femininity (e.g., breasts, hair, infertility, and sterility) and, among many additional stressors, the threat of death. The idea of "cancer world" conveys the vicissitudes, struggles, and life-changing experiences of adjusting to life with breast cancer. It reflects the liminal state of survivors, who may at times be asymptomatic and feel more like their pre-cancer selves, and at other times may be more severely affected by cancer treatments. "Cancer world" goes beyond the publicized pink ribbons to reveal the raw reality of having breast cancer and can only be grasped fully by breast cancer survivors.

Together in "cancer world," the survivors identified the physical and emotional stressors of breast cancer: worries over relationship changes with a partner, the future stability of one's family, the frustration of memory loss due to chemotherapy (or "chemobrain"), intense fatigue and related side-effects of chemotherapy, financial pressure, loss of sexuality or femininity, infertility and sterility, breast disfiguration, the heightened risk for genetic relatives to develop the disease, and the fear of death. It is important to note that while hair loss, fatigue, breast disfiguration, and chemobrain could be seen as humorous in some fashion, severed relationships, passing on heritable risk factors, and death were never interpreted as such; thus, survivors negotiate the boundaries of what can be considered humorous. This consensus over what is and is not humorous conveys a common understanding, sensitivity, and empathy among the survivors. This underlying understanding provides a common ground from which survivors can cope.

Survivors and staff members at the support center and clinic know one another by name and never walk by one another without offering a warm greeting. Survivors informed me that they often make private telephone calls to one another during off-hours, send cards of encouragement, and pray for one another. Solidarity through an array of connections is omnipresent and carries a great deal of significance when conducted out of the public view. These social links through which survivors grapple with stress are also part of "cancer world," and reinforce that it is an exclusive place intended for survivors alone to interact.

As mentioned, the transitions in public awareness, philanthropic efforts, and political attention have promoted the social spaces of support groups, breast care clinics, and community events for survivors, as well as the normalcy of the act of reaching out to one another. Survivors often drew from their own experiences to help new survivors with their transition into "cancer world" and to help one another. The youngest of the survivors met me in the support center on a day where she herself had no treatments or appointments, but was there to stay with a new patient throughout her chemotherapy session. She explained that the process of sitting in the chemotherapy center is particularly physically and emotionally stressful. She described her purpose there as "going to bat" for the other survivor, despite not knowing her before the cancer diagnosis.

The depiction of solidarity builds on why humor is a popular and effective means of coping. Fundamentally, the support center creates an environment where survivors do not have to face the burden of breast cancer alone, nor do they have to always be optimistic or cheerful, as is often expected of them in American culture (King, 2006). Knowing when to initiate a joke about a side effect of breast cancer treatment is rooted in the shared experience and mutual understanding of breast cancer that is present among survivors. This humor is also a means of excluding those who do not understand "cancer world." Non-survivors cannot consider things like hair loss and chemobrain funny. In fact, humor used by survivors was often met with confused silence, as was the case when a male colleague had not seen a survivor in months and complimented her on her "new hairdo," to which she replied, "Yeah, well, that's what happens when ya get CANCER! Ha!" The startled man began to apologize as the survivor continued to laugh, presumably thinking it would be inappropriate to laugh along. When she recounted the story at the support center, other survivors laughed endlessly along with her. Thus, the polarizing effects of humor are observed here: those survivors who can understand the humor bond closer together, while those who cannot grasp the humor realize they cannot understand the reality of breast cancer survivors. Rather than a means of exclusion, it is a way of asserting that the experience of breast cancer is complex and not easily understood by those who have not gone through it.

Another aspect of "cancer world" is the dimension of time and the realization that the use of humor is a process that does not occur instantaneously when a woman is diagnosed. One survivor, who later used humor comfortably, bluntly stated that at the start of her experience with breast cancer, "I didn't find humor, I found homicide!" She even reacted physically to medical personnel who performed her biopsy. Like movement across the borders of "cancer world," humor too is nonpermanent. This point suggests that the use of humor among breast cancer survivors is a fluid process rather than an inherent aspect of the illness experience; that is to say, humor does not occur overnight, nor is humor always appropriate in every situation. Rather than denial, the reality of the stress of cancer must be accepted before a survivor can begin to cope and, in turn, experience the cognitive-affective shift present within the use of humor. The social bonds present within the support center are a particular motivator of this shift and a key outlet for stress. Therefore, the following section explores how the use of humor reinforces social bonds and the mutual support that survivors provide each other.

Dealing with "Cancer World"

The reinforcement of social support through humor starts with mutual understanding among the survivors, which builds a more open and empathic network. This social support not only fosters deeper social bonds among the survivors, but also leads to stress relief and coping.

While "cancer world" is full of hardships, at the support center it is not a place of isolation and desolation. Informants expressed that as quickly as they were severed from their pre-cancer lives, they were received into an empathetic domain where others had endured a similar loss of stability and knew well the fears and worries of breast cancer. The survivors have both social and emotional bonds that are fostered by acknowledging the reality of struggling through breast cancer treatment and the empathetic desire to help others.

Psychological stress emerges foremost from a lack of control, followed closely by a lack of predictability (Sapolsky, 1994). Clearly, breast cancer evokes feelings of a loss of control and uncertainty. Informants would often describe their feelings of intense frustration and helplessness after their cancer diagnosis. One survivor emotionally recounted, "I went to the parking lot and shook my fists and yelled and cried," and others described not knowing where to turn to next. Psychological stress, in addition to being caused by a lack of control, also occurs when outlets for frustration are unavailable. One of the most influential modulators of this stress response is the outlet of social affiliation (Sapolsky, 1994), like that provided by the support center. This is where links can be drawn between psychosocial stress and the occurrence of peer assistance in emotional coping. Because the women share not only a specific and private location but also a social label, coping with breast cancer becomes a meaningful social act. The breast cancer support center shows how building relationships can lead to a better quality of life, and on a greater scale, reveals the importance of building positive emotions, even during traumatic times.

Psychoneuroimmunology has been used to demonstrate that increased social support is associated with lower mean concentrations of cortisol, suggesting that social support as a coping mechanism carries implications of stress buffering among physiological pathways and overall better health (Cohen & Wills, 1984; Dunn, Bruce, Ikeda, Old et al., 2002). Conversely, further supporting evidence for the importance of social support in the experience of breast cancer, Turner-Cobb, Sephton, Koopman, Blake-Mortimer et al. (2000) found that metastatic breast cancer patients who reported less social support had significantly higher mean cortisol levels.

Humor is linked to social support and coping because it reinforces social bonds, group connectivity, and common understandings within a social group (Kershaw, Northouse, Kritpracha, Schafenacker, et al., 2004). It builds on existing interpersonal relationships, heightens personal and group states of positivity, and influences others to also engage in the use of humor (Francis, 1994). Because it is socially acceptable, if not valorized, to be a breast cancer patient in the United States, these survivors can draw upon their social label as breast cancer survivors to join together in emotional release within the support center.

During ethnographic interviews, I asked about the cathartic effect of humor on the stress of breast cancer. The women explained by stating, "We all know what we're going through." Inside this world of mutual understanding and the security of knowing that others care, there is a certain degree of futility in expressing angst over the effects of the disease, namely breast disfiguration or reconstruction, hair loss, and memory loss. At the support center and during community outings, complaining about these fears and worries to another woman felt unproductive and antithetical to coping.

Being constantly confronted by the negative effects of breast cancer left survivors feeling "tired of being consumed by what-ifs." Over time, they realized that privileging stress and letting it take hold of the mind led to feelings of hopelessness. One survivor stated that stress "just doesn't work. It can't get you through it." During a participant-observation session, women sitting around a table, all with short hair from the effects of chemotherapy, said they felt socially connected to one another, and because of this, did not have to occupy themselves by explaining why breast cancer is so stressful. Thus, rather than restating the same fears among each other, survivors drew from their mutual understanding to express the cumbersome and laughable transitions into "cancer world." They used their time and energy to uplift and strengthen each other against the stress of breast cancer through laughter. Thus, social bonds provide an outlet for emotional release, while shared understanding allows for the promotion of positive emotions. The following section attends to how this promotion of positive emotions is enacted linguistically and how humor is an outward signifier of an internal, cognitive shift.

Language, Humor, and Meaning

Humor is a hallmark of the power of communication to transform and transcend meaning. By linguistic means, the use of humor within this group of breast cancer survivors outwardly projects an internal shift in the ways the women perceive the hardships of cancer. Although humor does not remove the hardships, it signifies that the women are not completely overcome by hair loss, chemobrain, breast deformation, and other negative side effects. With the use of humor, these stressful aspects are brought to the fore, but instead of succumbing to them and feeling hopeless, the women show they are taking control by laughing. Such a reinterpretation of the meaning of the stresses of breast cancer marks a cognitive shift in the women's external reality.

The potency of humor and the explanation for why it is a complex form of communication lie within its paradoxical nature. That is, for these cancer survivors, humor is a form of play that reflects something that is in fact quite serious, or "not play" (Bateson, 1972/2000). The seemingly lighthearted laughter actually transforms that which is serious and threatening into a more benign and less intimidating form. In this context, the negative effects of breast cancer never become easy, but humor shows that these effects will not consume a survivor and the woman she was before her diagnosis.

In this context, humor operates within cognitive and emotional states to linguistically project a change in a state of mind. But more than simply projecting this state of mind, humor also functions to reduce the stress response and to influence others to engage in the change in the cognitive state; that is, the use of humor among survivors influenced others to see aspects of "cancer world" as humorous. Said a patient succinctly, "I didn't see the humor in it all until receiving it from others around me here [at the clinic]." Humor, then, is a learned process, and like other learned processes, it impacts both neurological processing and social perception (Downey, 2010). This does not indicate that all the survivors endured the same experiences with breast cancer, but that this positive frame of mind is something that can be shared socially.

Portraying specific ethnographic examples of the use of humor is hindered by the fact that they are generally only understood as funny in the context of being a survivor. Nevertheless, the elements of stress relief and affective reward can be demonstrated by several examples of the humor of hair loss, which survivors termed "*hair* today, gone tomorrow," or "*wiggin'* out." Another middle-aged survivor burst into laughter as she told me that the hair on her head after chemotherapy looked like "a map of the United States with forty-eight states missing!" That which was threatening is projected as humorous and therefore is an outward expression of positive emotion and defiance.

These ethnographic examples are not exceptional, and the survivors shared them with me because they believe that humor lessens the burden of stress and anxiety that surrounds breast cancer. Although stress does not disappear, their level of anxiety is lessened by the camaraderie of the support system and the neutralization that humor produces. As one survivor explained, "Humor lets you take the seriousness out of knowing things will never be the same again." Thus, humor provides a way of acknowledging that, while breast cancer is incredibly difficult, this stress will not take complete control of a survivor's life. The ability to outwardly express this feeling is what allows survivors to do, as one woman stated, "*move* on."

This idiomatic phrase of "moving on," though brief and common, is rich with affective significance and metaphor. It also has a strong connection to the idea of "cancer world." Moving on, or away, from "cancer world" reflects what is suggested in the neurological literature regarding humor's ability to create a distance between oneself and a stressor, allowing the negative experience to be taken less seriously (Abel, 2002). In other words, through humor and moving on, the survivors are linguistically projecting their cognitive-affective shift from the stress of breast cancer. As opposed to maintaining a cheerful attitude during the struggle of cancer, this sometimes crude but always personalized humor operates outside of the societal expectations of relentless optimism from breast cancer survivors. Instead, humor represents a candid affirmation of the hope for survival that, on a woman's own terms, takes into account the hardships of breast cancer.

The metaphor of cancer as a place suggests that it is an illness in which fluctuation and movement can occur. In other words, "foreignization" of the state of *having* cancer represents it as a location that one can enter and depart. Moving on is the act of emotionally departing from the anxiety, uncertainty, and threat inherent in cancer. The labeling of a metaphorical but shared space builds on the idea of a more exclusive solidarity expressed by the survivors. The metaphor of cancer as a place also suggests that it is not part of the inherent makeup of the survivor. In other words, cancer is present in the woman's life, yet she cannot be defined solely on the basis of having breast cancer. It is this sense of control and power over breast cancer that relates back to the positive cognitive appraisal and the reduction of stress, thereby enhancing the ability of humor to act as a compelling coping mechanism. As one patient stated, "You feel stronger when you can laugh, adapt, and keep moving on."

The labeling and compartmentalization of cancer through "cancer world" is also an aspect of taking command of an otherwise uncontrollable disease and gives survivors a sense of control over the borders of this world. Sitting in the support center surrounded by photos of survivors in the community, one survivor related being struck by her worries over losing her femininity. She asked herself, "Aren't you more than breasts? Aren't you more than hair?" Another survivor was "traumatized by the loss of [her] hair, but now [realizes] how trivial the whole thing was." These women were able to remove the stress of the non-crucial aspects of survival and make light of the situation where they could. Thus, humor is a means of proving that personality and femininity are deeper than breasts and hair.

Making side effects humorous is also a means of acknowledging and deferring the inevitability of stress, while stripping away the power it carries over an individual. In other words, it is the outward communicative means of indicating that, although drastic side effects are being experienced by the body, an individual will not be overcome by stress and helplessness. The deeper meaning, then, is not just a temporary positive message displayed through humor, but a deep inner belief that the survivors can take control of the stress of breast cancer. In essence, as a saying common among survivors puts it, these women have cancer, but cancer does not have them.

Recess and Reward: The Positive Effects of Humor

When I asked the survivors to describe the feeling of laughter, they often repeated notions such as, "Laughter is just my way of exchanging all this *overwhelming* [sic] for something else. It's my mental recess," or "Why do I laugh? It's just so carefree. It releases everything, like I don't have any inhibitions," and furthermore, "Laughter is the opposite of stress. It's just what we *have* to do." Survivors described humor and laughter as a cognitive break from the stress. For instance, one woman stated, "Laughter lets me get out of this fear. It brings me peace, it's a stress-reliever, and I just go off to another place." These quotes reflect research findings that the cognitive elements of humor, both in its perception and the expression of laughter, activate reward centers within the brain (Moran, Wig, Adams, & Janata, 2004). In this case, the reward of humor comes from this idea expressed by the survivors as "mental recess"; but in fact, neurological research depicts strong evidence that the brain is in reality doing anything but taking a break.

Drawing direct links to humor as a reward, Mobbs, Greicius, Abdel-Azim, Menon, et al. (2003) provide strong evidence of an association between humor and a network of subcorticol structures that are crucial components of the mesolimbic dopaminergic reward system. In other words, humor is identifiable within neurological centers of positive emotion that allow the mind to perceive an emotional reward. Furthermore, fMRI data reveal that humor activates the medial ventral prefrontal cortex, another region that regulates reward-related behaviors (Azim, Mobbs, Jo, Menon, et al., 2005; Goel & Dolan, 2007). Along the psychoneuroimmunological pathways, the use of humor and laughter are linked with upregulated immune function (Bennett, Zeller, Rosenberg, & McCann, 2003). These data illuminate the bidirectional pathways of humor and neurological function, and show the physiological depth of humor as a coping strategy.

Neurological studies examining the sex-specific effects of humor show the activation of brain structures related to semantic knowledge and juxtaposition (temporal-occipital junction and temporal pole) and language-processing centers (inferior frontal gyrus) in both men and women; however, women are found to experience greater activation of the reward-response centers (nucleus accumbens of the mesolimbic region) (Azim et al., 2005). Lefcourt, Davidson, Prkachin, & Mills (1997) also determined that women who use humor to cope with stressful tasks have lower systolic blood pressure than women who do not, but that the reverse is true for men. The researchers hypothesize that humor among women involves considerably more self-directed storytelling than among men, who engage in joke-telling that includes attacks on others. Therefore, humor among men is more competitive, whereas humor among women invites more personal social connections and sensitivity. This is suggested to lead to more social connections and empathy (Lefcourt et al., 1997).

But because social cohesion, concern for the self and others, and mutual understanding are all included within the cognitive-emotional response of humor, the use of humor essentially cannot be reduced to its neurological elements, nor is it accurate to state that humor is simply part of a neurological reward. Lende (2005) explains how the activation of neural reward centers, such as the mesolimbic dopaminergic system, is inherently bound up in sociocultural contexts, social interactions, and personal meaning-making. Thus, describing neural correlates and the phenomenon of physiological stress relief is important for understanding that humor is a higher-order psychological and physiological phenomenon (Azim et al., 2005), but to reduce humor to a hard-wired neurological reward would underemphasize the socially meaningful experience of humor among the breast cancer survivors. The motivation for social cohesion and a shared concern for others are critical for grasping the meaning of humor.

Studies linking humor to positive effects in the body are limited by the methodological issue of examining humor out of its natural context and the difficulty of linking one cause to one effect within such a complex sociocultural and physiological interplay. A neuroanthropological model is constructive because it does not prioritize singular physiological pathways, but instead focuses on how the embodied experience of humor creates social meaning among survivors. In other words, survivors are not considering the activation of reward centers of the brain or the upregulation of immune function; instead, they laugh with each other because, as one survivor said, "laughter just makes me feel good." The importance of humor lies not in what it does neurologically and physiologically, but rather in the social engagement and interpretive meaning-making that it is part and parcel of the humor experience.

Conclusion

Examining humor from a neuroanthropological perspective reconceptualizes what it means to practically cope with a stressful illness. Humor is a compelling coping mechanism in the context of breast cancer survivorship, and one that requires both sociocultural contexts and neurological functions. Socially, breast cancer survivors can form and reinforce social bonds. Cognitively, the survivors can distance themselves from the immense stress of breast cancer. Linguistically, they can outwardly project their own strength in the face of this hardship and influence others to do the same. Neurologically, the survivors benefit from the mediation of reward centers and the reduction of stress. Thus, humor as a coping mechanism has multiple levels of meaning, and to privilege any one of them is to miss the full importance of this holistic coping mechanism.

Humor challenges assumptions of breast cancer and coping in three key ways: 1) it is a form of optimism that forces an acknowledgment of the true stress of cancer, and an ultimate cognitive departure from the crippling effects of stress; 2) it demonstrates that coping is not always a linear, clear-cut process, but rather can occur in spontaneous, fleeting moments; and 3) that the human ability to instigate humor and respond with laughter is part of a greater suite of traits related to stress relief and group bonding that deserves greater attention. It requires a new way of considering neurological responses to stress, such as humor, that accounts for the flexibility of sociocultural contexts in shaping coping mechanisms. Humor then can act as a lens through which to see the sociocultural influence of coping mechanisms, and how such coping mechanisms are simultaneously mediated by neurological emotional responses.

This emotional response to breast cancer must not be considered a universal phenomenon, nor should it be expected that there is a single model of breast cancer survivorship. The uniformity present in this ethnography is a product of the particular environment studied, which included a local, accessible breast care clinic and support center that is run by patient-centered clinicians and staff devoted to the well-being of survivors in the greater community. There is no doubt that this clinic is part of the previously discussed breast cancer advocacy context that is so prevalent in the United States. Therefore, I strongly warn against pathologizing cancer survivors outside of these kinds of environments who are not able to seek stress relief. In other words, there is nothing *wrong* with a survivor who does not find breast cancer, or any cancer, humorous. Additionally, to marginalize the warranted negative feelings that are also an inherent part of experiencing cancer, or to imply that positive thinking is morally right, would be a detrimental fallacy (De Raeve, 1997). Without the kind of social connection the women had in the support center, this unique and meaningful means of coping would likely not have been as prominent. Ethnography is a means of examining these and other structural and sociocultural determinants of health, such as access to medical care and support, the persistence of cancer, and additional, unrelated life stressors that can determine the degree to which a woman is able to cope.

In this ethnographic context, humor is a self-identified marker of de-stressing, yet humor must be considered as only one manifestation of social support, emotional coping, and the release of stress from breast cancer. This point is particularly important considering the rising rates of breast cancer in contexts outside of the United States (particularly in the developing world, where cancer is still stigmatized; Porter, 2008), and the fact that breast cancer is the leading cause of cancer death among women globally (Bray, McCarron, & Maxwell Parkin, 2004). Thus, it will be imperative to use ethnographic methods and neuroanthropological analyses to understand the various contexts of breast cancer survivorship and the well-being of women affected by the disease.

The use of humor in this context is situated within a dynamic flux of sociocultural context and neurological function that enacts stress-relieving properties. Both aspects are equally invested and inextricably tied in the production and usefulness of humor. This work has striven to demonstrate how sociocultural constructions of breast cancer are received, emotionally experienced, and coped with at the community-based level. Placing the neurological and psychological correlates of coping strategies into the ethnographic context can illuminate the sociocultural factors behind stress reduction, in turn giving a voice to those enduring hardships like breast cancer.

Pursuing applied neuroanthropology at the community level not only allows the possibility of making concrete contributions on the ground, but also aids in theoretical development by demonstrating the plasticity of the relationship between our neural, physiological, and sociocultural pathways. In this example, the use of humor as a coping mechanism for breast cancer is a call to stimulate dialogue and challenge biological determinism, while highlighting the vast importance of sociocultural context in shaping neurological and physiological processes. Ultimately, I draw from one informant, who stated, "Cancer has no boundaries." Perhaps we will strive to break down our own research boundaries in order to truly understand how people manage the stress and hardships of the illness experience.

Acknowledgments

This chapter is possible due to the guidance, patience, and encouragement from Daniel Lende and Greg Downey, as well as their major theoretical contributions to neuroanthropology. Special thanks to Daniel Lende, Margaret McKinney-Arnold, Caroline Nally, Stephanie Pelligra, and Kathleen Carroll, without whose collaboration, insight, and guidance this ethnography would not have been possible. I also thank the members of the "Encultured Brain" conference at the University of Notre Dame, Chikako Ozawa-de Silva, Bradd Shore, Peter J. Brown, and Bonnie Fullard, who offered invaluable critiques of this chapter.

Note

1. The term "survivor" is based on the United States National Cancer Institute definition. A person is a survivor from the time of a cancer diagnosis to the end of life (see Twombly, 2004).

References

Abel, M. H. (2002). Humor, stress, and coping strategies. *Humor: International Journal of Humor Research*, 15(4), 365–381.

Apte, M. L. (1985). *Humor and laughter: An anthropological approach*. Ithaca, NY: Cornell University Press.

Azim, E., Mobbs, D., Jo, B., Menon, V., & Reiss, A. L. (2005). Sex differences in brain activation elicited by humor. *Proceedings of the National Academy of Sciences of the United States of America, 102*(45), 16496–16501.

Bateson, G. (2000). *Steps to an ecology of mind*. Chicago: University of Chicago Press. (Original work published 1972)

Bennett, M. P., Zeller, J. M., Rosenberg, L., & McCann, J. (2003). The effect of mirthful laughter on stress and natural killer cell activity. *Alternative Therapies in Health and Medicine, 9*(2), 38–45.

Braun, S. (2003). The history of breast cancer advocacy. *Breast Journal, 9*(2), 101–103.

Bray, F., McCarron, P., & Maxwell Parkin, D. (2004). The changing global patterns of female breast cancer incidence and mortality. *Breast Cancer Research, 6*, 229–239.

Christie, W. (2005). The impact of humor on patients with cancer. *Clinical Journal of Oncology Nursing, 9*(2), 211–218.

Cohen, S., & McKay, G. (1984). Social support, stress and the buffering hypothesis: A theoretical analysis. In A. Baum, S. E. Taylor, & J. E. Singer (Eds.), *Handbook of psychology and health* (Vol. 4, pp. 253–267). Hillsdale, NJ: Lawrence Erlbaum.

Cohen, S., & Wills, T. A. (1984). Stress, social support, and the buffering hypothesis. *Psychological Bulletin, 98*(2), 310–357.

Coreil, J., Wilke, J., & Pintado, I. (2004). Cultural models of illness and recovery in breast cancer support groups. *Qualitative Health Research, 14*(7), 905–923.

Culver, J. L., Arena, P. L., Antoni, M. H., & Carver, C. S. (2002). Coping and distress among women under treatment for early stage breast cancer: Comparing African Americans, Hispanics and Non-Hispanic whites. *Psycho-Oncology, 11*(6), 495–504.

Culver, J. L., Arena, P. L., Wimberly, S. R., Antonia, M. H., & Carver, C. S. (2004). Coping among African-American, Hispanic, and Non-Hispanic white women recently treated for early stage breast cancer. *Psychology & Health, 19*(2), 157–166.

Darwin, C. (1955). *The expression of the emotions in man and animals*. New York: The Philosophical Library. (Original work published 1872)

De Raeve, L. (1997). Positive thinking and moral oppression in cancer care. *European Journal of Cancer Care, 6*(4), 249–256.

Dodson, D. L. (2005). Making a difference: Behind the scenes. In S. Thomas & C. Wilcox (Eds.), *Women and elective office: Past, present, and future* (2nd ed., 129–151). Oxford: Oxford University Press.

Downey, G. (2010). "Practice without theory": A neuroanthropological perspective on embodied learning. *Journal of the Royal Anthropological Institute, 16,* S22–S40.

Dunn, G. P., Bruce, A. T., Ikeda, H., Old, L. J., & Schreiber, R. D. (2002). Cancer immunoediting: From immunosurveillance to tumor escape. *Nature Immunology, 3*(11), 991–998.

Folkman, S., Lazarus, R.S., Dunkel-Schetter, C., DeLongis, A., & Gruen, R.J. (1986). Dynamics of a Stressful Encounter: Cognitive Appraisal, Coping, and Encounter Outcomes. *Journal of Personality and Social Psychology, 50*(5), 992–1003.

Francis, L. E. (1994). Laughter, the best mediation: Humor as emotion management in interaction. *Symbolic Interaction, 17*(2), 147–163.

Fredrickson, B. L. (1998). What good are positive emotions? *Review of General Psychology, 2*(3), 300–319.

Freud, S. (1959). Humor. In J. Strachey (Ed.), *Collected Papers* (J. Riviere, Trans., Vol. 5, pp. 215–221). New York: Basic Books. (Original work published 1928)

Gervais, M., & Wilson, D. S. (2005). The evolution and functions of laughter and humor: A synthetic approach. *Quarterly Review of Biology, 80*(4), 395–430.

Gibbs, R. W., & Franks, H. (2002). Embodied metaphor in women's narratives about their experiences with cancer. *Health Communication, 14*(2), 139–165.

Goel, V., & Dolan, R. J. (2007). Social regulation of affective experience of humor. *Journal of Cognitive Neuroscience, 19*(9), 1574–1580.

Golden-Kreutz, D. M., & Andersen, B. L. (2004). Depressive symptoms after breast cancer surgery: Relationships with global, cancer-related, and life event stress. *Psycho-Oncology, 13,* 211–220.

Goldstein, D. M. (2003). *Laughter out of place: Race, class, violence, and sexuality in a Rio shantytown.* Berkeley, CA: University of California Press.

Holahan, C. J., Moos, R. H., Holahan, C. K., & Brennan, P. L. (1997). Social context, coping strategies, and depressive symptoms: An expanded model with cardiac patients. *Journal of Personality and Social Psychology, 72*(4), 918–928.

Jemal, A., Siegel, R., Xu, J., & Ward, E. (2010). Cancer statistics, 2010. *CA: A Cancer Journal for Clinicians, 60*(5), 277–300.

Kasper, A. S., & Ferguson, S. J. (2002). *Breast cancer: Society shapes an epidemic.* New York: Palgrave.

Kaufert, P. A. (1998). Women, resistance and the breast cancer movement. In M. Lock & P. Kaufert (Eds.), *Pragmatic women and body politics* (pp. 287–309). New York: Cambridge University Press.

Kershaw, T., Northouse, L., Kritpracha, C., Schafenacker, A., & Mood, D. (2004). Coping strategies and quality of life in women with advanced breast cancer and their family caregivers. *Psychology & Health, 19*(2), 139–155.

King, S. (2006). *Pink Ribbons, Inc.: Breast cancer and the politics of philanthropy*. Minneapolis: University of Minnesota Press.

Kuiper, N. A., & Martin, R. A. (1998). Is sense of humor a positive personality characteristic? In W. Ruch (Ed.), *The sense of humor: Explorations of a personality characteristic* (pp. 159–178). Berlin, Germany: Mouton de Gruyter.

Kuiper, N. A., McKenzie, S. D., & Belanger, K. A. (1995). Cognitive appraisals and individual differences in sense of humor: Motivational and affective implications. *Personality and Individual Differences, 19*, 359–372.

Lazarus, R. A., & Folkman, S. (1984). *Stress, appraisal, and coping*. New York: Springer.

Lefcourt, H. M., Davidson, K., Prkachin, K. M., & Mills, D. E. (1997). Humor as a stress moderator in the prediction of blood pressure obtained during five stressful tasks. *Journal of Personality Research, 31*(4), 523–542.

Lefcourt, H. M., Davidson, K., Shepherd, R., Phillips, M., Prkachin, K. M., & Mills, D. E. (1995). Perspective-taking humor: Accounting for stress moderation. *Journal of Social and Clinical Psychology, 14*, 373–391.

Lende, D. H. (2005). Wanting and drug use: A biocultural approach to the analysis of addiction. *Ethos, 33*(1), 100–124.

Lerner, B. H. (2006). Power, gender, and pizzazz: The early years of breast cancer activism. In M. C. Rawlinson & S. Lundeen (Eds.), *The voice of breast cancer in medicine and bioethics* (pp. 21–30). Dordrecht, The Netherlands: Springer.

Levine, J. (1969). *Motivation in humor*. New York: Atherton Press.

Matthews, H. F. (2000). Negotiating cultural consensus in a breast cancer self-help group. *Medical Anthropology Quarterly, 14*(3), 394–413.

McMullin, J. M., & Weiner, D. (2009). Introduction: An anthropology of cancer. In J. M. McMullin & D. Weiner (Eds.), Confronting cancer: metaphors, advocacy, and anthropology (pp. 3–25). Santa Fe: School for Advanced Research Press.

Mobbs, D., Greicius, M. D., Abdel-Azim, E., Menon, V., & Reiss, A. L. (2003). Humor modulates the mesolimbic reward centers. *Neuron, 40*(5), 1041–1048.

Moran, J. M., Wig, G. S., Adams, R. B., Janata, P., & Kelley, W. M. (2004). Neural correlates of humor detection and appreciation. *NeuroImage, 21*(3), 1055–1060.

Nesse, R. M. (1990). Evolutionary explanations of emotions. *Human Nature, 1*, 261–289.

Palmer, J. (1994). *Taking humour seriously*. London: Routledge.

Porter, P. (2008). "Westernizing" women's risks? Breast cancer in lower-income countries. *New England Journal of Medicine, 358*(3), 213–216.

Radcliffe-Brown, A. R. (1940). On joking relationships. *Africa: Journal of the International African Institute, 13*(3), 195–210.

Sapolsky, R. M. (1994). Individual differences and the stress response. *Neurosciences, 6*, 261–269.

Sherwin, S. (2006). Personalizing the political: Negotiation the feminist, medical, scientific, and commercial discourses surrounding breast cancer. In M. C. Rawlinson & S. Lundeen (Eds.), *The voice of breast cancer in medicine and bioethics* (pp. 3–19). Dordrecht, The Netherlands: Springer.

Skott, C. (2002). Expressive metaphors in cancer narratives. *Cancer Nursing, 25*(3), 230–235.

Söllner, W., Maislinger, S., DeVries, A., Steixner, E., Rumpold, G., & Lukas, P. (2000). Use of complementary and alternative medicine by cancer patients is not associated with perceived distress or poor compliance with standard treatment but with active coping behavior. *Cancer, 89*(4), 873–880.

Sontag, S. (1979). *Illness as Metaphor*. New York: Vintage Books.

Spross, J. A., & Burke, M. W. (1995). Nonpharmalogical management of cancer pain. In D. B. McGuire, C. H. Yarbo, & B. R. Ferrell (Eds.), *Cancer pain management* (pp. 159–206). London: Jones and Bartlett.

Stahly, G. B. (1989). Psychosocial aspects of the stigma of cancer. *Journal of Psychosocial Oncology, 6*(3&4), 3–27.

Sulik, G. (2011). *Pink ribbon blues:* how breast cancer culture undermines women's health. New York: Oxford University Press.

Turner-Cobb, J. M., Sephton, S. E., Koopman, C., Blake-Mortimer, J., & Spiegel, D. (2000). Social support and salivary cortisol in women with metastatic breast cancer. *Psychosomatic Medicine, 62*, 337–345.

Twombly, R. (2004). What's in a name: Who is a cancer survivor? *Journal of the National Cancer Institute, 96*(19), 1414–1415.

Wild, B., Rodden, F. A., Grodd, W., Ruch, W. (2003). Neural correlates of laughter and humour. *Brain, 126*(10), 2121–2138.

9 Embodiment and Male Vitality in Subsistence Societies

Benjamin Campbell

Anthropologists should see the relevance of neurological fact to all cultural history.
—Weston La Barre (1984)

For anthropologists, embodiment was originally defined by Csordas (1990) as "the non-physiological experience of the body." It is unclear, however, exactly what Csordas meant by physiology in this statement. Intuitively, it does seem that the relatively slow and widespread physiological action of hormones on target organs is much too slow to account for the rapid, dynamic, and subjective emotions that we experience in our heads. Yet even at the time, the mind-body dualism inherent in Csordas's notion of embodiment had been challenged. The James-Lange theory of emotion (James, 1884) had long explicitly posited that emotion arose out of the perception of physiological changes in the body. In other words, the theory suggested that subjective emotion reflected the physiological experience of the body.

Since 1990, a whole variety of findings have made it clear that mind and body are related by myriad neurological, immunological, and hormonal mechanisms. The sheer variety of such connections undercuts the viability of any sort of mind-body dualism. Importantly for our purposes, Damasio (1999) updated the James-Lange theory, suggesting that a variety of somatic markers such as muscle tone, heart rate, endocrine release, posture, and facial expression represent the experience of emotion, which provides the context for cognitive decision-making.

More recently, Bud Craig (2002, 2009, 2010) has provided a concrete mechanism for both the James-Lange theory of emotions and Damasio's somatic markers in postulating what he calls the "sentient self," a global representation of bodily status within the right anterior insula. Such an "embodiment" is based on the integration of information about physiological conditions such as thirst, dyspnea, air hunger, sensual touch, sexual arousal, sensations of warmth or cold, bladder and gastric distention, and muscle metabolism throughout the body via alpha and delta c fibers that travel through spinal cord to the thalamus, and finally to the right anterior insula. In addition, the right anterior insula receives afferents from the amygdala, the center of emotional attention. Hence, the neural representation in the insula reflects bodily status in the context of emotion (i.e., as a feeling).

The existence of an afferent network relaying information about peripheral organs and tissue to the brain means that the physiological state of the body can be transmitted to the brain fast enough to play a role, along with more direct responses to environmental stimuli, in shifting emotional states. Furthermore, because bodily status as represented in the insula includes both physiological assessment and motivational information, Craig (2009) calls it "homeostatic." The global somatic representation in the insula is primarily concerned with bodily survival and whether action needs to be taken to maintain essential bodily functions as a whole. As long as bodily functions are within tolerable limits, and there are adequate physiological resources to meet threats in the environment, no action is called for. Such a state of no inherent need for change may be thought of as a feeling of well-being, and the energy it represents as "vitality."

At the same time, cognitive psychologists have argued for what they call embodied cognition (see Clark, 1997; Lakoff & Johnson, 1999; Varela Thompson, & Rosch, 1992, among others); that is, cognition based on the experience of the body. In practice, this means that cognition arises from sensory inputs and somatic sensations that the prefrontal cortex integrates and abstracts to generate structured thought. Such cognition need not always draw directly on current somatic sensation, but can be based on multimodel representations of past experience. This is what Barsalou (2009, p. 1281) calls a simulation, or the reenactment of perceptual, motor, and introspective states acquired during experience with the world, body, and mind.

Thus, rather than being the non-physiological experience of the body as Csordas suggested, "embodiment" appears to represent the neurophysiological experience of fundamental bodily processes centered around "well-being." The fact that the insula is active in intuition (Allman, Watson, Tetreault, & Hakeem, 2005) suggests that well-being may be thought of as a bodily sense (introception) or feeling. On the other hand, embodied cognitions, or "body thoughts" as Andrew Strathern (1996) refers to them, are the result of a cognitive process that builds on the experience of embodiment. However abstracted by introspection, culturally elaborated, or both, such thoughts ultimately reflect the underlying physiological state of the body and its well-being.

For a biological anthropologist, the notion of embodiment is intriguing because it suggests that variation in the physiological condition of the body, often considered in terms of varying ecological conditions and specific hormonal responses which can be objectively measured, can also be considered part of the individual subjective experience of well-being and vitality. As such, embodiment gets right to the heart of the biocultural nature of the human condition (Campbell & Garcia, 2009).

The notion of the sentient self provides a "bottom-up" pathway—that is, one that goes from the body to the brain—for understanding cultural variation in subjective well-being and vitality. Physiological functions such as the amount of lactic acid in muscles (a marker of physical activity), the degree of stomach distention (a marker of food digestion), and the function of the bowels (a marker of parasite load) provide meaningful information about well-being in and of themselves. To the extent that such markers reflect predictable environmental factors such as habitual exercise, nutrition, and disease burdens, to name some of the most obvious, across different societies, those societies may share general similarities in their experience of the body (i.e., embodiment).

In what follows, to simplify my exploration of the role of physiological status in subjective well-being and vitality, I will focus exclusively on men. While at some level the neurophysiological mechanisms of embodiment are similar across humans (that is what it means to be a single species), they are surely conditioned differently in men and women. Sex steroids, including both estrogen and testosterone, are known to have differing effects on mood and behavior (see van Wingen, Ossewaarde, Backstrom, Hermans et al., 2011, for a recent neuroendocrinological perspective) in addition to their role in the physical development of sexually dimorphic male and female bodies.

Thus, I start by considering evidence from clinical and population studies that testosterone, the male steroid hormone, is related to well-being among men. I then suggest that differences in the ecological conditions of industrialized and subsistence societies may lead to differing patterns of testosterone production over the life course and differences in the experience of embodiment. I provide evidence from a study of Ariaal pastoralists from northern Kenya that testosterone is related to well-being, thus extending the impact of testosterone on well-being to subsistence populations as well. Finally, I discuss how the subjective experience of androgen-dependent mood and bodily structures may underlie cultural beliefs about male vitality—that is, the role of embodiment in bodily thoughts—specifically focusing on the culturally widespread belief in a vital substance linking semen, spine, and the brain, referred to as *muelos* by Weston La Barre (1984).

Testosterone and Vitality

Over the past fifteen years, standardized trials have established the impact of testosterone on subjective well-being and libido among men. Administration of testosterone to hypogonadal men not only decreases depression scores, but leads to increased good mood (Jockenhovel, Minnemann, Schubert, Freude, et al., 2009; Wang, Alexander, Berman, Salehian et al., 1996; Wang, Berman, Longstreth, Chuapoco et al., 2000) and feelings of energy (Wang et al., 1996). In addition, reports of sexual motivation and erections also increase with testosterone administration (Wang et al., 2000; Jockenhovel et al., 2009).

Among aging men, low testosterone levels are associated with depression (Hintikka, Niskanen, Koivumaa-Honkanen, Tolmunen, et al., 2009; Eskelinen, Vahlberg, Isoaho, Kivelä, et al., 2007, Shores, Sloan, Matsumoto, Moceri, et al. 2004) as well as low energy, low mood, and reduced libido (Seidman, Araujo, Roose, Devanand, et al., 2002), effects that can be reversed by testosterone administration (Seidman, Orr, Raviv, Levi, et al., 2009). Epidemiological studies of middle-aged community-living men report that testosterone is positively related to libido (Travison, Morley, Araujo, O'Donnell, et al., 2006) and negatively related to anxiety (Berglund, Prytz, Perski, & Svartberg, 2011), demonstrating that the link between testosterone and male subjective well-being extends outside the clinic to the population at large.

The neurological basis for the documented association of testosterone and well-being is not clear. Based on hypogonadal men, Azad, Pitale, Barnes, & Friedman (2003) report that testosterone administration increases brain perfusion in the midbrain and superior frontal gyrus as well as the mid-cingulate gyrus. Activity in the superior frontal gyrus (Brodman's area 8) has been associated with uncertainty (Volz, Schubotz, & von Cramon, 2005), suggesting that testosterone may promote an optimistic assessment or positive outlook. However, the small number of subjects in these studies means that these results and any interpretation must be considered preliminary.

The study above does not speak to possible effects of testosterone on activity within the insula as the neurological source of well-being. However, testosterone could also affect well-being by its actions on peripheral tissues, which then are integrated at the level of the insula. For instance, in addition to promoting the development and maintenance of muscle, testosterone promotes red blood cell production (Shahidi, 1973; Molinari, 1982), increasing oxygen delivery to all tissues (not just muscle) and promoting their functioning. The increased bodily functionality would be incorporated in the global sense of well-being represented in the insula.

In industrialized societies characterized by abundant energy, little infectious disease exposure, and a sedentary lifestyle, the bodies of men are in positive energy balance, experience little energy drain from infections, and are in minimal demand for physical exertion. One consequence of these favorable conditions for somatic growth is that men exhibit relatively high testosterone levels by the end of adolescent development (Ellison, Bribiescas, Bentley, Campbell, et al., 2002). As a consequence, men experience robust changes in baseline testosterone levels from their peak in their 20s throughout the aging process (Muller, den Tonkelaar, Thijssen, Grobbee, et al., 2003; Feldman, Longcope, Derby, Johannes, et al., 2002). In addition, testosterone is inversely related to adiposity, a marker of energy storage, but only at higher levels of adiposity, thus largely decoupling testosterone from energetic status on a daily basis (see McDonald et al., 2009 for a recent meta-analysis).

In contrast, the ecological context of most subsistence societies can be characterized by habitual physical exercise, limited caloric intake, and high disease burdens. In other words, energy output is greater relative to energy intake while surplus energy is circumscribed. Compared to well-nourished industrialized societies, testosterone levels in subsistence societies are lower (Ellison et al., 2002) but decline little, if at all, with age (Ellison et al., 2002; Campbell, Gray, & Ellison, 2006a). Furthermore, contrary to Western populations, testosterone levels are positively related to nutritional status (Campbell, O'Rourke, Lipson, et al., 2003, ; Campbell, Leslie, & Campbell, 2006b), and men may experience short-term fluctuations in testosterone related to changes in food availability and work load (Campbell et al., 2006a; Vitzthum, Worthman, Beall, Thornburg, et al., 2009; Worthman & Konner, 1987).

Based on the association among testosterone, subjective well-being and libido documented in western populations, differing testosterone profiles between industrialized and subsistence populations suggest differences in temporal patterns of subjective well-being in men. Age-related hormonal declines among men in industrialized populations have been associated with declines in well-being (Hintikka et al., 2009; Eskelinen et al., 2007).

Coupled with increasing sedentary lifestyles and adiposity, the steep decline from peak physical capacity and testosterone levels contribute to the image of the grumpy old man, irritable because he can no longer do the things he used to do.

On the other hand, men in subsistence cultures do not experience the same clear decline in testosterone and physical activity with increasing adiposity (Ellison et al., 2002; Campbell et al., 2006a), making age a less salient dimension for changes in well-being. Instead, such men may experience changes in subjective well-being based on fluctuations in circulating testosterone on the order of days or weeks (Vitzthum et al., 2009). Below, I provide empirical support that testosterone is related to subjective well-being among the Ariaal, pastoral nomads from northern Kenya. The findings, though admittedly preliminary, are consistent with the association between testosterone and well-being previously documented in industrialized populations.

Case Study: Testosterone and Subjective Well-Being among Ariaal Men

The Ariaal are pastoral nomads of Marsabit District, Kenya. They form a cultural bridge between the Rendille camel herders of the lowland desert and Samburu cattle herders of the upland regions, following Samburu cultural traditions while herding camels in the lowland desert and cattle in upland regions (for a more detailed description of the Ariaal, see Fratkin, 2003). While the subsistence practices of the upland and lowland groups differ substantially, the two populations represented here are highly related. The settlement was started in the 1960s by individuals from the lowland group, who followed missionaries promising a source of water on the mountain. More generally, the Ariaal experience conditions are common to subsistence societies with low energy intake, habitual physical activity, and high disease burdens (Campbell et al., 2006a).

Based on the discussion above, I predict that higher levels of testosterone will be positively associated with greater energy levels, greater libido, and more subjective enjoyment of life among Ariaal men. I also expect that these three variables measuring quality of life will not decline with age. I include allelic variation in the DRD2 dopamine receptor in the statistical analyses because previous findings have associated genetic differences in DRD2 binding with greater reward-seeking (Comings & Blum, 2000).

As part of a larger study of aging I conducted with Peter Gray, we asked men to answer questions from the WHO quality of life questionnaire

(WHOQOL). From the WHOQOL we selected three specific areas that might be related to the subjective experience of testosterone: satisfaction with energy level, positive emotions, and satisfaction with sex. Each topic consisted of four separate questions, with answers on a 5-point scale ranging from never to always. All of the questions were asked, and answers recorded, through an interpreter. We used an event calendar to determine individual ages. Because of the uncertainty in this method, age is treated in 10-year age groups (20s, 30s, 40s, 50s, and 60+) rather than as a continuous variable in the statistical models presented below.

Participants in this study included 102 men from the nomadic encampment of Lowogosa and 103 from the settlement of Songa near the town of Marsabit. We collected saliva samples to assay for testosterone. Samples were collected within minutes of 9:00 a.m. to control for circadian variation in salivary testosterone. We obtained values for 201 of the men. We also collected hair samples to determine allelic variation in the Taq1 A1 allele of the DRD2 dopamine receptor gene. The Taq1 A1+ genotype has been associated with substance abuse, which Comings and Blum (2000) refer to as "reward-deficiency syndrome," or reduced well-being under conditions of normalcy. Dan Eisenberg was able to obtain genotypes for 87 settled and 65 nomadic men. A more detailed description of the methods is in Campbell et al. (2006a) and Eisenberg, Campbell, Gary, & Sorenson (2008).

To determine if testosterone is related to the three outcomes chosen above, I ran generalized linear models with residence and age group as fixed variables. Testosterone was included as a continuous covariate. DRD2 Taq1 A1+ was added to the model to determine if it changed the testosterone results. An interaction between testosterone and DRD2 Taq1+ was tested to determine if the impact of testosterone might be affected by dopaminergic sensitivity. Because of the difference in the number of men with testosterone and those with DRD2 Taq1 A1+ values, these models are shown separately in tables 9.1A and 9.1B.

As shown in table 9.1A, both residence and age group are strong predictors of satisfaction with energy (F = 30.3 and 7.5 respectively; p < 0.001), less so for satisfaction with sex (F = 5.3 and 4.2 respectively; p < 0.01), but not for positive emotions (F = 1.4 and 0.6 respectively). As can be seen in figure 9.1, for all three variables, the effects of residence reflect greater satisfaction among the nomads, reflecting the global value placed on being nomadic in a nomadic culture. For satisfaction with energy, the effect of age group reflects a linear decline with age starting with the youngest age group. In contrast, for satisfaction with sex, age-group differences are

Table 9.1A

Testosterone and subjective measures of male vitality

Variable	Satisfaction w/energy	Positive emotions	Satisfaction w/sex
N	201	201	200
Model R^2	0.24	0.005	0.09
Predictor			
Residence, Nomad = 1	30.7***	0.6	5.31*
Age Group, Old > Young	7.5***	1.4	4.2**
Testosterone	3.7+	0.6	0.7

+p < 0.10; *p < 0.05; **p < 0.01; ***p < 0.001

Table 9.1B

Testosterone, DRD2 Taq1, and subjective measures of male vitality

Variable	Satisfaction w/energy	Positive emotions	Satisfaction w/sex
N	152	152	151
Model R^2	0.22	0.04	0.05
Predictor			
Residence	20.7***	3.8*	6.1*
Age Group	5.5***	0.7	2.1+
Testosterone	3.0+	4.4*	1.2
DRD2 Taq1+	0.2	4.4*	0.5

+ p < 0.10; *p < 0.05; **p < 0.01; ***p < 0.001

associated with peak satisfaction in the 40s, before erectile function shows a decline (Gray & Campbell, 2005).

When controlled for age group and residence, neither satisfaction with sex (F = 0.7; p = 0.39) or positive emotions (F = 0.6; p = 0.43) show a significant relationship with salivary testosterone, while satisfaction with energy level (F = 3.7; p = 0.06) shows a trend. As shown in table 9.1B, when DRD2 Taq1+ was added to the model for satisfaction with energy, salivary T remains a trend (F = 3.0; p = 0.09), while DRD2 Taq1+ is not significant (F = 0.2; p = 0.61). However, salivary testosterone becomes a significant predictor of positive emotions (F = .4.4; p = 0.04), as well as DRD2 Taq1 (F = 4.4; p = 0.04). For satisfaction with sex, neither salivary testosterone (F = 1.2; p = 0.28) nor DRD2 Taq1+ (F = 0.5; p = 0.50) are significant predictors. A testosterone by DRD2 Taq1+ interaction term is not significant for any of the three outcome variables (results not shown).

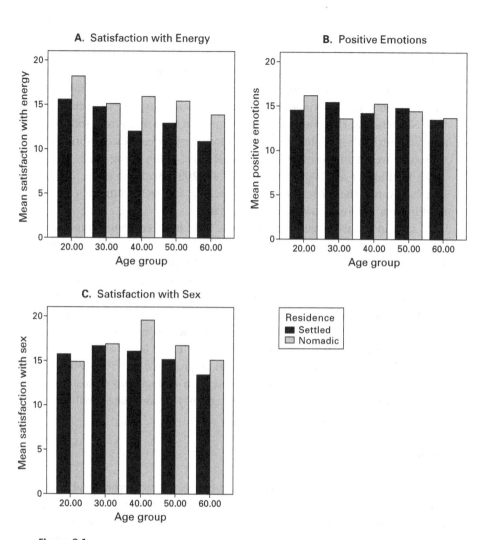

Figure 9.1

Ariaal male satisfaction by residence and age group. (A) Satisfaction with energy: note how nomadic males report more satisfaction at all age groups, as well as the linear decline across age groups from youngest to oldest. (B) Positive emotions: note the relatively small and inconsistent differences across both residence and age groups. (C) Satisfaction with sex: note the *u*-shaped distribution by age group, peaking in the 30s for settled men and 40s for nomadic males. See text for statistical analyses of the results for all three outcome variables.

The specific findings presented here must be considered suggestive. The analyses are post hoc, the sample size is small, and the statistical relationship between salivary testosterone and well-being is weak. In addition, the limited number of questions used to capture subjective mental states in the area of energy, enjoyment of life, and libido are susceptible to cultural bias. Nonetheless, the results are consistent with the results from industrialized countries that establish a clear relationship between testosterone and subjective well-being in men (Jockenhövel et al., 2009; Wang et al., 1996, 2000; Hintikka et al., 2009; Eskelinen et al., 2007; Seidman et al., 2002). In other words, the association of testosterone and well-being appears to hold in very different ecological circumstances.

In fact, in industrialized countries, the links between testosterone and well-being are among the strongest for hypogonadal men (Wang et al., 1996, 2000), including those who are hypogonadal as a result of aging (Seidman et al., 2002). Variation in testosterone is not so strongly related to differences in well-being among men with testosterone values in the normal range (Gray, Singh, Woodhouse, Storer, et al., 2005). Given the generally lower levels of testosterone among men in subsistence populations (Ellison et al., 2002), it is likely that testosterone levels among Ariaal men are at the lower end of the functional range. Thus, the cross-sectional results presented here suggest that variation in endogenous testosterone levels at the lower end of the normal range may be associated with individual variation in subjective well-being.

Cultural Beliefs about Male Vitality

The universal (or nearly so) relationship between testosterone and well-being in men, mediated by the brain, hinted at by these results supports a biocultural understanding of the subjective experience of male vitality. Instead of looking exclusively to cultural ideology for the origins of symbols and metaphors of masculinity, the neurophysiological experience that men have of their bodies needs to be considered as well. The subjective experience of ongoing and concurrent changes in energy level, libido, and sense of well-being may lead to their clear association, which in turn acts as the substrate for cultural metaphors of male vitality and well-being.

To illustrate this point, I start with an example population from Turkana who, like the Ariaal, are pastoral nomads from northern Kenya and with whom I have worked. The Turkana also live under the ecological conditions of habitual physical exercise, chronic undernutrition, and prevalent infectious disease (see Little & Leslie, 1999, for a comprehensive discussion). Furthermore, Turkana men show late reproductive maturation (Campbell, Leslie, Little, & Campbell, 2005), low body mass indexes and little decline in testosterone with age (Campbell et al., 2006b), similar to the Ariaal.

According to Broch-Due (1993), the Turkana believe that "the marrow of the spinal cord makes semen while the marrow of the bones contributes to the blood" (p. 64). Furthermore, she states that "in men the spinal column is connected to the phallus providing an outlet for the sperm." Broch-Due (1993) attributes this belief to the Turkana's observation of their cattle, in which "the specific linkage between the spinal column and the phallus is similarly based on anatomical observation pivoting on a peculiarity in the bulls' genitalia. The penis of this animal tightens through the movement of a muscle attached firmly to the last two vertebrae above the tail. This muscle appearing as an extension of the back, supports and enlarges the shape of the bull's penis and is seen as one single organ" (p. 65).

It is true that pastoral nomads base much of their theorizing about humans on the observation of their animals. For instance, my own discussions with Ariaal warriors uncovered that young warriors figured that just after the menstrual period was the best time to have sex with their girlfriends in order to avoid pregnancy, because that is when their bulls were the least likely to impregnate the cows.

Yet such specific observation of a bull's sexual anatomy (even if common to other pastoral nomads as well) cannot account for the fact that there is a widespread belief of a connection between spine, brain, and semen across subsistence societies (La Barre, 1984), many of whom do not raise cattle. Nor does it seem capable of accounting for related beliefs about connections between bone, blood, and semen found throughout classical literature (Heritier-Auge, 1989).

Furthermore, the belief linking semen, the spine, and the brain does not appear to rely on empirical observation. In his drawing *Studies of the sexual act and male sexual organ*, Leonardo da Vinci shows a connection, presumably the ureters, running from the tip of the penis all the way to the spinal cord, as well as to the bladder. Yet, anatomical observation reveals that the ureters do not connect to the spine, but to the testes and the bladder. Apparently, da Vinci perceived a connection between semen and the spinal cord that trumped his own, otherwise careful, observation of human anatomy.

Galen describes a similar connection based on the spermatic artery and vein in his book *On Semen* (DeLacy, 1992). So much for the scientific origins of Western medicine. As Levi-Strauss would put it, it is appears that the linkage between spine, semen, and brain is an idea that is "good to think" (1963, p. 89). In other words, connecting spine, semen, and brain is a metaphor that reflects men's experience of their bodies. Or, as Barsalou would have it, it is the result of a simulation that reenacts the multimodal experience that men have of particularly prominent parts of their bodies.

Muelos

In his book *Muelos: A Stone Age Superstition about Sexuality*, Weston La Barre (1984) defines the belief that the marrow of bones is associated with semen as "*muelos.*" He then goes on to describe a widespread belief among indigenous populations throughout the New and Old World that links semen to the spine and through the spine to the head or brain. Such an anatomical linkage is further associated with head-hunting. In many of these same societies, the taking of scalps in battle has the explicit goal of gaining the vitality of another man. Together, semen, the spine, and the brain constitute a uniform matter of male vitality transferrable across individuals.

One of the most dramatic examples of semen as a transferable form of vitality comes from the Sambia of New Guinea. Herdt (2005) describes the ritualized fellatio and ingestion of semen of older males by younger boys. Along with the administration of nose bleeding, this practice is thought to be essential to transfer the energy and substance that young boys need to develop into mature men. It is hard not to see this practice as a concrete manifestation of an embodied cognition about male sexual energy.

But male energy could be transferred in more cerebral ways as well. Among the Celts, as the Romans learned to their horror, the head of an opposing warrior was severed from his body and cleaned, the skull gilded and used as a drinking vessel. By drinking from such a cup, one imbibed the genius of a man and prolonged his own life (La Barre, 1984, p. 23).

La Barre goes on to trace the belief in the vitality of semen and its association with the spine and the brain from indigenous societies into classical times. For instance, the Greeks believed, like the Sambia, that the transfer of semen, in this case per anum, between older men and younger boys was important in the development of young males because it transferred something vital and noble (areté) directly from the elder (p. 78ff). Just as important for La Barre, classical literature also contains many references to bones and bone marrow as the source of male sexual energy. For instance, Pliny, like Galen, believed that marrow extended from the brain through the vertebrae and that such brain-marrow was the substance of semen (p. 83). In this sense the head is the seed of life. A similar belief that the skull is the source of seed is found in Hindu literature about the powers of Shiva (p. 94).

Confronted with what appears, by his own findings, to be a widespread, perhaps almost universal belief about the somatic source of male vitality, La Barre appears at a loss to explain it. He considers the belief an "archosis," defined in his words as "nonsense and misinformation so ancient and pervasive as to be seemingly inextricable from our thinking, a sacred ghost dance entrenched in the forgotten" (p. 10). In other words, La Barre considered the representation of male vitality as a substance connecting semen, spine, and skull to be built directly into our thinking. At the same time La Barre dismissed the notion, being at odds with our scientific understanding of the world. He might well have considered it an example of what Whitehead (1967 [1925]) would have called "the misplaced concrete," in which the abstraction (male vitality) is mistaken for its concrete parts (semen, spine, and head).

Embodied cognition, which in effect explains the basis for Whitehead's fallacy, gives us a way to understand a notion so apparently misplaced as *muelos*. As Lakoff and Johnson (1980) argue, the metaphor (vitality) contains elements of the original physical and emotional experience on which it is based (bodily sensations linked by testosterone). Or as Barsalou (2008, 2009) would have it, libido, vitality, and well-being are based on simulation; that is, reenactments of the sensory perceptions, motor output, and introspective states of the body. Under naturalistic settings, testosterone changes during both physical activity (Worthman & Konner, 1987) and sex (Escasa, Casey, & Gray, 2011) and plays a role in memory (Matousek & Sherwin, 2010; but see Young, Neiss, Samuels, Roselli, et al., 2010 for experimental results). Thus, changes in testosterone might be expected to bind feelings of energy and well-being together with sexual excitement and the physical sensations of sex, which involves sperm, the base of the spine, and the brain as a multimodal experience.

According to Barsalou (2008, 2009), such simulations or neural reenactments would be individually persistent and culturally widespread, important qualities for the development and maintenance of common beliefs about the male body and its vitality. Furthermore, I have suggested above that in societies where dietary intake is low, habitual exercise is common, and disease burdens are pervasive (i.e., every society prior to the advent of industrialization), male testosterone levels are low and fluctuating. The intermittent experience of low testosterone may be related in turn to fluctuations in the sense of vitality and libido, as well as contributing to beliefs that male energy is generally limited.

If the pervasive somatic and neurological effects of testosterone as the link between changes in well-being and libido and cultural representation of male vitality has an intuitive appeal, the association of the spine and head with sperm is more opaque. Clinical evidence suggests that the function of both the spine and the brain in males is androgen-dependent. Kennedy's disease, a condition in which a trinucleotide repeat in the androgen receptor results in the ineffective action of testosterone at the cellular level, may provide a clue. Men with the condition, which does not generally become apparent until middle age, experience degenerative changes in the spine, as well as the spinal bulbar muscles responsible for erections and, by extension, ejaculation (see Finsterer, 2010, for a recent review). Though a clinical condition, the symptoms of Kennedy's disease suggest that the function of both the spine and the bulbar muscles may be affected by low and fluctuating testosterone levels among men in energy-limited populations.

Perhaps more importantly, recent findings demonstrate a direct effect of testosterone on the amygdala (Manuck, Marsland, Flory, Gorka, et al., 2010; Derntl, Windischberger, Robinson, Kryspin-Exner, et al., 2009) and prefrontal cortices (Volman, Toni, Verhagen, & Reolofs, et al., 2011; Mehta & Beer, 2010). Higher endogenous testosterone levels are associated with greater amygdala activity (Manuck et al., 2010; Derntl et al., 2009; though see Stanton, Wirth, Waugh, & Schultheiss, 2010), as well as disruption of prefrontal amygdala connectivity (Volman et al., 2011; Mehta & Beer, 2010). The implication of these neurological patterns is that individuals with higher testosterone levels appear to show less prefrontal control of amygdala signals, consistent with less inhibited social behavior and increased expression of aggression.

At the level of embodied thoughts, such testosterone-dependent behaviors may help to account for the belief in taking the head of an enemy to obtain the source of vitality. If higher testosterone is related to the expression of successful aggression, then aggression may become associated through the effects of testosterone with somatic sensations of vitality. In other words, aggressive males may actually feel more alive as part of their aggressive behavior, a feeling they then begin to associate with the enemy heads they took as a trophy while in that high testosterone state.

My speculations about the subjective experience of the somatic and neurological impacts of testosterone as the basis for cultural metaphors of male vitality may seem fantastical to some readers. However, it is important to realize that testosterone is a prominent metaphor for male vitality in our society, not only as the result of our scientific knowledge, but its reductionist perspective as well. The subjective experience of male vitality is not testosterone itself, but involves the effects that testosterone has on the entire body and the brain. However, scientific progress means that we have grown accustomed to describing complex phenomena by their single most essential cause, and in the process often leave out the broader effects the original cause may engender.

When Dr. Brown-Sequard injected himself with extracts from animal testes back at the turn of the twentieth century, he reported increases in energy, muscle strength, and mental agility (Brown-Sequard, 1889), though it now appears that he was only (erroneously) demonstrating what was already thought to be true. Testosterone, as a thing in itself, was not isolated until 1935 (Freeman, Bloom, & McGuire, 2001). What has changed is not the male experience of vitality itself, but the focus of our explanations. La Barre may be right: the association of male reproductive substances and vitality appears inextricably stuck in our thinking.

Future Research

To say that the neuroanthropological study of embodiment, let alone its implications for male vitality, is almost entirely conceptual at this point would be an understatement (as well as a bad pun). As I understand it, neuroanthropology is focused on the impact of habitual cultural practices, such as meditation, skill acquisition, and prayer, in shaping the plasticity of the brain over the course of years. In this way, embodiment is not fundamentally different from skill acquisition or meditation. What does appear to be different is the relative importance of subcortical structures. The question to be addressed is, does testosterone have a sufficient enough impact on the brain and its development to play a role in embodiment?

Taking an fMRI scanner out into the bush among men in a subsistence society to address this question is unlikely to be practical now and in the immediate future. On the other hand, there are several separate investigations, easily pursued, that would test elements of my argument. One of the most obvious would be a more definitive whole brain scan of the impact of testosterone administration on the brain of hypogonadal men. Results would provide fundamental information about what parts of the brain are most impacted by testosterone, including whether testosterone administration has a demonstrable effect on activity within the insula. These basic findings could be followed up with an investigation of whether endogenous levels of testosterone in normal men are related to brain activity in regions of the brain other than the amygdala and orbital frontal cortex, as previously reported. Comparison of younger versus older men would demonstrate whether any such relationships change with aging, as I have suggested they might based on the decline in testosterone levels with age among Western men.

At the same time, field research might focus on cross-cultural variation in the relationship of well-being and testosterone among men in subsistence populations, using standardized methods to determine if the results from the Ariaal presented here are robust. In particular, by including a United States sample as one of the populations, the hypothesis that individual variation in testosterone plays a more direct role in male subjective well-being in subsistence populations relative to industrialized populations can be put to the test. The ability to collect both hormonal and genetic data from saliva means that variation in genes associated with testosterone action or well-being, such as the androgen receptor or the serotonin receptor, can be controlled for, providing better resolution of the effects of testosterone and the potential interaction with other factors.

The major issue for such a study would be to demonstrate measures of well-being that are culturally appropriate across cultures. Cross-cultural reliability has been demonstrated for the WHOQOL (World Health Organization, 1998). Still, we found that among the Ariaal we needed to simplify the questions further. Simple measures of well-being designed to measure the effects of testosterone among aging males, such as the ADAM scale, have been validated cross-culturally (Chu et al., 2008). The use of this scale, which includes questions such as "Do you feel grumpy?" and "Has your work performance deteriorated?" should be easily relevant across cultures. Functional interpretation of the results would be enriched by the understanding of the neurological effects of testosterone gained from the first set of investigations.

The second half of the question I have raised through the course of this chapter might be if the physical conditions of living play a role in the nature of cultural beliefs about male vitality. This question is not directly amenable to neurological and hormonal investigation. Instead, the cultural context of *muelos* might be approached using cross-cultural analysis based on the Standard Sample cross-cultural database. Data on calories, disease burdens, and amount of subsistence effort may not be available for a large sample, but variables such as subsistence type, degree of warfare, and types of parasites might serve as proxies. Data on the belief in *muelos* could be gathered from Le Barre's reports. If La Barre is correct and the belief is nearly universal, it should not be associated with any particular cultural conditions. Another approach would be to follow the more interpretive linguistic or hermeneutic approach taken by Weston La Barre, to ask if similarities in cultural beliefs about male vitality can be connected through the use of specific terms across related cultures.

The results of all three lines of investigation need not be directly related to each other. In other words, there need not be one specific analytic model to be tested—for example, one that includes both testosterone levels and words for the substance of male vitality, along with measures of brain activity. Instead, I see future research bringing together these different ways of knowing into a larger synthetic understanding of the male subjective experience of vitality and the factors the influence it. This is an example of what E. O. Wilson (1998) has referred to as consilience, or a jumping together of knowledge, in a book by the same name.

It also reflects what Bruno Latour in *We Have Never Been Modern* (1993) refers to as "hybrids," networks that link different societies, sciences, and individual experiences. Yet it is not possible to treat all three of the levels equally at one time. To paraphrase Latour, can anyone imagine a study that would treat male vitality as simultaneously naturalized, sociologized, and deconstructed? As a result I have not attempted to sociologize male vitality, leaving that topic for other scholars.

Conclusions

Testosterone has been related to male vitality and well-being. As such, it would seem to be a simple metaphor. After all, testosterone and its effects have come into focus only very recently. But, as Lakoff and Johnson (1980) suggest, metaphors are grounded in concrete sensory and somatic experience. And for men, the experience of vitality and libido is grounded in the body and neurology, touched at all levels by the effects of testosterone.

Should it surprise us if men in subsistence societies have elaborate beliefs about male energy and sexuality, given the centrality of physical activity and reproduction to their lives? Or that those beliefs may be similar across a wide variety of cultures, given how little the fundamental ecological conditions of human existence changed until the industrial revolution? What is new here is the argument that such beliefs are ultimately based on the introceptive experience of the body rather than a shared cultural ideology.

In this vein, it is worth considering that all humans, regardless of their culture, share Stone Age bodies and brains that stretch back some 250,000 years. What varies culturally are the socioecological conditions that set the priorities for the use of those bodies and influence their development from birth. The belief in *muelos* may be foreign to us not because we do not have equally elaborate beliefs about male bodies, but because differences in the ecological conditions of subsistence societies directly impact the experience of energy, well-being, and libido. No amount of cultural ideology can change that difference in experience.

Acknowledgments

I would like to thank Greg Downey and Daniel Lende for the opportunity to contribute to this volume, as well as their comments on an earlier version of this chapter. Data collection among the Ariaal was supported by the Wenner-Gren Foundation and carried out with the help of Peter Gray, Jason Radak, and Daniel Lemoille. Parts of this paper were presented during a talk at the Human Development Workshop at the University of Chicago.

References

Allman, J. M., Watson, K. K., Tetreault, N. A., & Hakeem, A. Y. (2005). Intuition and autism: A possible role for Von Economo neurons. *Trends in Cognitive Sciences, 9,* 367–373.

Azad, N., Pitale, S., Barnes, E. W., & Friedman, N. (2003). Testosterone treatment enhances regional brain perfusion in hypogonadal men. *Journal of Clinical Endocrinology and Metabolism, 88,* 3064–3068.

Barsalou, L. W. 2008. Grounded cognition. *Annual Review of Psychology, 59,* 617–645.

Barsalou, L. W. (2009). Simulation, situated conceptualization and prediction. *Philosophical Transactions of the Royal Society B, 364,* 1281–1289.

Berglund, L.H., Prytz, H.S., Perski, A., & Svartberg, J. (2011). Testosterone levels and psychological health status in men from a general population: The Tromsø study. *Aging Male, 14,* 37–41.

Broch-Due, V. (1993). Making meaning out of matter: Perception of sex, gender, and bodies among the Turkana. In V. Broch-Due, I. Rudie, & T. Bleie (Eds.), *Carved flesh/cast selves: Gendered symbols and social practices* (pp. 53–82). Oxford: Oxford University Press.

Brown-Sequard, C. E. (1889). The effects produced on man by subcutaneous injections of a liquid obtained from the testicles of animals. *Lancet, 2,* 105–107.

Campbell, B. C., & Garcia, J. R. (2009). Neuroanthropology: Evolution and emotional embodiment. *Frontiers in Evolutionary Neuroscience, 1,* e4.

Campbell, B. C., Gray, P. B., & Ellison, P. T. (2006a). Age-related changes in body composition and salivary testosterone among Ariaal Males. *Aging Clinical and Experimental Research, 18,* 470–476.

Campbell, B. C., Leslie, P. W., & Campbell, K. L. (2006b). Age-related changes in testosterone and SHBG among Turkana Males. *American Journal of Human Biology, 18,* 71–82.

Campbell, B. C., Leslie, P. W., Little, M. A., & Campbell, K. L. (2005). Pubertal timing, hormones and body composition among adolescent Turkana males. *American Journal of Physical Anthropology, 128,* 896–905.

Campbell, B. C., O'Rourke, M. T., & Lipson, S. F. (2003). Salivary testosterone and body composition among Ariaal males. *American Journal of Human Biology, 15,* 697–708.

Chu, L. W., Tam, S., Kung, A. W., Lam, T. P., Lee, A., Wong, R. L., et al. (2008). A short version of the ADAM Questionnaire for androgen deficiency in Chinese men. *Journal of Gerontology A: Biological Science and Medical Science, 63,* 426–431.

Clark, A. (1997). *Being there: Putting brain, body and world together again.* Cambridge, MA: MIT Press.

Comings, D. E., & Blum, K. (2000). Reward deficiency syndrome: Genetic aspects of behavioral disorders. *Progress in Brain Research, 126,* 325–341.

Craig, A. D. (2002). How do you feel? Introception: The sense of the physiological condition of the body. *Nature Reviews Neuroscience, 3,* 655–666.

Craig, A. D. (2009). How do you feel—now? The anterior insula and human awareness. *Nature Reviews Neuroscience, 10,* 59–70.

Craig, A. D. (2010). The sentient self. *Brain Structure & Function, 214,* 563–577.

Csordas, T. J. (1990). Embodiment as a paradigm for anthropology. *Ethos, 18,* 5–47.

Damasio, A. (1999). *The feeling of what happens.* New York: Harcourt Brace.

DeLacy, P. (1992). *Galen: On semen.* New York: John Wiley.

Derntl, B., Windischberger, C., Robinson, S., Kryspin-Exner, I., Gur, R. C., Moser, E., et al. (2009). Amygdala activity to fear and anger in healthy young males is associated with testosterone. *Psychoneuroendocrinology, 34,* 687–693.

Eisenberg, D. T., Campbell, B. C., Gary, P. B., & Sorenson, M. D. (2008). Dopamine receptor genetic polymorphisms and body composition in undernourished pastoralists: An exploration of nutritional indices among nomadic and recently settled Ariaal men in northern Kenya. *BMC Evolutionary Biology, 8*, 173.

Ellison, P. T., Bribiescas, R. G., Bentley, G. R., Campbell, B. C., Lipson, S. F., Panter-Brick, C., et al. (2002). Population variation in age-related decline in male salivary testosterone. *Human Reproduction, 17*, 3251–323.

Escasa, M. J., Casey, J. F., & Gray, P. B. (2011). Salivary testosterone levels in men at a U.S. sex club. *Archives of Sexual Behavior, 40*, 921–926.

Eskelinen, S., Vahlberg, T. J., Isoaho, R. E., Kivelä, S. L., & Irjala, K. M. (2007). Associations of sex hormone concentrations with health and life satisfaction in elderly men. *Endocrine Practice, 13*, 743–749.

Feldman, H. A., Longcope, C., Derby, C. A., Johannes, C. B., Araujo, A. B., & Coviello, A. D. (2002). Age trends in the level of serum testosterone and other hormones in middle-aged men: Longitudinal results from the Massachusetts male aging study. *Journal of Clinical Endocrinology and Metabolism, 87*, 589–598.

Finsterer, J. (2010). Perspectives of Kennedy's disease. *Journal of the Neurological Sciences, 298*, 1–10.

Fratkin, E. (2003). *Ariaal pastoralist of Kenya*. Reading, MA: Addison-Wesley.

Freeman, E. R., Bloom, D. A., & McGuire, E. J. (2001). A brief history of testosterone. *Journal of Urology, 165*(2), 371–373.

Gray, P. B., & Campbell, B. C. (2005). Erectile function and its correlates among the Ariaal of northern Kenya. *International Journal of Impotence Research, 17*, 445–449.

Gray, P. B., Singh, A. B., Woodhouse, L. J., Storer, T. W., Casaburi, R., Dzekov, J., et al. (2005). Dose-dependent effects of testosterone on sexual function, mood, and visuospatial cognition in older men. *Journal of Clinical Endocrinology and Metabolism, 90*, 3838–3846.

Herdt, G. (2005). *The Sambia: Ritual, sexuality and change in Papua New Guinea* (2nd ed.). Belmont, CA: Wadsworth Publishing

Heritier-Auge, F. (1989). Semen and blood: Some ancient theories concerning their genesis and relationship. In M. Feher, R. Naddaff, & N. Tasi (Eds.), *Fragments for a history of the body* (Vol. 3, pp. 158–175). New York: Zone Books.

Hintikka, J., Niskanen, L., Koivumaa-Honkanen, H., Tolmunen, T., Honkalampi K., & Lehto, S. M. (2009). Hypogonadism, decreased sexual desire, and long-term depression in middle-aged men. *Journal of Sexual Medicine, 6*, 2049–2057.

James, W. (1884). What is an emotion? *Mind, 9*, 188–205. Available online at http://psychclasses/york.ca/james/Emotion.

Jockenhövel, F., Minnemann, T., Schubert, M., Freude, S., Hübler, D., Schumann, C., et al. (2009). Timetable of effects of testosterone administration to hypogonadal men on variables of sex and mood. *Aging Male, 12,* 113–118.

Little, M. A., & Leslie, P. W. (Eds.). (1999). *Turkana herders of the dry savanna: Ecology and biobehavioral response of nomads to an uncertain environment.* New York: Oxford University Press.

La Barre, W. (1984). *Muelos: A Stone Age superstition about sexuality.* New York: Columbia University Press.

Lakoff, G., & Johnson, M. (1980). *Metaphors we live by.* Chicago: University of Chicago Press.

Lakoff, G., & Johnson, M. (1999). *Philosophy in the flesh: The embodied mind and its challenge to Western thought.* New York: Basic Books.

Latour, B. (1993). *We have never been modern* (C. Porter, Trans.). Cambridge, MA: Harvard University Press.

Levi-Strauss, C. (1963). *Totemism* (R. Needham, Trans.). Boston: Beacon Press.

Manuck, S. B., Marsland, A. L., Flory, J. D., Gorka, A., Ferrell, R. E., & Hariri, A. R. (2010). Salivary testosterone and a trinucleotide (CAG) length polymorphism in the androgen receptor gene predict amygdala reactivity in men. *Psychoneuroendocrinology, 35,* 94–104.

Matousek, R. H., & Sherwin, B. B. (2010). Sex steroid hormones and cognitive functioning in healthy, older men. *Hormones and Behavior, 57*(3), 352–359.

McDonald, A. A., Herbison, G. P., Showell, M., & Farquhar, C. M. (2009). The impact of body mass index on semen parameters and reproductive hormones in human males: A systematic review and meta-analysis. *Human Reproduction Update, 16,* 293–311.

Mehta, P. H., & Beer, J. (2010). Neural mechanisms of the testosterone-aggression relation: the role of orbitofrontal cortex. *Journal of Cognitive Neuroscience, 22,* 2357–2368.

Molinari, P. F. (1982). Erythropoietic mechanisms of androgens: A critical review and clinical implications. *Haematologica, 67,* 442–460.

Muller, M., den Tonkelaar, I., Thijssen, J. H., Grobbee, D. E., & van der Schouw, Y. T. (2003). Endogenous sex hormones in men aged 40–80 years. *European Journal of Endocrinology, 149,* 583–589.

Seidman, S. N., Araujo, A. B., Roose, S. P., Devanand, D. P., Xie, S., Cooper, T. B., et al. (2002). Low testosterone levels in elderly men with dysthymic disorder. *American Journal of Psychiatry, 159,* 456–459.

Seidman, S. N., Orr, G., Raviv, G., Levi, R., Roose, S. P., Kravitz, E., et al. (2009). Effects of testosterone replacement in middle-aged men with dysthymia: a randomized, placebo-controlled clinical trial. *Journal of Clinical Psychopharmacology, 29*, 216–221.

Shahidi, N. T. (1973). Androgens and erythropoiesis. *New England Journal of Medicine, 289*, 72–80.

Shores, M. M., Sloan, K. L., Matsumoto, A. M., Moceri, V. M., Felker, B., & Kivlahan, D. R. (2004). Increased incidence of diagnosed depressive illness in hypogonadal older men. *Archives of General Psychiatry, 61*, 162–167.

Stanton, S. J., Wirth, M. M., Waugh, C. E., & Schultheiss, O. C. (2009). Endogenous testosterone levels are associatedwith amygdala and ventromedial prefrontal cortex responses to anger faces in men but not women. *Biological Psychology, 81*, 118–122.

Strathern, A. (1996). *Body thoughts*. Ann Arbor, MI: University of Michigan Press.

Travison, T. G., Morley, J. E., Araujo, A. B., O'Donnell, A. B., & McKinlay, J. B. (2006). The relationship between libido and testosterone levels in aging men. *Journal of Clinical Endocrinology and Metabolism, 91*, 2509–2513.

Van Wingen, G.A., Ossewaarde, L., Backstrom, T., Hermans, E.J., Fernandez, G. (2011). Gonadal hormone regulation of the emotion circuitry in humans. *Neuroscience, 191*, 38–45.

Varela, F. J., Thompson, E. T., & Rosch, E. (1992). *The embodied mind: Cognitive science and human experience*. Cambridge, MA: MIT Press.

Vitzthum, V. J., Worthman, C. M., Beall, C. M., Thornburg, J., Vargas, E., Villena, M., et al. (2009). Seasonal and circadian variation in salivary testosterone in rural Bolivian men. *American Journal of Human Biology, 21*, 762–768.

Volman, I., Toni, I., Verhagen, L., & Reolofs, K. (2011). Endogenous testosterone modulates prefrontal-amygdala connectivity during social emotional behavior. *Cerebral Cortex, 21*, 2282–2290.

Volz, K. G., Schubotz, R. I., & von Cramon, D. Y. (2005). Variants of uncertainty in decision-making and their neural correlates. *Brain Research Bulletin, 67*, 403–412.

Wang, C., Alexander, G., Berman, N., Salehian, B., Davidson, T., McDonald, V., et al. (1996). Testosterone replacement therapy improves mood in hypogonadal men—a clinical research center study. *Journal of Clinical Endocrinology and Metabolism, 81*, 3578–3583.

Wang, C., Berman, N., Longstreth, J. A., Chuapoco, B., Hull, L., Steiner, B., et al. (2000). Pharmacokinetics of transdermal testosterone gel in hypogonadal men:

Application of gel at one site versus four sites: A General Clinical Research Center study. *Journal of Clinical Endocrinology and Metabolism, 85,* 964–969.

Wilson, E. O. (1998). *Consilience: The unity of knowledge.* New York: Knopf.

World Health Organization. (1998). The World Health Organization quality of life assessment (WHOGOL): Development and general psychometric properties. *Social Science & Medicine, 46,* 1569–1585.

Whitehead, A. N. (1967) [1925]. *Science and the modern world.* New York: Free Press.

Worthman, C. M., & Konner, M. J. (1987). Testosterone levels change with subsistence hunting effort in !Kung San men. *Psychoneuroendocrinology, 12,* 449–458.

Young, L. A., Neiss, M. B., Samuels, M. H., Roselli, C. E., & Janowsky, J. S. (2010). Cognition is not modified by large but temporary changes in sex hormones in men. *Journal of Clinical Endocrinology and Metabolism, 95,* 280–288.

III Case Studies on Human Problems, Pathologies, and Variation

10 War and Dislocation: A Neuroanthropological Model of Trauma among American Veterans with Combat PTSD

Erin P. Finley

I once saw a clinical psychologist and expert on post-traumatic stress disorder (PTSD) give a talk on combat PTSD to a group of several hundred medical students and faculty at the University of Texas Health Science Center in San Antonio. He began by engaging the audience with a series of photos projected on the auditorium's large central screen, most of which he had taken during his own deployment to Iraq several years previously. He showed us pictures of smiling Iraqi children, the behavioral health office where he spent much of his time providing care for service members struggling with combat stress, and the cafeteria where he normally ate his lunch. While pausing on this last image, he suddenly slammed his hand down on the podium with a resounding crack, flipped to a picture of the same cafeteria destroyed after a mortar attack, and began describing to the startled audience what can happen in the seconds and minutes after a mortar falls on an unsuspecting crowd.

Even as he did so, he began flipping rapidly through a series of images: clouds of white dust in the air; "You try to understand what just happened"; the face of a soldier with his eyes screwed up in terrible pain, his cheeks smeared with blood and soot; "You look around to see what you can do"; the disheveled wreck of a floor covered in debris; "If you're combat personnel, you go outside to secure the area"; the horribly mangled leg of a soldier on a gurney; "If you're a medic, you go to work trying to save whoever's in need." He flipped through slide after slide of damage and destruction for what could only have been a matter of seconds, but I and those around me were held still and breathless for the duration, shocked and stunned by the graphic images, the suddenness of the shift in tone, the dramatic disruption of what we had expected to be an ordinary morning presentation.

The psychologist's way of introducing his audience to combat PTSD was brilliant because the demonstration drove home a point too often left out of discussions of trauma: that all trauma begins with the sensory and perceptual experience of danger, with the physiological and cognitive awareness that something is very wrong. Trauma may come as a shock—to the ears, the eyes, the nerves of the skin feeling the heat of a blast, the mind as we begin to fathom that the predictable fabric of a normal moment has been irrevocably torn. Alternately, trauma may be perceived without coming as a shock at all; perhaps it arrives with a shudder of fulfilled dread at the creaking open of a bedroom door in the night, perhaps it rushes into a combat patrol that has held soldiers at a level of high-intensity alertness for many hours. But trauma is first and foremost a matter of sensation and perception, something akin to the breathless surprise that held the audience captive during the psychologist's presentation. Only later, after the experience is over, can its consequences begin to emerge in such a way that they may be labeled a disorder, a potential matter of mental illness. This chapter takes a closer look at the neuroanthropology of trauma, particularly in the context of male US veterans' experiences of combat and its aftermath. Combat PTSD among veterans of Iraq and Afghanistan provides an illuminating case study of how trauma results in cognitive and neurophysiological shifts that reveal complex interactions between the individual and his and her cultural environment.

It is worth noting that trauma is an extremely common experience, even in civilian populations. The National Comorbidity Study (NCS; Kessler, Sonnege, Bromet, Hughes, et al., 1995) surveyed a nationally representative sample in the United States and found that some 60% of men and 51% of women had experienced at least one traumatic event in their lives, whether in the form of physical or sexual assault, major injury, or witnessing these events happening to others. Although trauma is a necessary precursor to PTSD, only a minority of individuals exposed to trauma will ever develop PTSD. The NCS, for example, estimated a lifetime PTSD prevalence of 7.8% across the United States based on epidemiological assessments (Kessler et al., 1995). The risk of developing PTSD, however, increases with the number of traumas experienced, with the result that certain populations exposed to a greater number of cumulative life stressors—including inner-city residents, combat veterans, refugees, and so forth—typically exhibit higher rates of PTSD.

The process by which trauma evolves into PTSD has long been a source of debate across the continuum of scientists and care providers. Psychotherapists often view PTSD as a sign of identity loss or a kind of wound to the soul (e.g., Tick, 2005), whereas psychiatrists and clinical psychologists may instead describe PTSD as a stress disorder, focusing, respectively, on its neurobiological and cognitive underpinnings (e.g., Yehuda, 2002). Meanwhile, anthropologists and scholars in the humanities have attempted to illuminate trauma while challenging the idea that a diagnosis of PTSD is the only way to understand such suffering (e.g., Young, 1995). Social theorists point out that the violence causing trauma is often a direct result of social environment, reflecting structural inequalities and political economic conditions as much as one individual's distress (e.g., Summerfield, 1999). Surveying the largely separate literatures that have resulted from the study of trauma and PTSD in each of these disciplines leaves one with a lingering appreciation for the old story about blind sages trying to describe an elephant, with each man carefully listing the traits of one part of the pachyderm—flapping ears, long trunk, and skinny tail—while nonetheless failing to grasp the sum of its parts.

Efforts to conduct more integrative study of the complex processes involved in shaping PTSD risk are fairly recent. Yet a rapidly growing scientific literature on PTSD reveals that the etiology of the disorder relies upon interactions between social, cultural, biological, and psychological factors, although many of the specifics of these interactions are clear only in their broadest outlines. For example, while researchers widely agree that parental PTSD increases the likelihood of PTSD among offspring, a variety of pathways appear to explain this link. A parent with PTSD, for example, is likely to pass on certain patterns of hypervigilant behavior and attitudes about the safety of the world to his or her children (Dekel & Goldblatt, 2008). In addition, a PTSD-diagnosed parent may carry a genetic predisposition to the disorder, such that offspring who are exposed to traumatic events may be more likely to develop PTSD (Kilpatrick, Koenen, Ruggiero, Acierno, et al., 2007). Recent work by Rachel Yehuda and colleagues (2008) also indicates that epigenetic factors may shape PTSD risk. Their research suggests that maternal PTSD may shape a child's vulnerability as a result of prenatal endocrine conditions that influence the developing fetus's physical and emotional responsiveness to environmental stimuli.

Yet even as such integrative research has expanded the wider understanding of how PTSD reflects interactions between the individual and his or her environment, the difficulties inherent in developing and conducting truly multidisciplinary research have continued to inhibit the study of trauma and health. This may be due in part to a common tendency to conflate key terms. Too often, *trauma*, or exposure to a horrific event, is used interchangeably with *suffering*, which speaks more broadly to a state of generalized distress, and with *PTSD*, which is a biomedical category describing a cluster of symptoms that arise among a minority of those exposed to trauma, causing significant distress and impairment (Finley, in press). It is similarly easy to lose conceptual clarity when differentiating between *trauma* and *stress*, for both signal a critical challenge to the individual organism. It is perhaps best and simplest to note that while stress often does not include exposure to trauma, the experience of trauma inevitably imposes significant stress. Terms aside, anthropology, psychology, psychiatry, neuroscience, and epidemiology have all made significant contributions to the study of trauma, each bringing its own perspective to bear on the problem of what PTSD is, what it isn't, and what sort of responses and research it will require over the course of time. The primary insights of these streams of research can be synthesized into a framework that focuses on six variables implicated in the elicitation, experience, expression, and response to trauma-related distress: (1) cultural environment; (2) stress; (3) horror; (4) dislocation; (5) grief; and (6) cultural mediators.

The central goal of this chapter is to fuse these interwoven and mutually reinforcing variables into a neuroanthropological model of post-traumatic responses to aid in furthering the integrative study of trauma and health. The first part of this chapter examines each of the six model components as they emerge in the lives of male veterans of Iraq and Afghanistan who have been diagnosed with combat PTSD. Toward its conclusion, the chapter returns to the question of how the neuroanthropological model proposed here can move the study of trauma forward, suggesting three ways: first, by facilitating the effort to conduct comparative research on post-traumatic responses across heterogeneous cultural environments, populations, and types of trauma exposure; second, by pointing toward a research agenda that prioritizes unlocking the relationships between the cultural environment and the individual, with particular attention to evolving processes and interactions over time; and third, by suggesting a series of research foci and questions to aid in elucidating key cultural and experiential variables and developing preliminary hypotheses for future investigation.

Cultural Environment

In considering the neuroanthropology of trauma, we will return throughout this chapter to findings from research conducted during fieldwork in and around San Antonio, Texas, in 2007–2008 (Finley, 2011). The experiences and views of the veterans discussed here (all of whose names and other identifying features have been changed) provide a telling glimpse into both the universality and the local and individual uniqueness of experiences of trauma, and the role of cultural environment in shaping those experiences. The cultural environment in which a post-traumatic response emerges is taken here to include historical and political economic context; local social structures and patterns of kin and other relationships; the physical environment; and local worldviews and norms, particularly as they pertain to ethnotheories of trauma and expectations for a good life. It goes almost without saying that gender is a central variable in shaping an individual's experience of the cultural environment, so the focus here on male veterans should be acknowledged as incomplete in describing the full range of combat trauma responses among this generation of veterans. Even so, it is to be hoped that the model for neuroanthropological study proposed here can provide a meaningful starting place for examining the trauma responses of women veterans more closely in future research.

In the years since the attacks of September 11, 2001, more than two million US service members have been deployed to the conflicts in Afghanistan and Iraq, many returning home with significant health concerns related to the physical and psychological stresses of combat deployment. Epidemiological studies have identified rates of PTSD among Afghanistan and Iraq veterans of between 11% and 19% (Hoge, Auchterlonie, & Milliken, 2006; Hoge, Castro, Messer, McGurk, et al., 2004), also noting significant levels of new-onset depression among those exposed to combat (Cabrera, Hoge, Bliese, Castro et al., 2007; Wells, LeardMann, Fortuna, Smith, et al., 2010). By 2008, some 50,000 of these veterans had already been diagnosed with PTSD within the Department of Veterans Affairs (VA) health care system alone (Seal, Maguen, Cohen, Gima, et al., 2010).

Faced with such staggering numbers and veterans' growing need for health care and services, the American scientific and clinical establishments have leveled significant research attention at PTSD over the past decade. Authoritative understandings of combat trauma are driven primarily by an ethnopsychiatry of PTSD that, while far from homogenous, is centered around a largely agreed-upon definition and set of diagnostic criteria. PTSD is held to be a mental illness that results from exposure to a traumatic event and is characterized by three types of symptoms— hyperarousal, reexperiencing, and avoidance or numbing—that are severe enough to result in clinically significant impairment to normal function (American Psychiatric Association, 2000). PTSD as a formal diagnosis dates back to 1980, shortly following the close of the American war in Vietnam, and carries with it a broad array of cultural meanings related to the human costs of war; the stigma of mental illness; the honor attendant to military service and sacrifice; and the national benefits, including various forms of financial compensation and health care, owed to veterans in return for their service (Finley, 2011; Young, 1995). As seen in the discussion below, these and other features of the cultural environment have profound implications for shaping veterans' experiences of combat trauma.

Stress

In everyday American discourse, the word "trauma" is commonly used to discuss a wide range of profound and disturbing emotions, often without explicit or implicit recognition of the full physiological cascade involved (Finley, 2011). By contrast, emotion as conceptualized in this chapter borrows heavily from Paul Griffiths' (1997) discussion of affect program theory, which presumes that emotions are phenomenologically compelling, complex, and coordinated responses that include:

(a) expressive facial changes, (b) musculoskeletal responses such as flinching and orienting, (c) expressive vocal changes, (d) endocrine system changes and consequent changes in the level of hormones, and (e) autonomic nervous system changes ... [as well as] emotion feelings and cognitive phenomena. (77)

Griffiths points to research by Paul Ekman (1971) and others that suggests these affect programs are pancultural, occurring among individuals across every cultural environment (if not necessarily every individual), and that their emergence reflects both evolutionary predisposition and individual learning based on conditions in the local cultural and physical environment. For example, although the human affect program associated with fear contains features that are shared in common around the world, the fear experience is also responsive to individual variation and the specific threats of a particular setting, thus enabling adaptation that occurs in real time across the life course (Griffiths, 1997).

Placed in this context, it makes sense that much of the emotional experience of combat PTSD arises from what might best be thought of as a rewiring of the stress response in reaction to sustained periods of time spent under the intense and potentially life-threatening conditions of a war zone. Veterans of Iraq and Afghanistan who return home from deployment often describe a fundamental change in their sensory perception of and emotional response to the world around them, finding themselves reacting to stimuli with unprecedented vigilance.

Such a shift can be extremely adaptive to survival in a combat zone. Under life-threatening circumstances, humans have been shown to exhibit enhanced reaction time, responding to questions and accomplishing physical tasks more quickly (Marx, Brailey, Proctor, Macdonald, et al., 2009). At the same time, individuals under direct threat are likely to exhibit reduced powers of concentration, a muddled ability to learn, and little or no memory for tasks or information not associated with a direct threat (Arnsten, 1998; McEwen & Sapolsky, 1995; Sauro, Jorgensen, & Pedlow, 2003; Shors, 2006). In other words, a stress-related neurobiological process assists the human brain in focusing intently on surviving immediate threats. This may occur even to the point of sacrificing the ability to complete other cognitive tasks, tasks that must—at the moment of staring down an insurgent with a gun—seem of relatively low priority. For deployed service members living under threat for prolonged periods of time, evidence suggests that this neurobiological reshuffling persists after their return. Bryan Marx and his colleagues recently reported that, even a full year after returning from deployment, soldiers who had lived through intense combat experiences still had a *faster* reaction time than those with little combat experience (Marx et al., 2009). In other words, these soldiers had adapted to the need to respond quickly to potential threat, and this adaptation held up even after the return to a comparably low-threat environment.

These findings suggest that it is *normal* for those who live in a highly stressful environment to adapt to that environment in ways that linger even after returning to relative safety. This stress-related adaptation explains many of the reactions considered to be normal for combat personnel in the post-deployment period: reacting quickly to perceived danger, feeling naked without a firearm and exposed and vulnerable while driving. Most service members are likely to find that spending a year in a combat zone will result in noticeable changes in reaction and cognition, but these changes are not typically either permanent or debilitating.

At a more extreme level, as occurs when a service member begins to develop PTSD, the stress response can stimulate a level of physiological reactivity that interferes with the ability to live a normal life. Veterans with PTSD often describe being unable to engage in everyday activities like going to the grocery store, where the crowds may make them feel unreasonably anxious or where they may imagine assailants behind every corner. The discomfort these veterans report is mirrored in the findings of physiological research on sensory and information processing post-trauma. One recent study found that individuals with PTSD have a more dramatic increase in heart rate after viewing trauma-related images than do traumatized individuals without PTSD (Ehlers, Suendermann, Boellinghaus,

Vossbeck-Elsebusch, et al., 2010). Individuals with PTSD were also slower to recover, meaning that their heart rate stayed elevated for a longer period of time after the images were taken away. Another study reported that individuals with PTSD are less able to filter out unnecessary stimuli and select only relevant sensory information in the environment (Holstein, Vollenweider, Jacnke, Schopper, et al., 2010). Such an inability to distinguish between relevant and irrelevant information can be a profound liability, perhaps particularly so in an industrialized society like the United States, where tasks requiring attention and prolonged concentration are a vital part of many types of work. Many recent veterans return to work or school after separating from the military, and those with PTSD frequently describe having difficulty with homework, test-taking, remembering meetings, or paying bills. These neurophysiological shifts, in other words, can have direct implications for the ability to function in normative American society, often to veterans' profound frustration.

However, culture and neurophysiology collide long before these downstream effects begin to take their toll. The cultural environment may be inextricably implicated in an individual's cognitive and physiological priming for the perception of an event and his or her emergent response to it (Jonas, Graupmann, Kayser, Zanna, et al., 2009; Ng, Han, Mao, & Lai, 2010). Priming in this sense refers to the process by which exposure to a specific repeated stimulus influences how an individual responds to a subsequent stimulus. Studies of culture and priming have become more common in recent years as researchers have identified priming as a potential mechanism for understanding cultural difference (Bardi & Goodwin, 2011; Suh, Diener, & Updegraff, 2008). Much of this research has approached priming as purely a cognitive or perceptual task conducted in a psychological laboratory with the intention of examining how priming influences the expression of cultural attitudes, such as a collectivist vs. an individualist orientation. One can imagine, however, that repetitive experience within a given cultural environment might itself provide a sort of priming with the potential to shape how an individual is likely to respond to a given event. The following example illuminates how the individual experience of trauma remains profoundly subjective in a way that seems grounded as much in cultural worldview and environment as in an individual's physiology and personal history.

A veteran named Chris once told me, "The first time I was in a firefight ... I froze. I didn't move." Although a member of the US Air Force, he was deployed to Afghanistan in the early days of the war to provide logistical support to a Special Forces unit. Their outpost came under fire unexpectedly, and at the time, Chris was mostly aware of the noise and his own shock. "They were screaming, 'Return fire, return fire!!' And none of us were shooting—everyone was surprised." When he did begin firing some endless few seconds later, Chris says he went through magazine after magazine of ammunition, firing wildly at targets far out of range. He didn't even realize that he had wet himself until after it happened.

The memory of these few seconds of frozen time ultimately became part of his PTSD, what Chris called "the little monster inside." His account of the firefight is steeped in shock and immobility, although the activation of his sympathetic nervous system—indicated by his urinating on himself—would suggest that he experienced a level of fear that may or may not have risen to consciousness. His profound experience of stress at that moment, however, was shaped not only by his physiological reactivity but also by the fact that, as an airman in the early post-9/11 era, he was not trained for combat in the same way the Special Forces soldiers he was accompanying had been. Much of combat training in the Army or Marines is dedicated to helping service members dampen the kind of stress response that can inhibit prompt and effective action under exactly the circumstances Chris described. Training seeks to develop a set of deep-seated cognitive and muscle-memory skills related to combat so that a combatant can go through the necessary motions even when he may consciously still be saying to himself, "Wait, somebody's shooting at me?!" Although the Air Force has since retooled its program of basic training, in 2002 the airman's role as on-the-ground support staff for soldiers and Marines was a fairly new one, and the training had yet to catch up with changing demands. Neither Chris's sense of himself nor his nervous system were ready for that firefight, and this lack of preparedness likely shaped his initial response. (To be clear, there is no intention here of suggesting that Chris's reaction was somehow "natural" by virtue of being untrained, but only of recognizing that it would likely have been *different* with training.)

In consequence, for the next five years, Chris viewed this event as "humiliating" evidence of personal failure. "I didn't react. I didn't find a target and kill it. I pissed my pants." He compared his own actions with what he imagined others might have done, and was devastated by the disparity. "Some of them may have jumped on a grenade or made the 500-meter shot through the eye that killed the bad guy, made the one-shot-one-kill that never happens." He measured himself against an almost action-movie–like image of the perfect warrior, even as he was able to acknowledge, wryly, that the "one-shot-one-kill" is an almost mythical rarity. Both his initial and evolving experiences of the firefight, then, were

a product of his reaction to the event as well as his view of what his performance meant as a judgment on his military service, perhaps even his
manhood.

The moment of danger itself, Chris's immediate response, and his long-
term perspective on that response each reflect the interplay between his
cultural, cognitive, and neurophysiological experiences. Examining how
an individual or population experiences such moments of traumatic stress
has considerable potential for clarifying not only the immediate felt consequences of the event, but the cultural, environmental, and life-history
context shaping the perception and course of the unfolding post-traumatic
response.

Horror

According to the *Diagnostic and Statistical Manual of Mental Disorders* (2000),
two criteria must be present in order to define an event as a trauma. The
first is experiencing an event that involves "actual or threatened death or
serious injury" to either oneself or another (American Psychiatric Association, 2000). The second is horror, for an event is only considered to be
traumatic if it prompted feelings of "fear, helplessness, or horror" (Yehuda,
2002). As used here, however, the term "horror" glosses a wide variety of
distressing emotions that can be elicited by trauma: horror itself, but also
shame, disgust, terror, helplessness, and so forth. Horror connotes an
emotion so powerful as to overwhelm the individual's capacity for immediate sense-making or cognitive processing.

Horror defines those events that are most likely to reemerge in recurrent
nightmares and intrusive memories, elements of the second cluster of
symptoms required for a PTSD diagnosis, referred to as "reexperiencing."
Many such memories are marked for their gore, as when soldiers speak of
"pink misting," the ephemeral cloud of blood one sees when a body
explodes. One veteran described seeing the faces of corpses superimposed
on those around him even several years after his return from combat.
Other memories may be marked for how they seem to offend the natural
order, as in the case of children mutilated or killed in the course of a battle.
Many people experience the intrusion of horrific thoughts or images
during the process of recovering from difficult events (e.g., a traumatic
assault, the death of a loved one), only to find such occurrences diminish
over time as the sufferer finds a way to make sense of the event. For individuals with PTSD, this gradual recovery may not happen without outside

intervention; these memories and images may continue for decades without resolution.

How do we make neuroanthropological sense of the fact that certain memories are so intrusive? Clearly the content of trauma-related memories contains powerful messages regarding potential dangers to be avoided. Evolutionarily speaking, it makes sense to retain a visceral grasp of not only an acknowledged danger, but also of warning signals—for example, the sound of an incoming mortar heard the instant before a roof collapses— that may help to identify and avoid similar dangers in the future. The clarity of horror-related intrusions, then, likely results in part from an evolutionarily acquired bias toward remembering lessons with relevance to immediate survival.

Horror, however, should also be recognized as reflecting a challenge to deeply held beliefs about the self and world, an insight that is perhaps the most significant contribution to the understanding of PTSD from within psychoanalysis and the humanities (Caruth, 1995; Frankl, 1946/1984; Herman, 1992). Most immediately, trauma can pose a painful challenge to individual identity. Chris was shocked and dismayed to find that his first response to combat was primarily one of desperation and terror. The resulting disappointment haunted him for years afterward, providing continual fodder for self-disparagement and feelings of inadequacy. In his own assessment, he was not the man he had thought himself to be.

Horrific events can also force a more global reassessment, particularly when an individual struggles to bring the trauma into alignment with preexisting beliefs regarding a just world (Hafer & Begue, 2005). Elaine Scarry (1985) famously wrote of torture as "unmaking the world," for the intentioned infliction of suffering can be so morally anathema as to splinter an entire worldview. A veteran named Carlos was deployed to Iraq during the height of the war and faced violence day after day, but says that what made him "snap" was when he witnessed an Iraqi girl shot with a grenade launcher. "It was actually a misfire—a mistaken fire, I should say. They thought she was a terrorist, and it was a little girl bouncing a ball in the dark." Even after he came back home, he saw the little girl in flashbacks. She had survived the shooting, but when he saw her in his mind she came to him "more injured or dead or staring at me or pointing at me. Just totally unrealistic." In speaking of the shooting he described it as "senseless." He could not reconcile his notion of moral combat with the accidental shooting of a little girl, and this glaring disjuncture likely contributed to the persistence of this particular memory.

Dislocation

The notion of trauma-related dislocation draws upon the metaphor of physical dislocation, in which a sudden breach or jolt results in injury (Finley, 2011). The dislocations many veterans describe arise out of their experiences with combat deployment and the changes it has created in them. These changes often result in a series of uncomfortable distances for veterans: from others in their lives; from an earlier mode of perception; from feeling at home in the world; or from some prior, preferred sense of self.

Often dislocations arose out of the stress-related changes described above, changes that many veterans found had left them with an emotional landscape distinguished mainly by numbness, anxiety, irritation, and anger. Carlos, who otherwise appeared a gentle and loving family man, described himself after his snap as "very angry." Veterans commonly wondered aloud why they couldn't grieve a loved one who had died, or mused that they weren't sure if they loved a girlfriend, or expressed bafflement that they could get angry in the blink of an eye. They often spoke of a sort of emotional flattening, as though their affective response to the world had collapsed its range. One veteran in his mid-twenties told me he felt like a pregnant woman whose emotions were out of control: "If I'm not sad now, I'm pissed off." He said he will get "overly mad" in response to relatively minor conflicts, and added, "I don't know why I get like that."

One recent theory why individuals with PTSD often appear hyperreactive to perceived threat draws on animal and human imaging studies to propose that traumatic stress leads to an "altered architecture" of the brain and its information processing (Rockstroh & Elbert, 2010). Rockstroh and Elbert have observed that repeatedly stressed brains appear to use what they call a "low road" for sensory processing that speeds up the ability to configure a threat-response by bypassing the prefrontal cortex. In other words, instead of using a "high road" pathway that privileges the prefrontal cortex's ability to analyze complex data and regulate emotion accordingly, these brains have developed pathways that privilege a more rapid response, changing from "a careful analyzer of the environment to a rapid threat detector" (Rockstroh & Elbert, 2010, p. 14).

Even as theories and evidence accumulate to explain such stress-related neuroplasticity, however, it remains necessary to consider how the experience of the resulting shifts is framed by an individual's personal history and cultural expectations. For example, if a veteran has always prided himself on being a calm and easygoing guy, then finding himself continually on edge and on guard may prompt a crisis of identity. Chris struggled to make peace with himself as a man who felt debilitated by fear his first time under fire. In both of these examples, combat led to a shattering of pre-war notions of personal identity.

Noting how survivors talk about what has changed most significantly in their lives as a result of the trauma provides a flexible way in which to examine its downstream effects. Dislocations also provide an opportunity to consider the adaptability of emotional responses to local conditions. Paul Ekman was one of the first to recognize that while an emotional response is likely to exhibit significant commonalties across all cultures, the conditions eliciting that response may vary significantly, as do the approved means of expression (Ekman, 1971; Griffiths, 1997). Take, for example, the experience of anger. The source of an individual's anger may be culturally appropriate, although the manner of its expression might not. One older veteran sheepishly described losing his cool at the grocery store—not as the result of any direct threat, but because he felt the cashier had been rude to the elderly women ahead of him in line. He perceived that the young cashier had failed to show proper respect to a vulnerable elder. The source of his anger, therefore, was entirely justified within the cultural setting, although his means of expressing it—in this case, by yelling at the cashier—was not, and provoked considerable disapproval from his wife, who witnessed the exchange.

His wife's perception of the event sheds some light on two further means by which post-combat changes in veterans' affect and behavior may be culturally mediated: (1) by the interpretive framework in which these changes are understood (as rudeness, as PTSD, etc.); and (2) in the social responses to those changes deemed appropriate within that framework. In this case, the older veteran's wife viewed the conflict with the cashier as evidence of her husband's inability to control his behavior rather than, as other wives of veterans might, as evidence of his PTSD. Veterans often find that their friends and family are more or less willing to tolerate post-combat shifts depending on their understanding of what these changes represent (Finley, Pugh, & Jeffreys, 2010b).

Friends and family members who are not well-acquainted with PTSD may blame the veteran rather than the illness for unpleasant changes in behavior or attitude, often resulting in damaged relationships. The impact of PTSD on veterans' social networks has seen growing attention in recent years, with more and more studies reporting high levels of relationship failure and divorce and low levels of marital satisfaction among veterans and their spouses (Goff, Crow, Reisbig, & Hamilton, 2007; McLeland,

Sutton, & Schumm, 2008; Monson, Taft, & Fredman, 2009; Renshaw, Rodrigues, & Jones, 2008). Several studies have noted that a spouse's marital satisfaction appears to be closely associated with the perceived meaning of post-war changes in the veteran and relationship (Finley, Baker, Pugh, & Peterson, 2010a; Renshaw et al., 2008). For example, a wife who believes her husband is simply behaving badly, as in the case with the cashier, is much less likely to be tolerant than a wife who believes her husband's behavior reflects wartime suffering in service of a noble cause.

Drawing on examples like this, it seems clear that many of the disruptions that veterans describe result from the powerful ways in which experiences of stress and horror have challenged their ability to live in accordance with a culturally normative status quo. Returning to Griffiths, it is important to remember that while profound emotion reflects a complex of physiological responses, it can be triggered as powerfully by internal cognitive processes as by an external event. That being the case, it comes as no surprise that it is often a veteran's way of thinking about these personal, social, and cultural dislocations, rather than either stress or horror alone, that results in the third and perhaps most commonly underappreciated aspect of combat trauma for American veterans: grief.

Grief

Grief is part of the natural mourning process associated with the loss of fellow service members in combat. It is sometimes taken as a cultural truism that there is no intimacy like that between men who have been through battle together (Simons, 1997), and many of these men had lost close friends in the war, often during periods of intense fighting when survivors had little time to stop and mourn. Chris said of bidding farewell to his Special Forces mentor and role model, "His body's gone, and you might have the opportunity to salute his boots and his dog tags. Then you go right back out and you do it again and somebody else dies." Both the immediate and the long-term aftermath of these deaths could be haunting. One veteran described falling apart while still in the combat zone, sobbing uncontrollably and losing the ability to speak after picking up the pieces of his best friend's body. Another man, now returned back home, spoke of long nights spent crying over buddies who hadn't made it through the war. If a veteran feels some sense of personal responsibility for failing to prevent the death, or even for simply surviving the same events, his grief may be compounded by terrible guilt.

The 2010 documentary *Restrepo*, directed by Sebastian Junger and Tim Hetherington (who was himself later killed in a combat incident in Libya), captured the immediate aftermath of an American soldier's death in Afghanistan. Patrolling the mountainous countryside around their isolated base, the local unit came under attack while strung out in a long line along a hillside. During the course of the fighting, one of the men toward the front of the line was killed, and within a period of a few minutes, word of his death spread to other soldiers arriving on the scene. One soldier stood stunned at the news of his buddy's death, unable at first to believe the news and then overwhelmed by his emotional response. Watching his face on the screen, it is impossible to know what he was feeling, but he may have been experiencing something akin to the visceral swamping that Renato Rosaldo (1993) described after his wife's death in a tragic accident:

the deep cutting pain of sorrow almost beyond endurance, the cadaverous cold of realizing the finality of death, the trembling beginning in my abdomen and spreading through my body, the mournful keening that started without my willing. (9)

In the film, the moment does not last long. The unit is left in a precarious position on the hillside and must continue on toward safety; the soldier is given no more than a few moments to pull himself together and get moving again.

But this is not to say that his grief halted there. Descriptions of combat trauma going back to the 1940s pay homage to its frequent occurrence alongside grief-specific symptoms, particularly where the traumatic event included interpersonal loss (Pivar & Field, 2004). More recent work has identified key differences between the signs of PTSD and those of bereavement (Raphael & Martinek, 1997). Whereas the attempt to avoid reminders of traumatic event is a classic and central symptom of PTSD, those who are grieving may actively seek out reminders of those they have lost. Whereas PTSD is often typified by anger and fear, grief may be characterized by longing and sadness. Following this pattern, certain of the phenomena described by veterans in the study appear more consistent with grief than PTSD. One veteran told of sitting in the early hours of the morning in front of his computer, compulsively watching YouTube videos of combat scenes uploaded by service members in Iraq or Afghanistan, tears streaming down his face but unable to look away. Most illustrative of the longing thought to be more characteristic of grief than PTSD was the refrain heard from so many veterans: "I wish I could go back." In a study conducted among Vietnam veterans, Pivar and Field (2004) found that grief symptoms were common and often severe among those who had lost valued comrades during the war, even though those experiences were more than thirty years in the past. They also found that grief symptoms emerged as distinct from symptoms of PTSD and depression, even among those veterans who reported all three kinds of distress, and ultimately concluded that "unresolved grief will endure over time if it is not addressed by clinical intervention" (753).

Although bereavement appears to be under-recognized for the role it plays in shaping many veterans' experiences of post-deployment distress, grief also goes beyond the mourning of combat casualties. Veterans' testimonies suggest that grief may also result from losses that are less immediately tangible. Many of these men felt a sense of loss at passing out of the military and returning to the civilian world, often finding themselves uncomfortable with returning to a different set of norms after years of life in the service. Many expressed frustration that civilians seem to expend their energy on nonessential things. Combat veterans have lived in settings where every decision may mean the difference between life and death. Civilians expressing irritation about a broken cell phone or work-related stress can provide a jarring reminder that the stakes are very different in a war zone. Some veterans saw this disconnect between military and civilian priorities as a sign of their own personal growth post-deployment, as evidence that they had learned to keep life's priorities in perspective. But others found that the perceived disparity between military and civilian moral values posed an almost insurmountable barrier to closeness and understanding. For some, friends, coworkers, and even family members who could not understand the profound stakes of war came to seem as distant as if they lived in another world altogether.

It is likely that this sense of disconnection may be exacerbated by the fact that, since 1973, the American military has been an "all-volunteer force," with the result that there is a growing social disparity between those who serve and those who do not (Michalowski & Dubisch, 2001). There is also something of a cultural legacy that has been handed down from that same period, when political conflict over the conduct of the Vietnam War too often took the form of blame loaded upon the heads of the veterans who had served there (Michalowski & Dubisch, 2001). Many Vietnam veterans were left feeling like outcasts, with no honored or acknowledged way to talk about their war experiences. Although there is greater social recognition and respect for military service in the current era, the residual caveat that someone who hasn't been through combat "just can't understand" may serve to inhibit the kind of open discussion and sharing of those experiences that can be most effective in rebuilding post-war intimacy with family and friends (Finley, 2011; Finley et al., 2010b).

Grief, then, results from veterans' internal and external dislocations: bereavement and interpersonal loss; being rent from a previously cherished view of self; overwhelming distress at finding the world can be without order, sense, or safety; and gaps in experience and understanding that lead to feeling distant from loved ones. All of these changes can foster a profound sense of grief and loss, exhibited in the disheartenment expressed by so many veterans.

Cultural Mediators

There is one final element to be added to this exploration of post-traumatic responses, and that is the role of cultural tools and processes with the potential to mediate the relationships between other trauma-related variables. For example, the evidence-based cognitive behavioral treatments for PTSD, which have proven to be highly effective in reducing symptoms over the long term, rely primarily on reexposing veterans to their traumatic memories. Therapeutic methods require veterans to write or narrate these events in close detail, and then to read or listen to their own narration of the memory until the associated anxiety decreases to a manageable level. This process of gradually habituating to stressful memories is accompanied by cognitive therapy that enables an individual to reassess the meaning of that memory. For example, Chris, having been through therapy, can now say of his first firefight, "Nobody was shooting. I didn't freeze for that long—maybe five seconds at the most. My body was evacuating itself to give it greater speed. I didn't do so bad." He says he can now look at these memories "rationally," and considers himself "98% cured" of PTSD. In addressing the stress and horror associated with his memories through therapy, Chris has found that his anxiety and shame have diminished, and so too has his grief. The treatment has disrupted what had been a negative cycle of increasing frustration and despair.

Cultural mediators need not take the form of what can be easily identified in Western terms as "treatment," but may emerge in a variety of other culturally salient forms across populations. Ritual emerges as a prime candidate, of which two excellent examples include the post-war work of *curandeiros* engaged in removing the violence from former combatants in Mozambique (Nordstrom, 1998) and the death rituals of Mozambican refugees in Zaire (Englund, 1998). Approved ways of expressing emotion or telling stories around the experience of trauma may provide another means of diminishing dislocation and distress; forms of storytelling by veterans occurring in different contexts and to different ends has been described in the work of Theresa O'Nell (1999) and Edna Lomsky-Feder (2004).

Social support is widely recognized to play a critical role in mediating the impact of trauma on long-term physical and psychological health outcomes (Ahern, Galea, Fernandez, Koci, et al., 2004; King, King, Fairbank, Keane, et al., 1998; Ozer, Best, Lipsey, & Weiss, 2003), and may take a wide variety of forms depending on the cultural setting and type of trauma, ranging from a network of local kin relations to formalized groups of self-identified survivors. Judging by the veterans in this study, personal role models may also play an important role in influencing how individuals navigate their response to trauma. Many veterans spoke of drawing on the example provided by other veterans or family members in deciding upon a course of appropriate response once PTSD began to interfere with their lives.

Of course, the local ethnotheory of trauma itself may play a role, affecting not only the manifestation of a post-traumatic reaction in light of local idioms of distress (Nichter, 1981), but also the course of those manifestations over time. If the experience of trauma-related suffering is locally understood to result in long-term disability or, alternately, a path to personal growth, one might expect to see commensurate patterning in local trajectories of trauma and resilience. As with emotions, human pathways through sickness and healing have a way of following the scripts available to them (following what Moerman [2002] has called the "meaning" effect).

The presence of such cultural mediators points yet again to the interrelationship between the many variables implicated in a trauma response, while also highlighting the potential modifiability of the neuroanthropological processes involved. The task of identifying potential mediators for trauma in a given setting may be approached by a series of questions: What cultural processes, therapies, or other tools are brought to bear in the aftermath of trauma? How are these locally understood? How might they intersect with physiological and cognitive processes at the individual level to result in a modified trauma response over time? Does the presence or absence of a potential mediator appear to be associated with differential outcomes?

A Neuroanthropological Model of Post-Traumatic Response

In describing each of the elements listed above, the aim of this chapter has been to illustrate how a neuroanthropological model of trauma may have utility for a variety of purposes: to facilitate the comparison of trauma responses across cultural settings, populations, and types of trauma exposure; to foster a better understanding of the relationship between the cultural environment and cognitive and neurophysiological responses to trauma; and ultimately, to elucidate how interactions between each of these dimensions function at individual, cultural, and pancultural levels to shape the elicitation, expression, and experience of trauma across the full and highly variable range of human response.

Within this neuroanthropological model, cultural environment provides the sociocultural, material, historical, and political frame in which trauma takes place and responses to it emerge. Stress encapsulates the physical and psychological adaptations brought on by traumatic events. Horror describes the way in which an individual may be overwhelmed by the trauma, left at a loss for how to make meaning of the event amid the challenges to notions of personal safety, self, and world left in its wake. Dislocations arise when the consequences of the trauma event, as well as cognitive and physiological shifts related to stress and horror themselves, result in changes in an individual's sensory perception, neurochemistry, sense of self, or social status and relations that may be painful or distressing. Where such dislocations occur, they may also become a source of terrible grief, in the sense of mourning both interpersonal losses and prior notions of self or worldview. Based on the fieldwork among combat veterans with PTSD discussed here, it appears that the relationship between each of these variables moves primarily in the direction of the larger bolded arrows in the model shown in figure 10.1, with the smaller arrows representing hypothesized bidirectional effects or weaker relationships. Finally, the model includes arrows to highlight, first, the chronological component of trauma response, which is likely to evolve considerably over time, with the acute stress and horror of the initial event taking on new forms and consequences as time marches on. The second arrow acknowledges the potential role of cumulative stressors that may be ongoing; cross-cultural research has shown that such stressors are a major factor predicting vulnerability and resilience in the wake of trauma (King et al., 1998; Silove, 2005). For veterans, these stressors might take the form of multiple deployments to the combat environment, the challenges of returning home and readjusting to civilian life, or difficulties with unemployment or divorce.

Adopting a neuroanthropological view of PTSD offers a useful way of thinking simultaneously about both the etiology and the experiential meaning of PTSD. Many models of PTSD exist, most of which are sophisticated and well suited to the task of driving forward the understanding of specific mechanisms and treatment efforts. Where this neuroanthropological

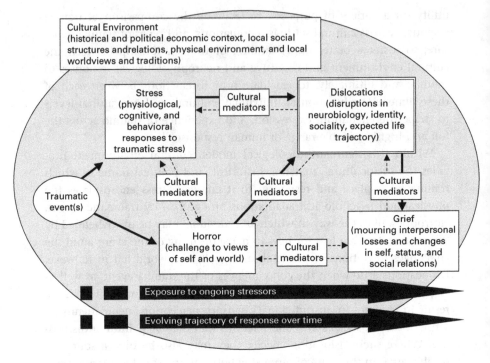

Figure 10.1
Neuroanthropological model of post-traumatic response.

model makes a contribution is by incorporating a broader view of the
interactions between multiple levels of human experience, thus providing
a more holistic perspective on the many factors that influence the emergence of varying trauma responses. For those examining PTSD from a
cross-cultural perspective, this model also provides a series of loci at which
to consider how culture, experience, and neurobiology come together.
Much ink has been spilled in debating to what extent PTSD can be generalized as a universal phenomenon and to what extent the diagnosis has been
overused as a way of understanding (and legitimizing) experiences of post-
traumatic suffering (e.g., Becker, 1995; Ingleby, 2005; Pederson, 2002; Summerfield, 1999, 2000, 2001; Young, 1995). The model proposes a means of
mapping both shared and distinctive experiences of trauma by offering a
template for gathering answers to organizing questions: how is traumatic
stress understood and manifest in this cultural environment? What is
deemed to be so horrifying about this particular traumatic event? What is
its cultural significance? What dislocations have occurred that may make

it more difficult to recover? What are sources of grief for trauma survivors in this environment, and what cultural mediators seem to play a role in easing or exacerbating the severity of their distress over time?

This neuroanthropological model can also be useful in developing hypotheses to probe the relationship between various mechanisms in creating and perpetuating disorder. For example, grief may be one outcome of the physiological, emotional, and social dislocations created by experiences of stress and horror, but that grief might also be hypothesized to play a role in exacerbating continued stress, thus driving increasing symptoms and dislocations in a cyclical pattern. Alternatively, one's sense of horror may continue to increase over time, rather than decreasing, if a particular memory becomes linked with awareness of the unfolding consequences of that event. Memory is known to be an evolving entity—not static like film, but imaginal and emotional, often with gaps that may be partially filled by assumptions or reinterpretations made after the fact (Engelhard, van den Hout, & McNally, 2008; Geraerts, Kozarix-Kovacic, Merckelbach, Peraica, et al., 2007). A memory may take on rising significance as it is swept along in the ever-flowing river of continued experience and reflection, and in so doing, may prompt a new intensification of the horror associated with the original event, thus resulting in a reinvigorated sense of stress, new dislocations, and fresh grief.

In their edited volume on trauma, Kirmayer, Lemelson, and Barad (2007) ultimately conclude that "although the neurobiological processes underlying an acute posttraumatic stress response have universal components, their temporal configuration and interaction is powerfully shaped by developmental, social, and cultural contexts" (470). Their point regarding the important roles of time and development across the life course is particularly well-taken. Understanding that experiences of trauma and PTSD evolve over time provides an invitation to follow and observe that evolution, and to identify psychological, physiological, social, and cultural features that may mean the difference between chronic suffering, some form of resolution and healing, and even post-traumatic growth.

Thinking in terms of a developmental perspective, some of the most interesting work on post-deployment health and well-being among veterans was conducted several decades ago by Glen Elder and colleagues (1987; 1994) among veterans of World War II. They found that veterans who were mobilized to war at an older age had greater difficulty returning home and reintegrating into society. Older recruits departed from the usual trajectory of American men's adulthood when they were forced to leave their marriages and children for years at a time and to disrupt the progress of their careers and education. Those who mobilized at a younger age, by contrast, had not yet achieved these markers of adult development and so were able to pursue them along a normative—if initially delayed—track after their return from the war. These findings point to the importance of examining dislocations from a life-course perspective, identifying where trauma may have interfered with expected life trajectories.

Readers familiar with the standard diagnostic criteria for the PTSD symptom cluster may note marked similarities between three of the core trauma model dimensions—that is, stress, horror, and grief—and the three clusters into which PTSD symptoms are organized, namely hyperarousal, reexperiencing, and avoidance. Certainly there are marked similarities, and yet there are conceptual distinctions worth noting. "Hyperarousal," for example, is used to describe symptoms of troubled sleep, irritability and anger, hypervigilance, and difficulty concentrating, all of which signal a level of stress-related arousal. Nonetheless, broadening the category to examine stress-related cognitive and neurophysiological phenomena not included in the diagnostic criteria for PTSD enriches the effort to describe how stress-related adaptations may vary or remain consistent across a variety of cultural and physical environments. Given the extensive body of research on human responses to acute and chronic stress (e.g., McEwen, 2002; McEwen & Wingfield, 2003; Sapolsky, 1998; Seeman, Singer, Ryff, Love et al., 2002), refocusing on *stress* rather than *hyperarousal* also provides added impetus to consider how stress-related responses are likely to evolve over time. Similarly, while "reexperiencing" as a category captures the intrusiveness of certain memory-related symptoms after trauma, the concept of "horror" speaks more directly to the task of identifying the cultural, social, and meaning-related cognitions that may exacerbate the event's importance for an individual. The third and final symptom cluster for PTSD, "avoidance," describes behavioral and cognitive attempts to avoid reminders of a traumatic event. The category includes emotional numbing and the suppression of feelings not immediately associated with survival. Although there are similarities between emotional numbing and certain kinds of powerful grief, the notion of "grief" leaves open a much wider field for understanding the role of bereavement and other kinds of mourning related to personal and interpersonal losses related to the trauma, including a range of grief-related symptoms that are not addressed within the current definition of PTSD. In summary, stress, horror, and grief are each included here in an attempt to push the field of comparative inquiry beyond the limits of the PTSD diagnostic category and to allow for a more processual view of how responses to trauma emerge as a product of continual interaction between the individual and his or her cultural environment.

A final benefit of this neuroanthropological model is that it provides an interdisciplinary framework with which to reexamine how best to help trauma survivors establish and maintain positive lives post-conflict. PTSD-diagnosed veterans actively respond to their post-deployment distress and other life challenges by making selective use of the resources at their disposal: social relations, treatment options, education, career-building, and economic resources, as well as particular narratives about their illness experience (Finley, 2011). Employing a model of PTSD that looks beyond psychological symptoms provides an opportunity to imagine responses that go beyond clinical treatment. As a nation, the United States and its military and VA institutions are doing a much-improved job in providing prevention, screening, and treatment services in support of the psychological health of service members and veterans. However, too few of these efforts have focused on responding to what are clearly key experiences driving distress—those of dislocation and grief. By broadening our understanding of combat PTSD to include a more holistic model of its implications for those affected, we can refine our efforts to be more meaningful, more focused on identifying those factors that interfere with post-traumatic growth, and more effective.

References

Ahern, J., Galea, S., Fernandez, W. G., Koci, B., Waldman, R., & Vlahov, D. (2004). Gender, social support, and posttraumatic stress in postwar Kosovo. *Journal of Nervous and Mental Disease, 192*(11), 762–770.

American Psychiatric Association. (2000). *Diagnostic and statistical manual of mental disorders* (4th ed., text rev.). Washington, DC: American Psychiatric Association.

Arnsten, A. (1998). The biology of being frazzled. *Science, 28*(6370), 1711–1712.

Bardi, A., & Goodwin, R. (2011). The dual route to value change: Individual processes and cultural moderators. *Journal of Cross-Cultural Psychology, 42*(2), 271–287.

Becker, D. (1995). The deficiency of the concept of posttraumatic stress disorder when dealing with victims of human rights violations. In R. J. Kleber, C. R. Figley, & B. P. R. Gersons (Eds.), *Beyond trauma: Cultural and societal dynamics* (pp. 99–114). New York: Plenum Press.

Cabrera, O. A., Hoge, C. W., Bliese, P. D., Castro, C. A., & Messer, S. C. (2007). Childhood adversity and combat as predictors of depression and post-traumatic stress in deployed troops. *American Journal of Preventive Medicine, 33*(2), 77–82.

Caruth, C. (Ed.). (1995). *Trauma: Explorations in memory*. Baltimore: Johns Hopkins Press.

Dekel, R., & Goldblatt, H. (2008). Is there intergenerational transmission of trauma? The case of combat veterans' children. *American Journal of Orthopsychiatry, 78*(3), 281–289.

Ehlers, A., Suendermann, O., Boellinghaus, I., Vossbeck-Elsebusch, A., Gamer, M., Briddon, E.,. (2010). Heart rate responses to standardized trauma-related pictures in acute posttraumatic stress disorder. *International Journal of Pathophysiology, 78*, 27–34.

Ekman, P. (1971). Universals and cultural differences in facial expressions of emotion. In J. K. Cole (Ed.), *Nebraska symposium on motivation*. Lincoln, NE: University of Nebraska Press.

Elder, G. H. (1987). War mobilization and the life course: A cohort of World War II veterans. *Sociological Forum, 2*(3), 449–472.

Elder, G. H., Shanaham, M. J., & Clipp, E. C. (1994). When war comes to men's lives: Life-course patterns in family, work, and health. *Psychology and Aging, 9*(1), 5–16.

Engelhard, I. M., van den Hout, M. A., & McNally, R. J. (2008). Memory consistency for traumatic events in Dutch soldiers deployed to Iraq. *Memory, 16*(1), 3–9.

Englund, H. (1998). Death, trauma and ritual: Mozambican refugees in Malawi. *Social Science & Medicine, 46*(9), 1165–1174.

Finley, E. P. (2011). *Fields of combat: Understanding PTSD among veterans of Iraq and Afghanistan*. Ithaca, NY: Cornell University Press.

Finley, E. P. (In press). The chaplain turns to God: Negotiating post-traumatic stress disorder in the U.S. military. In A. Hinton & D. E. Hinton (Eds.), *Rethinking trauma: Memory, symptom, and recovery after genocide and mass violence*. Durham, NC: Duke University Press.

Finley, E. P., Baker, M., Pugh, M. J. V., & Peterson, A. (2010a). Patterns and perceptions of intimate partner violence committed by returning veterans with post-traumatic stress disorder. *Journal of Family Violence*.

Finley, E. P., Pugh, M. J. V., & Jeffreys, M. (2010b). "Talking, love, time": Two case studies of positive post-deployment coping in military families. *Journal of Family Life*. Retrieved from http://www.journaloffamilylife.org/militaryfamilies.html.

Frankl, V. E. (1984). *Man's search for meaning*. New York, NY: Washington Square Press. (Original work published 1946).

Geraerts, E., Kozarix-Kovacic, D., Merckelbach, H., Peraica, T., Jelicic, M., & Candel, I. (2007). Traumatic memories of war veterans: Not so special after all. *Consciousness and Cognition, 16*, 170–177.

Goff, B. S. N., Crow, J. R., Reisbig, A. M. J., & Hamilton, S. (2007). The impact of individual trauma symptoms of deployed soldiers on relationship satisfaction. *Journal of Family Psychology, 21*(3), 344–353.

Griffiths, P. E. (1997). *What emotions really are*. Chicago: University of Chicago Press.

Hafer, C. L., & Begue, L. (2005). Experimental research on just-world theory: Problems, developments, and future challenges. *Psychological Bulletin, 131*(1), 128–167.

Herman, J. (1992). *Trauma and recovery: The aftermath of violence—from domestic abuse to political terror*. New York: Basic Books.

Hoge, C. W., Auchterlonie, J. L., & Milliken, C. S. (2006). Mental health problems, use of mental health services, and attrition from military services after returning from deployment to Iraq or Afghanistan. *Journal of the American Medical Association, 295*, 1023–1032.

Hoge, C. W., Castro, C. A., Messer, S. C., McGurk, D., Cotting, D. I., & Koffman, R. L. (2004). Combat duty in Iraq and Afghanistan, mental health problems, and barriers to care. *New England Journal of Medicine, 351*(1), 13–22.

Holstein, D., Vollenweider, F. X., Jacnke, L., Schopper, C., & Csomor, P. A. (2010). P50 suppression, prepulse inhibition, and startle reactivity in the same patient cohort suffering from posttraumatic stress disorder. *Journal of Affective Disorders, 126*, 188–197.

Ingleby, D. (2005). Editor's introduction. In D. Ingleby (Ed.), *Forced migration and mental health: Rethinking the care of refugees and displaced persons* (pp. 1–27). Utrecht, The Netherlands: Springer.

Jonas, E., Graupmann, V., Kayser, D. N., Zanna, M., Traut-Mattausch, E., & Frey, D. (2009). Culture, self, and the emergence of reactance: Is there a "universal" freedom? *Journal of Experimental Social Psychology, 45*, 1068–1080.

Kessler, R. C., Sonnege, A. C., Bromet, E., Hughes, M., & Nelson, C. (1995). Posttraumatic stress disorder in the national comorbidity study. *Archives of General Psychiatry, 52*(12), 1048–1060.

Kilpatrick, D. G., Koenen, K. C., Ruggiero, K. J., Acierno, R., Galea, S., Resnick, H. S., et al. (2007). The serotonin transporter genotype and social support and moderation of posttraumatic stress disorder and depression in hurricane-exposed adults. *American Journal of Psychiatry, 164*, 1693–1699.

King, L. A., King, D. W., Fairbank, J. A., Keane, T. M., & Adams, G. A. (1998). Resilience-recovery factors in post-traumatic stress disorder among female and male Vietnam veterans: Hardiness, postwar social support, and additional stressful life events. *Journal of Personality and Social Psychology, 74*(2), 420–434.

Kirmayer, L. J., Lemelson, R., & Barad, M. (Eds.). (2007). *Understanding trauma: Integrating biological, clinical, and cultural perspectives.* Cambridge: Cambridge University Press.

Lomsky-Feder, E. (2004). Life stories, war, and veterans: On the social distribution of memories. *Ethos, 32*(1), 82–109.

Marx, B. P., Brailey, K., Proctor, S. P., Macdonald, H. Z., Graefe, A. C., Amoroso, P., et al. (2009). Association of time since deployment, combat intensity, and posttraumatic stress symptoms with neuropsychological outcomes following Iraq war deployment. *Archives of General Psychiatry, 66*(9), 996–1004.

McEwen, B. S. (2002). The neurobiology and neuroendocrinology of stress: Implications for post-traumatic stress disorder from a basic science perspective. *Psychiatric Clinics of North America, 25*(2), 469–494, ix.

McEwen, B. S., & Sapolsky, R. M. (1995). Stress and cognitive function. *Current Opinion in Neurobiology, 5*(2), 205–216.

McEwen, B. S., & Wingfield, J. C. (2003). The concept of allostasis in biology and biomedicine. *Hormones and Behavior, 43*, 2–15.

McLeland, K. C., Sutton, G. W., & Schumm, W. R. (2008). Marital satisfaction before and after deployments associated with the global war on terror. *Psychological Reports, 103*, 836–844.

Michalowski, R., & Dubisch, J. (2001). *Run for the wall: Remembering Vietnam on a motorcycle pilgrimage.* New Brunswick, NJ: Rutgers University Press.

Moerman, D. (2002). *Meaning, medicine, and the "placebo effect."* Cambridge: Cambridge University Press.

Monson, C., Taft, C. T., & Fredman, S. J. (2009). Military-related PTSD and intimate relationships: From description to theory-driven research and intervention development. *Clinical Psychology Review, 29*, 707–714.

Ng, S. H., Han, S., Mao, L., & Lai, J. C. L. (2010). Dynamic bicultural brains: fMRI study of their flexible neural representation of self and significant others in response to culture primes. *Asian Journal of Social Psychology, 13*, 83–91.

Nichter, M. (1981). Idioms of distress: Alternatives in the expression of psychosocial distress: A case study from south India. *Culture, Medicine and Psychiatry, 5*, 379–408.

Nordstrom, C. (1998). Terror warfare and the medicine of peace. *Medical Anthropology Quarterly, 12*(1), 103–121.

O'Nell, T. (1999). "Coming home" among Northern Plains Vietnam veterans: Psychological transformations in pragmatic perspective. *Ethos, 27*(4), 441–465.

Ozer, E. J., Best, S. R., Lipsey, T. L., & Weiss, D. S. (2003). Predictors of post-traumatic stress disorder and symptoms in adults: A meta-analysis. *Psychological Bulletin, 129*(1), 52–73.

Pederson, D. (2002). Political violence, ethnic conflict, and contemporary wars: Broad implications for health and social well-being. *Social Science & Medicine, 55,* 175–190.

Pivar, I. L., & Field, N. P. (2004). Unresolved grief in combat veterans with PTSD. *Anxiety Disorders, 18,* 745–755.

Raphael, B., & Martinek, N. (1997). Assessing traumatic bereavement and posttraumatic stress disorder. In J. P. Wilson & T. M. Keane (Eds.), Assessing psychological trauma and PTSD (pp. 373–395). New York: Guilford Press.

Renshaw, K. D., Rodrigues, C. S., & Jones, D. H. (2008). Psychological symptoms and marital satisfaction in spouses of Operation Iraqi Freedom veterans: Relationships with spouses' perceptions of veterans' experiences and symptoms. *Journal of Family Psychology, 22*(3), 586–594.

Rockstroh, B., & Elbert, T. (2010). Traces of fear in the neural web—magnetoencephalographic responding to arousing pictorial stimuli. *International Journal of Pathophysiology, 78,* 14–19.

Rosaldo, R. (1993). Introduction: Grief and a headhunter's rage. In *Culture and truth: The remaking of social analysis* (pp. 1–21). Boston: Beacon Press.

Sapolsky, R. M. (1998). *Why zebras don't get ulcers: An updated guide to stress, stress-related disease, and coping.* New York: W.H. Freeman.

Sauro, M. D., Jorgensen, R. S., & Pedlow, C. T. (2003). Stress, glucocorticoids, and memory: A meta-analytic review. *Stress, 6*(4), 235–245.

Scarry, E. (1985). *The body in pain: The making and unmaking of the world.* New York: Oxford University Press.

Seal, K. H., Maguen, S., Cohen, B. E., Gima, K. S., Metzler, T. J., Ren, L., et al. (2010). VA mental health services utilization in Iraq and Afghanistan veterans in the first year of receiving new mental health diagnoses. *Journal of Traumatic Stress, 23*(1), 5–16.

Seeman, T. E., Singer, B. H., Ryff, C. D., Love, G. D., & Levy-Storms, L. (2002). Social relationships, gender, and allostatic load across two age cohorts. *Psychosomatic Medicine, 64,* 395–406.

Shors, T. J. (2006). Stressful experience and learning across the lifespan. *Annual Review of Psychology, 57,* 55–85.

Silove, D. (2005). From trauma to survival and adaptation: Towards a framework for guiding mental health initiatives in post-conflict societies. In D. Ingleby (Ed.), *Forced migration and mental health: Rethinking the care of refugees and displaced persons* (pp. 29–51). New York: Springer.

Simons, A. (1997). *The company they keep*. New York: Free Press.

Suh, E. M., Diener, E., & Updegraff, J. A. (2008). From culture to priming conditions: Self-construal influences on life satisfaction judgments. *Journal of Cross-Cultural Psychology, 39*(1), 3–15.

Summerfield, D. (1999). A critique of seven assumptions behind psychological trauma programmes in war-affected areas. *Social Science & Medicine, 48*, 1449–1462.

Summerfield, D. (2000). War and mental health: A brief overview. *British Medical Journal, 321*(7255), 232–235.

Summerfield, D. (2001). The invention of post-traumatic stress disorder and the social usefulness of a psychiatric category. *British Medical Journal, 322*, 95–98.

Tick, E. (2005). *War and the soul: Healing our nation's veterans from post-traumatic stress disorder*. Wheaton, IL: Quest Books.

Wells, T. S., LeardMann, C. A., Fortuna, S. O., Smith, B., Smith, T. C., Ryan, M. A. K., et al. (2010). A prospective study of depression following combat deployment in support of the wars in Iraq and Afghanistan. *American Journal of Public Health, 100*, 90–99.

Yehuda, R. (2002). Post-traumatic stress disorder. *New England Journal of Medicine, 346*(2), 108–114.

Yehuda, R., Bell, A., Bierer, L. M., & Schmeidler, J. (2008). Maternal, not paternal, PTSD is related to increased risk for PTSD in offspring of Holocaust survivors. Journal of Psychiatric Research, 42, 1104–1111.

Young, A. (1995). *The harmony of illusions: Inventing post-traumatic stress disorder*. Princeton, NJ: Princeton University Press.

11 Autism as a Case for Neuroanthropology: Delineating the Role of Theory of Mind in Religious Development

Rachel S. Brezis

One of the central concerns of psychological anthropology since its inception has been to determine how individual beings, with particular neuro-psychological foundations, become cultural beings (Sapir, 2002; Hallowell, 1955; Shweder & LeVine, 1984). To address this question, most psychological anthropologists have examined the psychological differences between individuals of different cultures. This chapter will present a different approach—one that looks not just at cultural diversity, but at the neurological diversity underlying every interaction between individuals and culture. Specifically, we will look at the ways in which individuals with autism, equipped with a particular neuropsychological foundation, interpret and interact with one aspect of their cultural surroundings, viz., religious practice and belief.

Autism is a neurodevelopmental condition characterized by impairments in social interaction, language, and communication, and a tendency for routinized behaviors. As a condition that fundamentally disrupts the juncture between self and others, or self and culture, autism provides a unique lens onto the process of acculturation. Just as Jakobson (1971) used the study of linguistic aphasia[1] to advance our understanding of normal linguistic functioning, autism may serve as a lens through which to view a certain "cultural aphasia." To use a different metaphor, because of the unique pattern of abilities and disabilities associated with autism, the condition may be seen as a refractive prism through which the white light of culture is broken down into its myriad color components. While persons with autism are exposed to the same cultural input as us all, they may appropriate certain aspects of culture more intensely than others, giving them new meanings and uses, thus helping us to discern the various components of acculturation that are normally orchestrated so smoothly we cannot tease them apart.

Of all aspects of culture, the degree of religious understanding among autistic persons provides an especially pertinent case study for acculturation, as several theories of religious development argue that religious belief emerges from a foundation of interpersonal understanding (Fowler, 1981; Hay & Nye, 2006). More specifically, Bering (2002) suggests that our ability to understand others' thoughts and intentions (theory of mind) lies at the basis of our ability to understand intentionality writ-large; and that lacking a theory of mind, autistic individuals would express an impersonal, mechanistic understanding of the universe.

Seeking the psychological foundations for religion, most researchers have compared the variety of religious practices across different religions, among experts or full lay practitioners (James, 2004; Whitehouse & McCauley, 2005). Instead, the present chapter aims to delineate the extent of religious belief in a population where researchers have predicted its absence. Given autistic persons' difficulty in inferring others' thoughts, would they be capable of conceiving of the world as directed by a spiritual agent? To answer this question, I conducted in-depth interviews and participant-observations in Bar Mitzvah ceremonies of Jewish children and adolescents with high-functioning autism in Israel. In brief, my respondents largely contradicted Bering's predictions, demonstrating belief in an agentive God who gives meaning to events in the world. These surprising results led me to a revaluation of Bering's hypothesis, and, drawing from emerging neuropsychological studies of self-understanding in autism, suggested a renewed focus on how deficits in self-understanding in autism are conducive to religious development.

Thus, rather than impeding autistic persons from accessing certain cultural elements, their neuropsychological constraints pushed some of them to appropriate and adapt those cultural elements to further their own self-development. As a model for future neuroanthropological research on the neural bases of acculturation, the present research demonstrates the ways in which an ethnographic study of autism can provide a test for theories of religious development, illuminate the study of autism from new perspectives, and at the same time generate new hypotheses regarding the neuropsychological foundations of culture.

What is Autism? Outline of a Prism for Neuroanthropology

Autism is a neurodevelopmental condition characterized by a spectrum of impairments; autistic individuals demonstrate a wide range of functioning, ranging from profoundly retarded to highly intelligent. The diagnostic criteria for autism include: (1) impairment in social interaction; (2) impairment in language and communication; and (3) a tendency for routinized, stereotyped behavior and interests, and a related lack of varied, spontaneous make-believe play (American Psychiatric Association, 2000). My research focused on individuals with high-functioning autism and Asperger's syndrome, who have spared language abilities and normal to above-normal intelligence, alongside their social and communicative difficulties (Kasari & Rotheram-Fuller, 2005).

While the etiology of autism is disputed, twin and family studies in the past thirty years have confirmed that autism is highly heritable (with estimates as high as 64%; see Abrahams & Geschwind, 2008, for a review), though the precise genetic basis for these heterogeneous conditions has yet to be determined (Bauman & Kemper, 2005). On a neurobiological level, converging evidence suggests that autism involves abnormalities in the cerebellum, the superior temporal sulcus, the medial temporal lobe (including the amygdala), and the frontal lobe (Penn, 2006). Though many attempts have been made to trace the complex behavioral manifestations of autism to particular brain regions, recent anatomical and functional imaging studies are pointing to an overarching deficit in brain connectivity (Muller, 2007), meaning that the behavioral deficits in autism may not emerge from abnormalities in particular brain regions, but rather, from decreased connectivity between them.

On a psychological level, the great diversity of abilities and disabilities which characterize the autism-spectrum conditions has been tied by researchers to several different core, underlying processes. One of the dominant paradigms in autism research claims that the core impairment in autism is in theory of mind, or the ability to infer others' thoughts and intentions (Baron-Cohen, Tager-Flusberg, & Cohen, 2000). Without theory of mind, autistic persons may lack the ability to detect emotion in others, engage in joint attention, and further develop language and other social skills. Bering (2002) relies on this paradigm to predict that persons with autism would not search for an intentional agent to give meaning to events in the world.

Another explanatory core for autism is a cognitive style that prioritizes details over gestalt perception, resulting in "weak central coherence," or an inability to organize details into higher-level patterns or structures (Happe & Frith, 2006). This deficit may make autistic persons' perception of themselves and the world incoherent and fragmented. Further, autism has also been framed as an overarching executive function disorder, resulting in compromised problem-solving abilities and difficulty generating new and flexible responses without external scaffolding (Russell, 1997). Both of these deficits may hinder individuals with autism from formulating and expressing their desires in complex social and spiritual situations. Finally, autism has been conceived as a pervasive difficulty in engaging with pretend play and symbolic, non-literal communications (Harris & Leevers, 2000), which may in turn hinder persons with autism from imagining meanings and beings beyond the here and now.

An emerging theory of autism links many of these existing paradigms to a central deficit in self-understanding. Just as persons with autism may be impaired in understanding others' thoughts and feelings, they may have a difficulty in conceiving of their own mental states (Frith & Happe, 1999; Hobson, Chidambi, Lee, & Meyer, 2006). In line with the underconnectivity hypothesis of autism, recent studies of functional connectivity in autism (Cherkassky, Kana, Keller, & Just, 2006; Kennedy & Courchesne, 2008) suggest that autistic persons may not properly activate the network of regions associated with self-reflection (viz., the "default network" encompassing the medial prefrontal cortex, posterior cingulate, precuneus, and angular gyrus). As we shall see below, such deficits in self-understanding may lead autistic persons to use cultural scripts creatively, relying on their narrative structures as scaffolds for self-development.

While these theories provide important insights into the core deficits of autism, by studying them in isolation we may lose sight of the autistic person's experience. Adding to the growing group of ethnographic studies of autism (Ochs, Kremer-Sadlik, Sirota, & Solomon, 2004; Nickrenz, 2007; Sirota, 2010), the present approach aims to provide a better understanding of the ways in which autistic persons orchestrate their unique set of strengths and weaknesses to provide meaning to themselves in the world.

Religious Development and Theory of Mind: Predictions for Autism

What happens when individuals with a neurodevelopmental disorder such as autism are raised in a religious setting? One of the recurring themes in theories of religious development is the idea that an individual's relationship to the divine is modeled upon social relationships (Fowler, 1981; Boyer, 1994; Bering, 2002; Hay & Nye, 2006). Yet in autism, both the ability to relate to others, and consequently also the ability to fully appropriate cultural scripts, may be disrupted; thus, at both these levels, the emergence of religious belief could be impeded.

Bering (2002) proposes that our ability to infer others' thoughts and intentions (theory of mind) served as the evolutionary basis for our automatic search for meaning and agency behind events in the world (existential theory of mind). Just as we search behind words, gestures, and facial expressions to uncover the speaker's intent, we are prone to search for meaning behind certain life experiences (for instance, "I was in a bad car accident *because* I needed to learn that my life is fragile"; Bering, 2002, p. 4). Existential theory of mind is thus a certain kind of preparedness that leads us to represent some nondescript agency as the cause of experiences. Importantly, the nature of this sense of agency is shaped by cultural convention, and any religious or nonreligious entity may be cast in its frame.

Given the particular difficulty that persons with autism have inferring others' thoughts and intentions, alongside their relative talent for physical inference, Bering predicts that their view of the world would be patently mechanistic, focusing on *how* things work, not *why* they do so. Bering quotes several autobiographical reports by persons with high-functioning autism or Asperger's syndrome in support of his prediction, such as the following quotation from Temple Grandin, in which God appears more as a physical principle than a complex psychological agent:

I came to the conclusion that God was an ordering force that was in everything. ... In nature, particles are entangled with millions of other particles, all interacting with each other. One could speculate that entanglement of these particles could cause a kind of consciousness for the universe. This is my current concept of God. (Grandin, 1995, as cited in Bering, 2002, p. 14)

Similarly, Edgar Schneider reports: "The only thing that has deeply moved me [about religion] is the *reasonableness* of it all" (Schneider, 1999, as cited in Bering, 2002, pp. 14–15). Similarly, a man with Asperger's syndrome writing on an Internet bulletin board reported that he is "conscious of no feedback from the divine" (Bering, 2002, p. 15).

For Bering, these examples highlight the lack of a sense of deep interpersonal relation between the worshipper and God, or a sense of emotional dependency on an intentional agent who has control over the existence and experiences of the individual. In these quotations, autistic persons' view of the supernatural is mechanistic and impersonal. Bering goes on to predict that persons with autism would not automatically engage in meaning-making when confronted with life-altering events. They would attempt to engage in ritualistic activity (e.g., prayer), but would be unable to read any symbolic device through which a supernatural agent would "respond." Most persons with autism would show little interest in spiritual matters, and those who do, would do so primarily to learn how to engage in acceptable behavior within their community.

Bering's predictions imply a complex set of characteristics, which may or may not appear in concert among different individuals. In order to test Bering's thesis against my data, I classified this complex picture into a list of variables, designating different domains of religious experience according to which specific individuals might be categorized. These aspects of religious belief include (1) behavioral adherence to religious rules; (2) a sense of an agentive God giving meaning to events in the world; and (3) a deep interpersonal relationship with God manifest in the ability to pray and "read" events in the world as responses to the prayer. Importantly, I do not intend to place any inherent valuation on the presence or absence of belief, but rather to observe the varieties of religious beliefs reflected through the autistic "prism".

According to Bering's predictions, a person with autism may have ritualistic behavior, but would lack the affective, representational, and interpersonal aspects of the existential search for meaning, or the religious belief in God. Going beyond Bering's anecdotal predictions, my research examined first-person accounts of religious belief in autism in order to test the hypothesized relation between interpersonal and religious understanding. Before proceeding to the results of the study, let me briefly describe the Jewish setting in which the respondents were raised.

The Setting: Why Judaism?

Judaism provides a unique setting to explore the religious development of children with autism due to its special emphasis on the behavioral performance of 613 biblical commandments and their derivations, alongside its lack of an explicit credo. As such, Judaism may be especially appealing to persons with autism, who seek structure and routine in their lives and struggle with more abstract theological and spiritual content. According to Jewish tradition, though spiritual intent (*kavana*) is encouraged, most commandments may be rewarded solely on the merit of their performance. Thus, Jewish-Israeli children and adolescents who choose to focus mostly on behavioral, rather than spiritual, aspects of their religious identity may nevertheless do so in a culturally legitimate manner.

Those who choose to believe in the spiritual realm may use several available Hebrew terms to frame their belief (though these terms do not appear in the respondents' quotations below, they were implicit in our conversations). Bering's notion of a higher agent directing events in the world can best be translated as *kavana elyona*, literally, "Higher intent." Another term used by some of my respondents is *hashgaha pratit*, or "personal providence," i.e., the idea that God is watching over each particular creature It has created. Different from the Anglo-Christian terms *fate* and *determinism, hashgaha pratit* and *kavana elyona* may hold either positive or negative connotations, and may allow for a variety of responses in the face of one's destiny, as my informants' responses illustrated.

In line with its emphasis on behavioral over spiritual performance, Jewish law treats children with mental disabilities, those with autism among them, according to their behavioral ability. Low-functioning, non-verbal children with autism are labeled with the Talmudic term *shoteh* ("imbecile"), and may be barred from reading the Torah to the community during a Bar Mitzvah ceremony (Amial, 2007). Certain rabbis, however, base inclusion on a case-by-case evaluation and may allow higher-functioning children (including all of my informants) to perform a modified Bar Mitzvah ceremony (Merrick, Gabbay, & Lifshitz, 2001).

Previous anthropological studies of autism and Judaism in Israel have focused on the perceptions of autistic children in the ultra-orthodox community as mediators to the spiritual world (Bilu & Goodman, 1997) or as the parents' spiritual trial (Shaked & Bilu, 2006).[2] Importantly, my study moves beyond the perceptions of autism in religious communities and toward a person-centered ethnography of the religious experience of the autistic individuals themselves. Ultimately, by speaking with the children, I hope to recognize and privilege their thoughts and feelings, and further provide them with a key to self-understanding that goes beyond objective-level accounts of their development.[3]

Methods: Participant Characteristics and Interview Procedures

Participants in the study included sixteen autistic individuals (four girls and twelve boys) aged between 9 and 26 (mean = 15.3; SD = 5.34). Participants were recruited through local support groups for parents of autistic children in and around Jerusalem. All participants were Jewish, and their degree of practice ranged from ultra-orthodox to secular. All participants had previously received a diagnosis of Asperger's syndrome or high-functioning autism from a clinical psychologist or a psychiatrist. Nevertheless, their level of functioning varied greatly. Some had severe speech impairments and could only communicate in writing; others had more subtle pragmatic impairments, noticeable in their difficulty in maintaining a common topic or making eye contact during conversation. The majority of the interviews were conducted in the subjects' homes, with or without the parents present; three of the adult subjects came to my home independently. Two subjects were excluded from the analyses as their responses were too incoherent or too minimal to be coded; one subject was excluded due to equipment malfunction.

The interviews were conducted in Hebrew and lasted between one and one and a half hours. They began with several background questions and simple descriptive questions regarding the respondents' and their parents' degree of religious practice (keeping kosher, Shabbat, holidays). They then proceeded to trace their religious education, in an attempt to understand whether their religious practices and beliefs had changed over the years, and whether the change was driven by personal choice. The interviews then proceeded to more personal questions of belief and the nature of their relationship with God as manifest, for instance, in their personal prayers and their perceived effect, and in the young people's understanding of personal destiny and fate. Most of the interviews proceeded smoothly as subjects were very eager to speak and readily shared their private thoughts, beliefs, and doubts, though some had to write out their responses due to communicative difficulties. Feedback from several parents and caretakers confirmed that I had been able to gain their child's trust and elicit from them sincere responses. The interviews were transcribed in full and coded thematically based on Bering's domains of religious experience.

Varieties of Religious Experience: A Critique of Bering's Existential Theory of Mind

The first observation found in the interviews was the overwhelming diversity and richness of responses I received, going beyond participants' individual backgrounds to unique idiosyncratic views of religious practice. I will begin by a discussion of the subjects' responses according to the criteria advanced by Bering and then offer possible explanations of the results.

Religious Adherence

As predicted by Bering, all of my informants who were raised in religious environments adhered to the religious behaviors practiced by their families and communities, which may fit with their attraction to structure and routine. Furthermore, none of my respondents had become (or planned to become) more zealous in their religious practice compared to those in their religious environments; nor did those who were exposed only to religious settings reject the behavioral performance of commandments. Thus, their diagnosis of autism did not seem to affect the behavioral aspect of their religious participation, as Bering predicts. At the same time, however, I found a variety of interpretations given to these habitual acts. Some of the respondents viewed their behavioral acts as the ultimate performance of their Jewish identity and their sole vehicle of connection with God; while for others, religious acts served as a basis for further spiritual meaning-making.

Sense of Agentive God

According to Bering's predictions, given their difficulty with theory of mind, autistic persons would not read events in the world as directed by an intentionality writ large, adopting instead an impersonal mechanistic interpretation. Yet, of the thirteen interviews I analyzed closely, the majority of respondents—eight—expressed an agentive understanding of God (age and gender did not affect the distribution of responses). The remaining five expressed neither an explicitly agentive framework nor a deep mechanistic understanding, choosing instead to respond negatively with "I don't know," with silence, or with a simple change of topic. While these responses serve as certain glimpses of the participants' views, and may index a confusion or active avoidance of the topic, they cannot be equated with the explicitly impersonal and mechanistic views ascribed by Bering to Grandin and Schneider.

An example of agentive understanding was given by Moriah (pseudonym), an 18-year-old girl from a religious home, who describes her belief in God as the causal agent for her life experiences:

I believe there is a God in the world. There is no chance occurrence in the world. It can't be that I randomly appeared here, in this interview, or that I randomly went to school where I did, or grew up where I did. ... It's hard to think that everything is man's deed. Clearly there is something beyond our understanding that causes these things to be. I may have watered the tree, but it doesn't mean that thanks to that it grew; I could've watered it and it would've wilted.

Moriah's strongly theistic approach is indicative of a belief in *kavana elyona*, the idea that various events in her life were predetermined by God, including the current interview session. Yet despite the rigid predetermination this excerpt implies, she does not face her fate with resignation. Instead, she frames her personal will within the cosmological structure, as the following excerpt suggests:

I believe I'm part of a greater structure. I'm not the machine; I'm a small screw in the machine. ... We are all, in the end, part of a greater process that's been going on for 2000 years: the construction of the Temple, God willing. And, in order for Him to come, I need to ... act in the optimal way. ... The best things to do right

now are the commandments: to study Torah, to guard my tongue, not to gossip. ...
We perform commandments and He speeds the coming of the Messiah.

Interestingly, though Moriah uses a mechanical metaphor to explain her understanding of the world, rather than being impersonal and random, Moriah's machine is imbued with agency both in her part as a "small screw" performing the commandments and in God's response of hastening redemption.

These views depart significantly from Bering's predictions, suggesting that some autistic persons may have an agentive view of God. These beliefs may be common among the respondents' age group and background (and future research should determine their extent). Nonetheless, their very presence, where their absence was predicted, is striking.

Still, a possible interpretation of these results might be that the respondents are simply repeating the cultural scripts available to them. While Moriah's choice of scripts is telling, it may not reflect the spontaneous religious meaning-making which Bering postulates as being dependent on theory of mind. I now turn to examples of respondents who used their cultural structures to give novel meaning to their lived experiences and who created unique interpersonal relationships with God.

Interpersonal Relationship with God
Of the thirteen interviews I have analyzed, two respondents went beyond simple repetitions of existing religious scripts to "read" God into events in their lives. Twenty-one-year-old Amichai, who was raised in a variety of religious settings, brought God into his life by foregoing his financial responsibility and trusting God to intervene:

RSB: So do you feel the presence of God?
Amichai: Yeah, sure.
RSB: How do you feel it? Have you ever asked God for something?
Amichai: Yes, for instance I just bought a bike that cost me 5000 shekels [approximately $1200] and suddenly I realized I had no money in my bank account. So I said to myself: if it comes it comes, and if not, we'll figure it out. As soon as I spoke, within a week, after several months of being unemployed, I got two jobs offers [and earned 6000 shekels]. ... This shows God's presence in the world.

In the context of Jewish-Hassidic tradition, Amichai's lack of action and resignation can be interpreted as a way of exhibiting his trust in God, which is then rewarded by His speedy and accurate response. Going beyond a passive repetition of cultural scripts, Amichai frames his personal dilemma and inner speech as perceived by God, and is thus prepared to view the resolution of his crisis as His answer. Thus, Amichai uses the religious frame to give meaning to his moments of crisis and their resolutions.

Similarly, Yoav, a 16-year-old boy who grew up in a traditional home and has recently followed his uncle in becoming more religious, brings God into his life by negotiating a particular agreement:

RSB: So do you wear *tefilin* on weekdays? [*Tefilin* are phylacteries—a ritual object which observant men are required to wear during weekday prayers.]
Yoav: No. The truth is ... I have an agreement with God that if he fulfills my wishes, then I put them on.

In the context of a tradition in which the performance of basic commandments, such as putting on *tefilin*, is seldom questioned, Yoav's decision to condition his own acts on God's responses is striking. Indeed, implicit in his agreement is the expectation that God communicates with him on a daily basis, allowing Yoav to decide whether or not he should put on his *tefilin*. Yoav clearly demonstrates an interpersonal relationship with God, a feeling not only that an agent directs events in the world, but that human beings have a say in this relationship. In both of these examples, the spiritual realm does not operate simply in a distant, mechanistic fashion; the supernatural permeates everyday life, interacting with Amichai and Yoav's thoughts and desires. Such views are indeed remarkable in the context of Amichai and Yoav's Jewish background, which does not mandate a belief in the spiritual effects of prayer, and allows for a religious practice based only on the performance of commandments if they so choose.

These findings are difficult to reconcile with the social difficulties alleged in autism, which, according to several theories of religious development (Fowler, 1981; Bering, 2002; Hay & Nye, 2006), would preclude persons with autism from expressing a belief in God as an agent directing events in the world. The first possible explanation, suggested above, is that the majority of respondents may have been repeating religious scripts which they had heard, and thus their religious beliefs can be attributed more to their cultural background or age than to their autistic way of thought. Indeed, the same critique may be voiced in relation to Schneider and Grandin, whose views may have been shaped more by their cultural surroundings—academic physics and engineering—than by their clinical condition of autism. Nonetheless, the presence of agentive, and even interactive, views of God, especially in a Jewish context that emphasizes behavioral practice over belief, warrants further inquiry.

The second possible explanation, which the study was not designed to refute, is that the group of children who exhibited agentive views of God was part of the minority of autistic children who can pass theory of mind tests (Ozonoff, Rogers, & Pennington, 1991). Having determined in this case study that Jewish autistic children can in fact hold agentive views of God, future studies should continue to investigate the necessity of theory of mind for religious development by incorporating theory of mind tests into studies of religious development in both autistic and typically developing populations across different religions.

My current data cannot conclusively refute the possibility that religious belief arises from interpersonal understanding. Yet the findings may suggest that there is also an alternative route to religious belief, which some persons with autism may use in order to develop beliefs such as the ones manifest in my interviews. Importantly, the use of such a route need not be limited to persons with autism, but may provide an account of a parallel route to religious development employed also by typically developing children.

In the following section I suggest, based on further evidence from my interviews, that such a parallel route to religious development can be achieved through narrative appropriation. Moreover, based on recent evidence from autism research, I suggest a shift of focus from the difficulty persons with autism have in understanding others, to their more primary difficulty in understanding *themselves*. If indeed one of the deficits in autism is a difficulty in stringing autobiographical events into a coherent sense of self, we could explain their religious beliefs as a coping mechanism that provides structure and meaning to their past and future selves. More broadly, despite their impediments in accessing cultural scripts (due to their interpersonal difficulties), autistic persons would nevertheless be more inclined to incorporate such scripts into their sense of self to compensate for their delayed self-development. This third, alternative, explanation is a tentative sketch for religious development in autism, and its importance lies in providing new, testable hypotheses for the phenomenon of religious acculturation.

Religious Narratives and Self-Understanding in Autism

The theoretical insight for my alternative explanation of religious development in autism emerged from a story told to me by Nethanel, an ultra-orthodox 10-year-old boy. Rather than providing a simple answer to my question, "Will you continue being religious when you grow up?" Nethanel began recounting an elaborate story that took about ten minutes to unfold and at first seemed completely irrelevant. In brief, the protagonist was a boy who had been cursed before his birth: on the day of his Bar Mitzvah, he would become Christian. His adoptive father (who, through a complex set of events, happens to be the Hassidic rabbi who issued the original curse) unknowingly sets the boy's engagement and plans the boy's wedding in conjunction with the day of his Bar Mitzvah. When the rabbi discovers the horrible turn of fate, a great spiritual combat ensues, and at the last moment, thanks to the rabbi's fasts and prayers, the child is saved from the "impure force" which has invaded his body. Nethanel did not answer my question directly, nor did he refocus his narrative when I interjected to ask about the relevance of the story to my question. Yet as the subsequent conversation between us clarified, Nethanel intentionally used this Hassidic story to convey to me that he does not know whether he will stay religious when he grows up, as *some* external force might determine the course of his life before his thirteenth birthday.

Nethanel's story expressed a strong sense of *kavana elyona* (Higher intent) and a weak sense of agency both in content and in form. First, similar to the agentive views of God presented in the previous section, the story expresses a certain degree of resignation in face of "external forces," relegating even his own belief to external factors. Moreover, this passivity is reflected in the very *structure* of his discussion: using a story he has heard in order to reflect on himself and refusing to answer my direct questions, rather than engaging in simple introspection and future projection to think about whether he would like to stay religious. Expanding on Nethanel's use of narrative, and reviewing the emergent literature on delayed self-development in autism, I suggest that autistic individuals may use cultural narratives to compensate for their weak sense of self. By appropriating religious narratives as proxies for their personal identity, they may thus situate themselves within a particular cultural landscape and at the same time articulate their own unique path.

Several studies of narrative ability in high-functioning autism have noted that while autistic individuals are capable of repeating existing narratives in great detail, they have difficulty composing coherent autobiographical narratives (Happe, 1991; Bruner & Feldman, 1993; Losh & Capps, 2003). These experimental data were further echoed in Solomon's (2004) naturalistic ethnographic study, which found that while autistic children were just as likely to launch narratives into conversation as their non-autistic interlocutors, their narratives were more likely to be fictional and preorganized, compared to the personal narratives their non-autistic interlocutors introduced. Rather than reflecting a general impairment in narrative ability, these findings may be readily explained by recent findings on memory abilities and self-understanding in autism.

Recent research on memory in autism has consistently found a discrepancy between autistic individuals' poor autobiographical memory relative to their normal (or above-normal) memory for semantic facts and given narratives (Boucher & Bowler, 2008; Lind, 2010). These data further add to accumulating research on the development of the self in autism, suggesting that individuals with autism may have difficulty encoding self-experienced events and tying them to a coherent sense of self (Hobson et al., 2006; Lind & Bowler, 2008). On a neurobiological level, autistic persons may have difficulty activating several midline structures of the brain during resting states, which in typical individuals are associated with routine self-reflection, memory, and future planning (Cherkassky et al., 2006; Kennedy & Courchesne, 2008; for further neuropsychological studies of the self in autism, see Uddin, Davies, Scott, Zaidel et al., 2008; Lombardo, Chakrabarti, Bullmore, Sadek et al., 2010). If autistic persons do not spontaneously turn to self-reflection and social reflection during resting states, this may impede their ability to incorporate experiences onto a coherent sense of self in relation to others (Iacoboni, 2006).

Autistic persons' difficulties with personal memory and self-understanding may also be reflected in their reduced sense of agency: without a coherent sense of past and present selves, they may have difficulty planning and projecting on to the future. Together, these studies are creating a shift in focus within autism research from the development of social understanding and theory of mind to the perhaps more basic development of the self. Yet how does this emerging paradigm of autism fit together with research on the religious lives of autistic individuals?

Though I did not originally set out to examine how religious beliefs might be influenced by a weak sense of self, a rereading of my transcripts confirmed this unique pattern of findings on self-understanding and narrative abilities in autism. On the one hand, many of my respondents exhibited limited autobiographical recall and a weak sense of agency. For example, one mother told me of how her 16-year-old boy was unable to choose what to eat for dinner and would roll dice every evening to decide. At the same time, many respondents had a strong attraction to prestructured narratives such as Harry Potter, The Transformers, Dungeons and Dragons, and Hassidic tales, depending on their cultural background.

Importantly, if we were to examine these autistic abilities in isolation, we might come to the conclusion that persons with autism are capable of repeating prestructured narratives, but that they are incapable of using them flexibly and creatively as scaffolds or models for self-construction (Bruner & Feldman, 1993). However, ethnographic studies provide evidence to the contrary. In their work on identity construction in autism, Sirota (2010) and Nickrenz (2007) demonstrate the ways in which personal interests in finance and dinosaurs or narratives shared at dinnertime conversations serve as frames for the construction of identity. Together with my own findings, they point to ways in which autistic individuals may use their talent (and passion) for repeating prestructured narratives to creatively construct their sense of self in relation to the world.[4]

While for most of my respondents the fictional and personal realms remained independent, some respondents chose to weave the two realms together, as we find in Nethanel's use of a Hassidic story to reflect on his own religious identity. A further example of interplay between personal and fictional narratives was presented to me by Nir, a 16-year-old secular boy, who made creative use of his immersion in Dungeons and Dragons (D&D) to express his sense of agency in the world:

Nir: I'm an avid D&D player. I mostly ... control games. I am the Dungeon Master. ... This whole game is like playing within a book and writing it at the same time.

RSB: Interesting. So you feel that this gives you control?

Nir: Yes, it gives me control over the story, and to the players it gives control over their future fate. ... That is, I can invent towns, and I can invent plains, and I can invent monsters, and I can invent professions, and I can invent anything I want! ... Gods? The Dungeon Master plays them!

Rather than treating the game of D&D as a predetermined narrative and reacting to it passively, Nir uses his knowledge of the game to exert his agency. He uses different tactics to convey his control over other players and even reenacts gods. Akin to Nethanel's use of religious narratives to express his thoughts about his personal future, Nir uses the cultural scripts available to him to construct his personal identity by playing out different agentive beings. Though their approach to the spiritual realm differs markedly—Nethanel expressing passivity and resignation before God and Nir expressing his God-like power over others—their means of expression ultimately serves the same function. Rather than speaking separately about their passions for fiction and their personal experiences and goals, the two become merged in their natural discourse, scaffolding their personal identity.

At the same time, we cannot tell whether Nethanel and Nir have achieved a degree of flexibility and fluency in their personal storytelling expected of their developmental stage (Miller, Hengst, Alexander, & Sperry, et al., 2000). One therapeutic application of this neuroanthropological research may be that caretakers can help autistic persons by co-narrating fictional narratives with them, pointing to ways in which their lives parallel fictional characters' lives, and thus helping autistic individuals to gain a better understanding of themselves.

In sum, the interviews point to possible ways in which cultural scripts served children with autism as bootstraps for self-construction and acculturation. Rather than maintaining the domains of fictional and personal knowledge separate, as experimental studies of memory and narrative ability in autism may suggest, some autistic individuals can harness their immersion in fictional and religious narratives to help shape their sense of self. Though I did not directly examine the degree of self-understanding and agency among my participants, by ascribing agency to God and appropriating religious narratives, my respondents may have compensated for a weak sense of self-understanding and personal agency. Thus, children who grew up in religious settings may cast their interpretations in religious narratives, providing them an alternative route to religious understanding which does not rely solely on interpersonal inferences; and those who grew up in a secular setting may use the narratives accessible to them to construct visions which extend beyond reality, giving transcendent meaning to events in the "here and now."

Conclusion

The present neuroanthropological study examined the ways in which persons with autism, equipped with a unique neuropsychological foundation, interact with their cultural surroundings. Specifically, given their difficulty inferring intentionality in others, Bering (2002) predicted that autistic persons would fail to develop agentive views of God. A series of in-depth interviews with Jewish children and adolescents with high-functioning autism in Israel revealed instead that autistic persons can hold an agentive view of God as directing events in the world, and a minority can even engage in a more personal exchange with God. Together, these findings point to the necessary existence of alternative routes to religious development. Drawing from further interview data and a review of new findings on impaired self-development in autism, I propose that rather than being incapable of developing a full religious understanding due to their neuropsychological limitations, some autistic persons may in fact be attracted to religious narratives as scaffolds for self-development.

What was most striking about the variety of religious responses exhibited by the Jewish autistic children in this study was not their *lack* of cultural competence, but rather, the degree of cultural adaptations they were capable of making given their neuropsychological limitations. Rather than remaining passive or apathetic in face of the religious materials they were exposed to, some of them actively interacted with those materials, struggling to interpret and live through the scripts they had heard, and even harnessing them to further their own self-growth.

On a broader level, these findings point the remarkable potential of the human brain to adapt and relate to cultural materials. To return to the analogy of autism as a "cultural aphasia," just as deaf children who have never been exposed to language may develop their own idiosyncratic sign language (Senghas & Coppola, 2001), and brain-injured children may make use of non-linguistic regions of the brain to acquire language (Feldman, 2005), in the same way, and despite their neuropsychological limitations, autistic children may use any of their available neuropsychological resources to satisfy their propensity for cultural learning. To return to the analogy of the prism, rather than blocking certain elements of culture from being exhibited, the neuropsychological constraints of an autistic mind may push the autistic person to use their mental and cultural resources creatively, coloring the cultural scripts that they appropriate with new hues. It remains to be determined how and why certain autistic persons succeed in appropriating cultural scripts to further their own development; and perhaps by examining this process we may find new therapeutic ways of bringing out the cultural potential in children and adolescents with autism who do not yet exhibit it.

This study has important implications for the two domains that inform it—religious development and autism—while at the same time setting the stage for further neuroanthropological inquiries. First, through an observation of the religious development of children and adolescents with autism, this study contributes to the ongoing debate regarding the neuropsychological foundations of religious belief, and in particular, the role of theory of mind (Whitehouse & McCauley, 2005). Though my findings cannot refute the role of theory of mind in typical religious development, the fact that religious belief may exist among autistic persons, even in a cultural context that would allow otherwise, suggests that there are potential compensations for lack of theory of mind. Further studies should continue to explore the relationship between theory of mind, self-understanding, and narrative ability in children's religious development. By measuring and comparing these different variables in both typically developing and autistic children, in different religious settings, we may determine the relative contribution of each to their religious development, and further explain the vast individual differences in patterns of cultural appropriation.

Second, while most studies of autism are experimental in nature, breaking down the complex profile of autistic persons into discrete abilities and disabilities, such an approach inherently prevents us from examining the ways these skills may (or may not) become integrated. The current study contributes to the growing corpus of anthropological studies of autism (Ochs et al., 2004), providing a more naturalistic account of the ways in which autistic persons orchestrate, rather than isolate, their various strengths and weaknesses. More importantly, while most psychological and anthropological studies of autism focus on individuals' negotiations with the "here and now," the present research demonstrates how expanding our lens of research onto autistic children's relation with the spiritual realm may in fact illuminate ongoing debates regarding the relative importance of theory of mind and self-development as core deficits of autism.

Finally, by observing the interaction between the mental architecture of the autistic way of thought and the Jewish environment in which it is immersed, the current research sets the stage for the expansion of psychological anthropology into neuroanthropology: from an exploration of personhood across cultures (Hallowell, 1955; Shweder & LeVine 1984), to an exploration of acculturation across different neurological capacities. Drawing on the established tradition of cultural psychiatry (Devereux, 1970; Kirmayer, 1989; Kleinman, 1988), and complementing the burgeoning field of cultural neuroscience (Chiao, 2009; Han & Northoff, 2008; Kitayama & Park, 2010), neuroanthropological research on different clinical or brain-damaged populations would thus provide the necessary testing ground for particular neural-cultural hypotheses.

For instance, some of the same anatomical structures that have been implicated in autism—the amygdala, the medial prefrontal cortex, and the superior temporal sulcus (Amaral, Schumann, & Nordahl, 2008; Pelphrey, Adolphs, & Morris, 2004)—are also found to mediate culturally varied behaviors (Chiao, Adams, Tse, Lowenthal et al., 2008; Kitayama & Park, 2010). Moreover, the underconnectivity hypothesis of autism may likely become a useful model for explaining the neural bases for cultural development (empirically, for instance, we may find that individuals such as Nethanel and Nir, who use cultural scripts creatively to compensate for their reduced sense of self, have greater connectivity in self- and social-networks than other autistic persons).

As we continue to use autism as a case study for acculturation, it is also important that we expand the study of neuroanthropology to other clinical populations. First, it is important to examine the breakdown of cultural abilities at different points in the lifespan: while autism provides a compelling case for early-onset cultural impairment, in order to tease apart abilities from compensations it is important to examine adults who have functioned as fully cultural beings, and are struck by a late-onset disorder such as schizophrenia or dementia to determine which cultural capacities are maintained despite the brain damage, and which are lost.

Second, it is important to complement research on persons with an all-encompassing brain disorder such as autism with research on persons with closely delineated brain damage. Following in the tradition of classic neuropsychology (Harlow, 1848; Broca, 1861; and see Anderson & Phelps, 2001, and Mendez, Lauterbach, Sampson, & ANPA Committee on Research, 2008, for modern-day approaches), neuroanthropologists may shed new light on the limits of cultural abilities exhibited by persons with lesions to the amygdala, the superior temporal sulcus, or the prefrontal cortex. By bringing the well-honed tools of ethnographic research to such clinical populations, neuroanthropologists will thus provide major contributions to the field of neuropsychology, while at the same time generating and testing new hypotheses regarding the neural bases of culture. Together, the triangulation of neurological, clinical, and ethnographic research will pave the road for an interdisciplinary study of neuroanthropology.

Acknowledgments

I would like to thank Daniel Lende and Greg Downey for inviting me to join this unique forum for discussion of neuroanthropology, and for their deep engagement as editors. This work was conducted as part of my graduate studies in the Department of Comparative Human Development at the University of Chicago, and supported by the Gianinno Foundation. I would like to thank Rick Shweder, Micere Keels, Elinor Ochs, Jesse Bering, Harvey Whitehouse, Yoram Bilu, Yehuda Goodman, Ofer Golan, Gila Vogel, Tami Yona, Ayelet Schnur, Yael Lehman, Jeff Helmreich, Kristen Gillespie, and Emmy Goldknopf for their help at various stages of data collection and writing. Finally, I would like to thank the many families and individuals with autism who shared with me their thoughts on God and religion. I hope that I have done justice to their words in my analyses.

Notes

1. A language disorder caused by brain damage.

2. For related views within the Anglo-Christian world, see Stillman (2006) and Isanon (2001); for an in-depth ethnographic study of Muslim-American parental views on autism, see Jegatheesan et al. (2010).

3. At the same time, I am acutely aware that any academic research aiming to "unravel" the autistic way of thought and "give voice to the hidden child behind the fortress," including my own work, may be influenced by the Romanticist idea that meaning can be gained from studies of the mentalities of children with disabilities. It is interesting to trace the historical progression of ideas about mentally disabled persons, and to view our current academic pursuit as another variant of the religious search for "truth" in the mute and retarded (following Foucault, 1961). For comprehensive accounts of the construction of autism within Western psychiatry and other cultures, see Nadesan (2005) and Grinker (2006).

4. Importantly, not all persons with autism may be equally disposed to create such integrated constructions, and further research should determine what individual and situational constraints can be removed to help autistic individuals integrate their strengths and weaknesses.

References

Abrahams, B., & Geschwind, D. (2008). Advances in autism genetics: On the threshold of a new neurobiology. *Nature Reviews. Genetics, 9*, 341–355.

Amaral, D. G., Schumann, C. M., & Nordahl, C. W. (2008). Neuroanatomy of autism. *Trends in Neurosciences, 31*(3), 137–145.

American Psychiatric Association. (2000). *Diagnostic and statistical manual of mental disorders* (4th ed., text rev.). Washington, DC: APA.

Amial, M. (2007). Aliyat Torah le-Autist? [May a Child with Autism Read from the Torah?]. Retrieved from www.moriya.org on April 27, 2007.

Anderson, A. K., & Phelps, E. A. (2001). Lesions of the human amygdala impair enhanced perception of emotionally salient events. *Nature, 411*(6835), 305–309.

Baron-Cohen, S., Tager-Flusberg, H., & Cohen, D. J. (Eds.). (2000). *Understanding other minds: Perspectives from developmental cognitive neuroscience*. Oxford: Oxford University Press.

Bauman, T., & Kemper, M. (2005). *The neurobiology of autism*. Baltimore: Johns Hopkins University Press.

Bering, J. (2002). The existential theory of mind. *Review of General Psychology, 6*(1), 3–24.

Bilu, Y., & Goodman, Y. (1997). What does the soul say? Metaphysical uses of facilitated communication in the Jewish ultraorthodox community. *Ethos, 25*, 375–407.

Boucher, J., & Bowler, D. (2008). *Memory in autism*. Cambridge, MA: Cambridge University Press.

Boyer, P. (1994). *The naturalness of religious ideas*. Berkeley, CA: University of California Press.

Broca, P. (1861). Remarque sur le siège de la faculté du langage articule, suivie d'une observation d'aphémie (perte de la parole) [Remarks on the seat of the faculty of spoken language, followed by an observation of aphemia (loss of speech)]. *Bulletins de la société anatomique de Paris, 36*, 330–356.

Bruner, J., & Feldman, C. (1993). Theories of mind and the problem of autism. In S. Baron-Cohen, H. Tager-Flusberg, & D. J. Cohen (Eds.), *Understanding other minds: Perspectives from autism* (pp. 267–291). Oxford: Oxford University Press.

Cherkassky, V., Kana, R., Keller, T., & Just, M. (2006). Functional connectivity in a baseline resting-state network in autism. *Neuroreport, 17*, 1687–1690.

Chiao, J., Adams, R.B., Jr., Tse, P.U., Lowenthal, W.T., Richeson, J.A., & Ambady, N. (2008). Cultural specificity in amygdala response to fear faces. *Journal of Cognitive Neuroscience, 20*(12), 2167–2174.

Chiao, J. Y. (2009). Cultural neuroscience: A once and future discipline. *Progress in Brain Research, 178*, 287–304.

Devereux, C. (1970). *Essais d'ethnopsychiatrie générale* [Essays on general ethnopsychiatry]. Paris: Gallimard.

Feldman, H. M. (2005). Language learning with an injured brain. *Language Learning and Development, 1*, 265–288.

Foucault, M. (1961). *Histoire de la folie à l'age classique* [Madness and civilization: A history of insanity in the age of reason]. Paris: Plon.

Fowler, J. (1981). *Stages of faith*. New York: Harper & Row.

Frith, U., & Happe, F. (1999). Theory of mind and self-consciousness: What is it like to be autistic? Mind and Language 14, 1–22.

Grinker, R. (2006). *Unstrange minds: Remapping the world of autism*. New York: Basic Books.

Hallowell, A. I. (1955). *Culture and experience*. Philadelphia: University of Pennsylvania Press.

Han, S., & Northoff, G. (2008). Culture-sensitive neural substrates of human cognition: A transcultural neuroimaging approach. *Nature Reviews Neuroscience, 9*, 646–654.

Harlow, J. M. (1848). Passage of an iron rod through the head. *Boston Medical and Surgical Journal, 39*, 389–393.

Happe, F. (1991). The autobiographical writings of three Asperger syndrome adults. In U. Frith (Ed.), *Autism and Asperger syndrome* (pp. 207–242). Cambridge, MA: Cambridge University Press.

Happe, F., & Frith, U. (2006). The weak coherence account. *Journal of Autism and Developmental Disorders, 35*(1), 5–25.

Harris, P. L., & Leevers, H. J. (2000). Pretending, imagery and self-awareness in autism. In S. Baron-Cohen., H. Tager-Flusberg, & D. J. Cohen (Eds.), *Understanding other minds: Perspectives from autism and cognitive neuroscience* (2nd ed., pp.182–202). Oxford: Oxford University Press.

Hay, D., & Nye, R. (2006). *The spirit of the child* (Rev. ed.). London: Jessica Kingsley Publishers.

Hobson, P. R., Chidambi, G., Lee, A., & Meyer, J. (2006). *Foundations for self-awareness: An exploration through autism*. Boston: Blackwell Publishing.

Iacoboni, M. (2006). Failure to deactivate in autism: The co-constitution of self and other. *Trends in Cognitive Sciences, 10*(10), 431–433.

Isanon, A. (2001). *Spirituality and the autism spectrum*. London: Jessica Kingsley Publishers.

James, W. (2004). *The varieties of religious experience*. New York: Barnes and Noble. (Original work published 1902)

Jakobson, R. (1971). *Fundamentals of language*. The Hague, The Netherlands: Mouton.

Jegatheesan, B., Miller, P., & Fowler, S. (2010). Autism from a religious perspective: A study of parental beliefs in South Asian Muslim immigrant families. *Focus on Autism and Other Developmental Disabilities, 25*, 98–109.

Kasari, C., & Rotheram-Fuller, E. (2005). Current trends in psychological research on children with high-functioning autism and Asperger disorder. *Current Opinion in Psychiatry, 18*(5), 497–501.

Kennedy, D., & Courchesne, E. (2008). Functional abnormalities of the default network during self- and other-reflection in autism. *Social Cognitive and Affective Neuroscience, 3*, 177–190.

Kirmayer, L. J. (1989). Cultural variations in the response to psychiatric disorders and emotional distress. *Social Science & Medicine, 29*, 237–239.

Kitayama, S., & Park, J. (2010). Cultural neuroscience of the self: Understanding the social grounding of the brain. *Social Cognitive and Affective Neuroscience, 5*(2–3), 111–129.

Kleinman, A. (1988). *Rethinking psychiatry: From cultural category to personal experience.* New York: Simon and Schuster.

Lind, S. (2010). Memory and the self in autism. *Autism, 14*(2), 1–27.

Lind, S., & Bowler, D. (2008). Episodic memory and autonoetic consciousness in autistic spectrum disorders. In J. Boucher & D. Bowler (Eds.), *Memory in autism* (pp. 166–187). Cambridge: Cambridge University Press.

Lombardo, M., Chakrabarti, B., Bullmore, E., Sadek, S., Pasco, G., Wheelwright, S., et al. (2010). Atypical neural self-representation in autism. *Brain, 133*, 611–624.

Losh, M., & Capps, L. (2003). Narrative ability in high-functioning children with autism or Asperger's syndrome. *Journal of Autism and Developmental Disorders,* 33(3) 239–251.

Mendez, M. F., Lauterbach, E. C., Sampson, S. M., & ANPA Committee on Research. (2008). An evidence-based review of the psychopathology of frontotemporal dementia: A report of the ANPA committee on research. *Journal of Neuropsychiatry and Clinical Neurosciences, 20*(2), 130–149.

Merrick, J., Gabbay, Y., & Lifshitz, H. (2001). Judaism and the person with intellectual disability. *Journal of Religion, Disability & Health, 5*(2–3), 49–63.

Miller, P., Hengst, J., Alexander, K., & Sperry, L. (2000). Versions of personal storytelling/versions of experience: Genres as tools for creating alternate realities. In K. Rosengren, C. Johnson, & P. Harris (Eds.), *Imagining the impossible: Magical, scientific, and religious thinking in children* (pp. 130–156). Cambridge: Cambridge University Press.

Muller, R. (2007). The study of autism as a distributed disorder. *Mental Retardation and Developmental Disabilities Research Reviews, 13*(1), 85–95.

Nadesan, M. (2005). *Constructing autism: Unraveling the "truth" and understanding of the social.* London: Routledge.

Nickrenz, E. (2007, March). *Compelling structures: The "special interest" in the lifetellings of young adults with Asperger's syndrome.* Paper presented at the Society for Psychological Anthropology biennial meeting, Manhattan Beach, CA.

Ochs, E., Kremer-Sadlik, T., Sirota, K., & Solomon, O. (2004). Autism and the social world: An anthropological perspective. *Discourse Studies, 6*(2), 147–183.

Ozonoff, S., Rogers, S., & Pennington, B. (1991). Asperger's syndrome: Evidence of an empirical distinction from high-functioning autism. *Journal of Child Psychology and Psychiatry*, *32*(7), 1107–1122.

Pelphrey, K., Adolphs, R., & Morris, J. P. (2004). Neuroanatomical substrates of social cognition dysfunction in autism. *Mental Retardation and Developmental Disabilities Research Reviews*, *10*(4), 259–271.

Penn, H. E. (2006). Neurobiological correlates of autism: A review of recent research. *Child Neuropsychology*, *12*(1), 57–79.

Russell, J. (Ed.). (1997). *Autism as an executive disorder*. Oxford: Oxford University Press.

Sapir, E. (2002). *The psychology of culture*. J. Irvine (Ed.). Berlin: Mouton de Gruyter.

Senghas, A., & Coppola, M. (2001). Children creating language: How Nicaraguan sign language acquired a spatial grammar. *Psychological Science*, *12*, 323–328.

Shaked, M., & Bilu, Y. (2006). Grappling with affliction: Autism in the Jewish ultra-orthodox community in Israel. *Culture, Medicine and Psychiatry*, *30*, 1–27.

Shweder, R., & LeVine, R. (Eds.). (1984). *Culture theory: Essays on mind, self and emotion*. Cambridge, MA: Cambridge University Press.

Sirota, K. G. (2010). Narratives of distinction: Personal life narrative as a technology of the self in the everyday lives and relational worlds of children with autism spectrum conditions. *Ethos*, *38*(1), 93–115.

Solomon, O. (2004). Narrative introductions: Discourse competence of children with autistic spectrum disorders. *Discourse Studies*, *6*(2), 253–276.

Stillman, W. (2006). *Autism and the God connection*. Naperville, IL: Sourcebooks.

Uddin, L., Davies, M., Scott, A., Zaidel, E., Bookheimer, S., Iacoboni, M., et al. (2008). Neural basis of self and other representation in autism. *PLoS ONE*, *3*(10), e3526.

Whitehouse, H., & McCauley, R. (2005). *Mind and religion: Psychological and cognitive foundations of religiosity*. Walnut Creek, CA: AltaMira Press.

12   Collective Excitement and Lapse in Agency: Fostering an Appetite for Cigarettes

Peter G. Stromberg

Students learn many things at American colleges and universities. Many of them, for instance, learn to smoke cigarettes. In spite of gradually declining smoking rates in the nation as a whole, smoking remains widespread on college campuses, with close to a third of students reporting that they have smoked in the last 30 days (Wetter, Kenford, Welsch, Smith, et al., 2004). And many students transition from occasional to regular tobacco use during their college years (Wetter et al., 2004; Colder, Flay, Segawa, Hedeker et al., 2008; Patterson, Lerman, Kaufmann, Neuner et al., 2004), a fact that is especially pertinent for the processes I will address here.

This situation entails at least a mild irony, in that we assume that the brightest of our young people attend college, and by now the evidence suggests to any clear-headed observer that acquiring a smoking habit is deeply unwise. This mild irony, in turn, reflects something more momentous, a culture-wide confusion about situations in which people feel they do things they have not fully decided to do. This confusion is an area where our deeply rooted conviction that human action is typically freely chosen runs up against the fact that people not infrequently become involved in courses of action that seem to be impelled by something beyond the person's will. This is an example of the "lapse in agency" of my title.

The term "agency" needs to be approached with some delicacy. The word is used in many different ways in contemporary debate, and ultimately touches on a range of political, theoretical, and philosophical issues. For example, agency has been an important concept in social science theory for several decades, and here the term is often tied to discussions about the extent to which individuals and groups are able to resist political domination. In part, the term is intended to atone for earlier social scientific approaches that often stereotyped members of non-Western groups. "For anthropologists and historians," writes Webb Keane (2007, p. 3), "the quest for local agency is often portrayed as an antidote to earlier assumptions about tradition-bound natives and timeless structures or to triumphalist narratives of empire and modernity."

Somewhat more abstractly, the concept of agency has emerged as a central concern in social theory, as sociologists and others have sought to explore how human beings are embedded in social structures. Linguistic anthropologists, to take a single example of this sort of work, have made considerable progress in detailing the ways in which the social apparatus of language facilitates and constrains agency. Such accounts make it clear that agency is typically a communally constructed accomplishment, not under the complete control of any individual.[1]

In this paper I use the term agency in a more basic sense than those I have just referred to. Human beings can execute motor programs through which they pursue desired goals, as when they grasp a piece of fruit or run away from danger (David, Newen, & Vogeley, 2008). In this, they are similar to other animals. But in addition, human beings are self-aware, and are therefore capable of conceptualizing their movements as action, as linked to their projects (see Synofzik, Vosgerau, & Newen, 2008, p. 222). Thus, I use "the sense of agency" in the way that it is usually employed in the cognitive neuroscience literature (following Gallagher, 2000), as referring to the basic awareness that one is author of one's own actions.[2] To say that I am author[3] of my actions is to say that I acknowledge responsibility for them.[4]

The sense of agency has decisive implications for the character of our social interaction. Not only do we experience ourselves as having projects, we are highly skilled at discerning the projects of others. One compelling account of how such "mind-reading" develops is provided by psychologist Michael Tomasello (Tomasello, 1999; Tomasello, Carpenter, Call, Behne, et al., 2005). Humans develop, toward the end of the first year of their lives, an ability often called "joint attention" that sets them decisively apart from nonhuman primates. Joint attention is the capacity to attend simultaneously to a person and whatever it is that person is attending to. Over time, this capacity entails discerning and potentially participating in projects together with others. Tomasello (1999, p. 68ff.) suggests that joint attention develops alongside the child's growing sense of herself as an agent: as she begins to understand herself as an agent, she understands other persons in this way as well. She understands that others are acting in pursuit of goals, and at around nine months of age, she begins to have the ability to grasp the goals of others and join in their projects.

Not everyone will agree with Tomasello's take on joint attention; however, he is but one of many prominent neuroscientists who stress that among the most important features that make our species unique is not that we pursue goals, but rather that we understand that *others* pursue goals. In contrast, as he formulates it, "non-human primates are themselves intentional and causal beings, they just do not understand the world in intentional and causal terms" (Tomasello, 1999, p. 19). Human mind-reading facilitates enormously powerful new forms of social cooperation, and in Tomasello's view is perhaps the most important mental ability that, over time, is responsible for the differences between human and animal ways of life.[5]

Summing up, I have argued that all human beings necessarily have a "sense of agency," an understanding that the actions they initiate and execute are linked to their projects, and that they understand other human beings in the same way. Understanding ourselves and others as agents—as creatures pursuing goals—is not a cultural idiosyncrasy, but rather something essential to human social life. The fact that our activity is embedded in a hierarchy of goals that can be grasped by (in principle) all members of a group is one of the most important features that distinguishes human activity from that of other primates (Taylor, 1985).

Yet we know from numerous ethnographic reports that it is not at all uncommon for people to report lapses in the sense of agency, situations in which their choices seem to be being controlled by something beyond themselves. For example, persons sometimes become possessed by spirits, and do and say things that they do not feel are self-authored. There are many cultural variants of this situation, in which people carry out actions but deny full responsibility for them. In contemporary American society, addiction can be considered an example of a lapse in the sense of agency, and here my goal is to look at one way in which this idea may develop.[6]

Lapses in the sense of agency—like the sense of agency itself—are also reflective of basic aspects of the human mind and modes of social interaction. Human nature reflects our heritage as a hypersocial species. Human beings evolved over millions of years in social groups wherein coordinated responses were a basic survival mechanism. These selective pressures yielded a creature with finely tuned mental equipment for coordinating activity with social partners and groups. Individual humans, and human ancestors, had to be highly sensitive to ripples of aggression or fear, for example; they had to be able to discern and quickly join in developing responses to opportunities and threats. Indeed, according to the explanation I have developed here, agency is itself fundamentally a social strategy, a way to closely integrate individuals into cooperative projects. It is not surprising, then, that people often report lapses in the sense of agency, situations in which they feel controlled by something beyond the self. In fact, our mental and interactive capacities entail that our actions are inextricably intertwined with the communities in which we live.

Human communities make full use both of the sense of agency and of human hypersociality in governing social life. As I have noted, one of the unique advantages of human beings as a species is that we evolved both a capacity for flexibility and a means whereby that flexibility could be captured in collective projects. (Michael Tomasello, 1999, calls this evolutionary dynamic of imitation the "ratchet effect.") But being so flexible and adaptable also has a downside, namely that it's hard to keep humans in line. Social life depends upon close coordination of activity, and thus we would expect that along with increasing flexibility human societies developed increasingly sophisticated means of controlling that flexibility. Among these means of control, I propose, are ideas and practices that create lapses in agency by fostering a subjective sense that at certain times we are directed by forces that are beyond the will of the individual.

Lapses in Agency in American Society

To this point I have introduced the sense of agency, and lapses in that sense, as possibilities that grow out of the basic characteristics of the human mind and human sociality. I turn now to an elaboration of this point in the particular society that frames my case study here. As is often commented, Western societies—and perhaps American society in particular—have struck out along a cultural path that highlights the capacity of individuals to control their actions. That path arguably begins at an ethical and religious system wherein human beings will be judged by God for their choices. Understanding humans to be free agents is, in addition, an important ideological component of Western legal systems and an economic system—free market capitalism—that has dominated these societies for centuries.

The strong individualism embedded in Western culture adds a particular twist to the more universal characteristics I have described. In the first place, it may well be that a view of human beings as free and autonomous agents is easily naturalized because there are some ways in which a level of agency is in fact built into our experience. Of course, it does not follow that the peculiarities of Western individualism reflect aspects of human nature. But it does mean, in these societies, that lapses in the sense of agency take on a particular cultural salience. Evidence that human beings are not necessarily free and autonomous actors is potentially a threat to foundational cultural assumptions, and therefore particularly disturbing.

As a result, in American society lapses in the sense of agency are typically the site of confusion, political conflict, and even illness. The point is not that lapses in the sense of agency are completely denied. As Claudia Strauss (2007) has recently pointed out, even in the cultural context of individualism, we who live in Western cultures still retain alternative models of agency that emphasize the fact that individuals do not always make free and autonomous choices. These models can be applied to understand certain kinds of situations where individuals seem to be compelled to act; in some situations, people in our culture are likely to understand human acts as being caused not by the will of the individual but by forces beyond his or her control.

However, the application of these models usually entails some sort of controversy and uncertainty. The situations discussed in this chapter are ones in which we—both as a society and as individuals—seem unsure about the causes of action. One of the most frequent versions of this occurs in situations in which a person has a subjective sense of being pulled along by an activity, so much so that the everyday sense of agency is to some extent compromised.

Such lapses in agency may take positive or negative forms. The negative forms entail situations in which people do something they would have a general preference not to do, or in which they do not do something they would have a general preference to do. Such problems may be sufficiently severe that they are assigned to the spectrum of pathology: severe compulsions or addictions, multiple personality disorder, hallucinations, and so on. The positive form of a lapse in agency may manifest itself as enchantment with some activity, so that a person feels lost in the activity, as when one reads a compelling novel.[7] More broadly, human beings seem to find considerable pleasure in a wide range of activities that induce a sense of not being completely in control of one's acts, as with intoxication, spiritual ecstasy, and losing oneself in play.

In contemporary American society, some of the most obvious examples of lapses in agency can be found in the domains of entertainment and consumption. Many in our society surrender themselves to the widely available fictions of the entertainment industry and find great enjoyment in a sense of being carried away by these fictions, an experience that is not entirely under their voluntary control. At the same time, people—perhaps the same people—may testify that they watch more television than they wish to, that they find certain foods irresistible in spite of their goal of controlling their weight, that they spend too much money even though they strive to exercise restraint. These negative forms of the lapse in agency raise some of the same theoretical questions that have been debated in the popular and professional literature on addiction, and these phenomena constitute a region of cultural uneasiness.

In the following, I will describe one example of the development of what may become an uncontrollable appetite.[8] In particular, this chapter will focus on the way certain characteristics of the human nervous system intersect with routines of tobacco use to produce a sense among novice cigarette users that they lack complete control of their actions when they smoke. In particular, I will look at situations of excitement and emotional arousal, during which individuals can be swept into currents of collective feelings and projects. Because—as Americans—these young people are likely to have little awareness of the ways in which their actions are conditioned by social factors, and to consider most of what they do as a reflection of their own autonomous choices, they are likely to construe lapses of the sense of agency as being due to a mysterious power. Following culture-wide assumptions, they come to understand this power as the addictive potency of tobacco.

Example: Routines in Early Cigarette Use

In the following, I take an activity that often creates a lapse in the sense of agency and ask what specific practices and psychological processes might create this feeling. The goal is not to produce a comprehensive account of the lapse in the sense of agency but rather simply to conduct a sort of reconnaissance, an exploration of the landscape of such lapses from a particular cultural starting point. That point of departure is the early-phase tobacco-use routines among a group of youthful cigarette smokers.[9] Drawing on a recent qualitative study, I will look at some routines of self-administration that likely contribute to the user's developing sense that tobacco consumption is not entirely under his or her control.

At its base, the argument is a simple one. Some of the activities that people engage in when they are starting to smoke, at least among the young men and women I studied, are much like those observed by Emile Durkheim (1912/1995) in his pioneering study of religious ritual, *The Elementary Forms of the Religious Life*. Durkheim proposed a general theory of ritual based on a study of descriptions of the religious practices of Australian aborigines. Although the ethnographic material Durkheim used is often of questionable accuracy, the theory he constructed has for the most part stood the test of time. Specifically, Durkheim argued that the excitement generated in collective rituals fostered, for the participants, ecstatic feelings and a sense that they had transcended the mundane character of day-to-day life. Although the actual source of these feelings was the "collective effervescence" that can be generated by intense social interactions, the Australians attributed these intense feelings of transcendence to the sacred objects that were always present at these rituals.

Thus Durkheim pointed to a fundamental confusion, or perhaps a socially adaptive misdirection, at the foundation of religious practices. The collective movements and interactions of ritual generate excitement in a group, and religious ideas name deities and other key values as the source of that excitement. Sociologist Randall Collins (2004) has recently pointed out that this technique of energizing values with emotional energy is not confined to religious rituals, but occurs frequently in daily life. To take a simple example, an interesting conversation may well generate a sense of social excitement, which then is typically attributed to one's conversational partner. Interesting conversations sustain social relationships, and conversations that produce no feelings kill social relationships.

Early smoking experiences typically occur in effervescent social gatherings marked by a high level of excitement and highly rhythmic activities, such as conversation and dancing. For this reason, one suspects that some of the observations Durkheim made about religious objects may also hold for the salient objects often present at these activities, especially consumables such as cigarettes, alcohol, and other drugs. These substances benefit from the same sort of transfer of social energy that Durkheim observed for the sacred objects of the Australian aborigines—that is, participants assume that these objects are the source of powerful transformative feelings they have experienced in the gathering.

Here, a detailed look at several different sorts of activities undertaken in conjunction with early-phase routines of tobacco use will show that these activities may in themselves generate thoughts and emotions that contribute to the subjective sense of power that may override one's sense of agency and foster a feeling that one is unable to control one's appetite. These behaviors can be grouped into three general categories: imitation and rhythmic entrainment, pretend play, and emotional arousal. The focus will be on trying to better understand, in light of recent advances in cognitive neuroscience, the mental mechanisms that work together to produce a lapse in the sense of agency. This analysis will thereby provide an updated and more detailed account of the processes that Durkheim grouped together under the label "collective effervescence." What seemed in his day to be the somewhat mysterious effects of social groups on individual minds can today be understood to be not mysterious at all, and instead to result from a combination of several cognitive and emotional factors that have been closely studied and documented by psychologists and cognitive neuroscientists.

The Social Character of Early-Phase Tobacco Use

The first year of campus life is a time when many young men and women who have had some exposure to cigarettes make the transition to being more regular smokers (Wetter et al., 2004; Colder et al., 2008). Several years ago I undertook—in cooperation with my colleagues Drs. Mark and Mimi Nichter—an ethnographic and qualitative study of college freshmen in an attempt to understand how this happens. This was a short-duration (approximately 16-month) longitudinal interview study of early-phase tobacco users on two college campuses (Stromberg, Nichter, & Nichter, 2007).

The first point to be made in this context is that the early-phase tobacco use we observed is highly social. This conclusion is based on considerable evidence from interviews in which our subjects, who were only occasional smokers, repeatedly told us that they smoked mainly or exclusively at parties, while more established smokers spoke of smoking at parties or in informal gatherings of smokers. One quantitative measure of this tendency is that I had a group of research assistants classify all narratives concerning tobacco in a group of 27 interviews with early-phase and more established smokers. Working together, they isolated 407 such stories and then divided these stories into ten groups based upon themes. By far the largest of these ten groups was the category "social aspects of smoking," comprising 112 (about 28%) of the 407 narratives. (In comparison, the next-largest category of narratives comprised only about 11% of the total.) About 75% of these 112 stories concern smoking in collective situations. In sum, then, when the tobacco users in this sample talked about smoking, they were more likely to talk about smoking in groups than about any other theme—and all this talk reflects patterns of actual behavior. Among the first-year college students we studied, parties and other social gatherings are far and away the most frequent context in which early-stage tobacco use occurs.

"Other social gatherings" refers primarily to informal gatherings of friends and acquaintances for purposes of conversation and smoking. Many young persons are introduced to smoking in these sorts of gatherings, and they come to associate cigarettes with the camaraderie generated in these groups. Here, for example, is a young woman discussing how she started smoking in high school:

We used to be able to go off campus during lunch and we would all go down to the smoke hole and stuff, and it isn't necessarily that —I mean I do admit that there was a chain reaction whenever somebody lights up, but we mainly met because of the common personalities that we had plus the fact that we all smoked; it was kind of a radiating force to hang out with people that smoke. So, yeah, whenever we would get together I noticed I smoked more because everybody else was smoking, and so you thought well, what's one more cigarette.

And here is a student describing a group of male friends, all established smokers:

With us together, it tends to be a very social event. Like, we have stoop times. We'll be hanging out in the dorm, and we'll drop everything and just go have a cigarette. And those are the best conversations you ever have, when you're all smoking. I don't know, it helps conversation. It really does, it's definitely a social thing, you know what I mean.

Both of these descriptions of informal social gatherings stress the excitement and sociality of the gatherings, features that also characterize descriptions we heard of more formal gatherings. We are not the only researchers who have noticed the importance of social gatherings as a context in which tobacco use (like alcohol use) is likely to occur. The fact that much cigarette smoking occurs at parties, or in other sorts of energized collective interactions, is often more or less assumed (see Silverstein, Kozlowski, & Schacher, 1977), and may be one of those observations that is so obvious that we do not need confirming evidence from academic studies.

However, for those who remain skeptical, the academic studies are available. In an enormous survey sponsored by the Harvard School of Public Health and conducted in the late 1990s, 17,592 college students returned questionnaires on smoking behavior (Emmons, Wechsler, Dowdall, & Abraham, 1998). One of the predictors of smoking in this study was found to be "endorsement of parties as important or very important" (Emmons et al., 1998, p. 105). Other recent studies (for example, Colder et al., 2008; Acosta, Eissenberg, Nichter, Nichter et al., 2008; Rigotti, Moran, & Wechsler, 2005) establish direct links between attendance at parties and smoking for young persons.

Imitation and Rhythmic Entrainment

Sociologist Randall Collins (2004) has recently published a lengthy history of what he calls tobacco rituals, and among the most prominent of these rituals are those centered on carousing, sexually charged, high-energy group interactions. The basis of this association between social interaction and early-stage smoking is, of course, complex and difficult to sort out. But probably the most important factor in creating this association is the simple fact that some people have a desire to imitate others whom they see smoking. As one neophyte smoker expressed the point in our study: "When you see someone else light a cigarette, you get this urge to do the same" (Stromberg, 2008, p. 441). A more established smoker who plans on trying to quit when he leaves campus for the summer makes the same point: "I'm hoping that maybe when I go back ... this summer, I'll be able to cut back a little bit ... Although then, it's just like, I'll be with my friends all the time, and they smoke like chimneys. So I don't know. When I'm around people that smoke a lot, it's really hard."

Why is this the case? It's a crucial question for this analysis. Building upon Collins's work and my own earlier research, I will suggest two central reasons that imitation is such a prominent feature of smoking behavior, again without meaning to suggest that this will constitute a complete explanation of the matter. The first is that tobacco use has a social history in Western culture, and that social history has imparted a certain kind of symbolic valence to cigarette smoking. Even today, after decades of declining adult smoking rates and increasingly intolerant public attitudes about smoking, cigarette smoking retains potent associations with sexual desirability and coolness. Collins argues that these associations were created in the early and middle decades of the twentieth century when, for a time, a premium was placed on the capacity to demonstrate elegance and social facility in public space. The precipitous economic growth of the twentieth century created more upper-level career positions than could be filled by traditional economic elites. Facility with cigarettes became one way of demonstrating, in public space, social facility more broadly. Cigarettes became an effective way to assert status claims in public gatherings, in that men who could convey a sense of elegance in their handling of a cigarette really were, for a time, more likely to rise in the class system and therefore were more attractive to women. For various reasons, this system began to break down after World War II; one of the most important reasons was the growth of a credentialing system that began to function as the effective system of allocating people to upper-level positions in business. However, the symbolic associations of cigarettes change more slowly than the structural situation of social mobility. As readers will surely recognize, in certain contexts the ability to handle a cigarette remains a powerful indicator of what could be called "erotic prestige."

Thus, among first-year college students, who cannot rely on previous family and secondary school identities in the new campus setting, powerful incentives encourage imitating those who are effectively using the symbolism of cigarettes to establish prestige in collective gatherings. (This is all an academic way to establish a point that some might regard as obvious: smoking is a way to look cool, especially if cool people are already doing it.)

However, status considerations may not be the only factor that cause some students to imitate smokers. As anyone who has ever been in a conga line can attest, we humans can be strongly motivated to entrain with rhythmic activities, even if those activities might be judged as unappealing in other contexts. A recent study by Kirschner and Tomasello demonstrated that even very young children are drawn into joint drumming activities in social groups, and argue for "a specific human motivation to synchronize movements during joint rhythmic activity" (2009, p. 299). Of course, we typically do not think of cigarette smoking as being a rhythmic activity such as dancing or drumming, but in fact it is probably the complex, rhythmic quality of cigarette-smoking practices that makes them so effective as an erotic display.

As an outside observer, my impression is that research on social coordination has often focused on breathing, speech, and manual movements. This emphasis is presumably because rhythmic entrainment—while by no means limited to these systems—is especially likely to appear in the realms of behavior where it served a strong adaptive purpose over millions of years of evolution. Manual and oral behavior, of course, are the basic mechanisms for feeding among primates. Thus it is no coincidence that mirror neuron systems were first observed in (quoting now from a review), "the rostralmost sector of the ventral premotor cortex that controls hand and mouth movements" (Pascherie & Dokic, 2006, p. 102).[10] In sum, the often unstated assumption is that mirror neuron systems evolved to facilitate more effective food acquisition among social primates. Further evidence for the association of mirror neuron systems and social coordination was recently provided by a study of entrainment in spontaneous finger tapping (Tognoli et al., 2007, p. 8192). The investigators associate this coordination with mirror neuron activity through brain wave patterns, as measured by electroencephalogram.

Thus, a rhythmic oral-manual activity such as cigarette smoking can to some extent provoke entrainment in the same way a musical rhythm does. People smoke at parties, at least in part, for the same reasons they dance. If we conceptualize this matter in broader terms, we can say that any joint rhythmic activity is a basic form of playful social cooperation. The point is that a natural human motive to entrain in rhythmic activity may be one reason for participation in early-phase cigarette smoking.

To briefly summarize, part of the reason that young persons on college campuses begin to smoke cigarettes is that when they encounter smoking in collective gatherings, they imitate this behavior. We have examined two possible reasons for this, one being a desire to increase status (primarily, in this context, sexual desirability) by emulating other high-status persons, and another being a general desire to engage in joint rhythmic play. Considering these two points together suggests a further observation: novice smokers learn to handle cigarettes in a form of apprenticeship that is not unlike learning an athletic skill from a coach or a teacher. They imitate the moves of the high-status individuals who have already mastered smoking as a form of display (see Downey, 2010). In doing so the novice illustrates something of her history in the grace with which she manipulates a cigarette, and potentially communicates ("from body to body," as Bourdieu, 1977, p. 2, says) her past association with other "cool" people. As others are pulled into the play of smoking, her strong "sense of the game" (another phrase from Bourdieu, via Downey, 2010) may allow her to emerge as a dominant figure.[11]

Pretend Play

The mention of play brings me to another form of imitation that occurs in these contexts. Our interviews provide strong evidence that beginning smokers find cigarettes to be a useful prop in facilitating social interaction, and particularly for manipulating their own images in a form of identity play (Stromberg et al., 2007; Stromberg, 2008, 2009). For example, the cigarette is often used as a prop in performances of pretending, as is expressed in the following quote from a young man in our study: "I started smoking just to watch and witness how being a smoker changed me into all these different groups and people." Or, a young woman: "Sometimes I do it [smoke] to kinda, you know, announce myself, like, 'Hey, here I am.' I have fun with that ... I feel different ... it makes me feel a little sexier" (both quotes from Stromberg et al., 2007, p. 16).

Such pretending is considered a form of extended imitation by simulation theorists such as Alvin Goldman (2005, pp. 91–92), who suggests that any form of role play is "an extension of the more primitive phenomenon of imitation." This sort of pretending, it should be noted, entails a series of restrictions on action. To the extent I wish to pretend to be, for example, a sultry temptress smoking a cigarette, I will try as much as possible to conform to the sultry temptress model, making sure to avoid any action that will undermine the identity I seek to project. However, such a situation is not merely one of restriction, for like many forms of play, this identity performance is also a medium for creative improvisation and freedom. That is, we enjoy play in no small part because it offers us wide-open opportunities.[12]

If I am playing in a basketball game, I must make split-second decisions: is the defense slightly unbalanced, opening up the possibility of driving to the left? Or should I pass it to Sam, who is open on the perimeter? Nine other players will have to adjust instantaneously to whatever I decide. Ironically, this spectacular openness in the game is created by a set of highly restrictive rules about what one may and may not do while playing basketball. The same thing could be said about playing with a cigarette— the prop involves at once rules and improvisational opportunities. In fact, the rules can be said to create the opportunities.

This tension is precisely why this situation must be understood as a lapse in the sense of agency. In play of this sort, flexibility and restriction work together to create a potentially stimulating experience that has features of both improvisation and control. We do not notice the control because we surrender so willingly to the conventions, as we do when we become caught up in a fiction, ritual or sporting context. Yet at its heart, the game or pretend play are lapses in the sense of agency no different in principle from those situations wherein one feels drawn into activities one does not value. The lapse in the sense of agency is created by any experience of being led by something beyond the self. Play and games are in this sense opportunities for ecstasy, for creating the subjective sense that one has stepped outside the structure of the everyday.

The broader point is that playful activities can create a sense of being drawn into a world, the world of the game, without having fully intended to exit the day-to-day world. As such, playful activities are apt to give us the feeling that something beyond our will is impelling us to continue. Pretend play is likely to be enjoyable in no small part because of the opportunities it presents for creative improvisation, but paradoxically this enjoyment also can contribute to a sense that one is not fully the author of one's action.[13] As is the case with imitation more broadly, to the extent that cigarettes enter the context of pretend play, they can come to be associated with a lapse in the sense of agency. In the strongest version of this association, the cigarette comes to be understood as the reason behind the compelling nature of the play in which it is used as a prop.

Emotional Arousal

Of course, at parties and other exciting social gatherings, people do not just imitate smoking and join in the collective play around the cigarette; they are likely to join into lots of other collective processes as well. This more widespread imitation is probably a central component of Durkheim's "collective effervescence," the powerful sense of excitement that frequently occurs in lively collective gatherings. The most important factor underlying the heightened sense of excitement people can feel in crowds is primitive emotional contagion, a process that has been much studied by psychologists in recent decades (Hatfield, Cacioppo, & Rapson, 1992; Hatfield, Cacioppo, & Rapson, 1994; Neumann & Strack, 2000). One theme that emerges from this work is that emotional mimicry is, to quote Hatfield and her colleagues, "a phylogenetically ancient and basic form of intraspecies communication" (1994, p. 79). These authors cite research in support of the contention that "primates are prewired to respond to emotional faces with a strong ANS [autonomic nervous system] response" (Hatfield et al., 1994, p. 19).

Imitation is only one factor that produces collective effervescence; the experience is also exacerbated by rhythmic entrainment. The music, dance, and conversation that typify parties are highly rhythmic activities in which, as has already been noted, people seem to have a natural tendency to entrain. Randall Collins (2004, p. 62) points out that consumption at parties also fits into this category of behavior, and collective consumption is an effective way of creating entrainment and, consequently, emotional energy:

The physical substance ingested—the alcohol, coffee, tea, soft drinks, the party cake, the shared dinner or, in older times, a shared smoke—of course have some sensory character of their own. But they are not solitary pleasures. ... The ingestion of food and drink is part of the bodily coparticipation; these are ritual substances when they are consumed together in the atmosphere of a sociable occasion.

In a number of different ways, then, collective gatherings tend to generate unusual states of emotional arousal. This arousal is likely to be interpreted as coming from outside the individual, for in fact it is. And that interpretation, of course, lends further strength to the impression that forces are working to compromise one's accustomed responsibility for one's own mental states and actions.

Participants in collective gatherings often report, in addition to emotional arousal, unusual cognitive states such as altered perception of self and time. A number of our subjects referred to such states in discussing parties. The most frequent comments have to do with amnesia or "lost time" experiences. For example, consider the following statement from an experienced smoker:

When I was, like, really hard-core into the rave scene, I would go for nights when I would have two packs of cigarettes in my pocket when I walked into the rave. I'd leave the next morning, and I wouldn't have any cigarettes. I'd have no idea where they went, [but] I knew I smoked them all.

Of course, this amnesia is undoubtedly augmented by the drug ecstasy, but it is by no means unusual for people to report similar amnesias from other sorts of parties. Several times we heard stories along the lines of "time flies when you're having fun," an experience people can have simply with a lively conversation: "Whenever I go outside in the evenings especially, there are always other people that I know out there lighting up too, and we tend to get into these big, long discussions about stupid things, and before I know it an hour has gone by" (Stromberg, 2009, p. 134).

Social scientists have repeatedly documented amnesias and other experiential discontinuities such as a loss of awareness of self or the environment in association with the sort of strong emotional arousal that may occur in excited crowds. And of course one does not need to review the literature to know that such effects may occur at an exciting party or among the crowd at a rock concert. Typically, such mental effects are referred to as dissociation (Hollan, 2000). This term is often broadly used; "dissociation" might designate, on one end of the spectrum, the profound depersonalization of a fugue state, or on the other end, normal daydreaming (Seligman & Kirmayer, 2008).[14] For this reason, controversies swirl around the concept of dissociation, controversies that are peripheral to the argument here. In this context, the point is simply that the contexts in which college students begin to smoke may help to induce striking mental discontinuities.

There are a number of theories about why and how such responses are generated in the nervous system. Recently, Rebecca Seligman and Laurence Kirmayer (2008) have suggested that dissociation is based in an adaptive capacity to suspend awareness of peripheral aspects of everyday consciousness in order to focus conscious information processing on some immediate problem. They argue that this ability derives from a cortical capacity to inhibit connections with those parts of the nervous system that under normal conditions provide an ongoing stream of emotional and sensory material. The result is feelings of disconnection within the stream of mental experience. Particularly worth noting, in the present context, is an often-mentioned disruption in the sense of linkage between volition and execution (2008, p. 47), resulting in "experience and behavior that does not feel self-directed" (2008, p. 49).

In a way, Seligman and Kirmayer's treatment of dissociation parallels that which I have outlined for lapses in the sense of agency: they see dissociation as a mental process that may be appropriated, in different cultural contexts, to serve various social and individual ends (and indeed, what they call dissociation overlaps considerably with what I have termed a lapse in the sense of agency[15]). The college party is, it could be argued, a culturally elaborated routine in which dissociative modes of awareness are cultivated and valued.

In summary, intense collective interaction has often been observed to produce unusual mental states that create a lapse in the sense of agency. The neural mechanisms producing such states need no longer be regarded as mysterious; rather, they are beginning to be understood. Across human cultures and throughout history, these states have been exploited to support experientially grounded arguments that powerful forces exist, forces that go beyond everyday experience and can overwhelm our taken-for-granted sense of agency. In the present case, many students attribute these powers to the effects of drugs such as alcohol and tobacco. Of course, the chemicals in these potent substances undeniably do affect subjective experience. However, the force, direction, and interpretation of these experiences are strongly shaped by cultural factors such as widespread assumptions about the powers and effects of drugs and the resultant expectations of users.

Conclusion

I have argued that there is a sense of agency that is fundamentally human, more basic than the ideological elaborations we find in, say, contemporary American society. No surprise, then, that we also find many societies mark out for special notice experiences in which people report they are not, or not fully, the author of their own activities. Such lapses in agency are often harnessed by societies to regulate certain aspects of social life. One of the ways this happens is illustrated in Durkheim's theory of ritual. Lapses in agency create a subjective sense of being controlled by something that transcends the self. The evocation of this transcendent power can be used to valorize key objects, ideas, and practices, to make them seem to have powers that go beyond those encountered in the day-to-day world.

Until recently, the mechanics of these lapses in the sense of agency have remained mysterious, but advances in our understanding of the brain now make it possible to isolate some of the key neural processes that contribute to these situations. It turns out that there are fairly straightforward reasons at the neural level why persons in high-intensity social gatherings are likely to feel excited, and to have little or no sense of having willed this condition. I have pointed, for example, to the central role of imitation, a process that occurs more or less constantly in interaction, usually without conscious awareness. As we imitate others, we may begin to share in their emotions or to be drawn into complex rhythmic play with them. These, and other processes, may well create a subjective sense that everyday expectations of agency are compromised.

It is not controversial to observe that human ways of life depend upon ideas and practices that must be learned by members of functioning communities. It then follows that all communities must also have ways of both establishing and, over time, revising these ideas and practices. The lapse in the sense of agency is one way that many communities accomplish this. Ideas and practices can be infused, through the processes I have described, with the authority of a power that seems to reach beyond the expected and everyday sense of agency.

This happens not only in momentous rituals, but in other forms of social interaction that go on around us all the time. For example, this paper has examined a set of practices whereby one segment of novice tobacco users are encouraged to develop an embodied conviction that their appetite for cigarettes is not completely under their control. Cognitive and emotional processes such as imitation, entrainment, emotional contagion, and mild forms of dissociation are provoked in the social experiences that often occur in the early months of college life, and in themselves provoke strong mental responses. A prominent feature of many of these responses is a lapse in the sense of agency, a sense of not having fully authored one's own actions. Thus if a student smokes in this context, he or she is well positioned to interpret such feelings, and the feeling of emotional arousal in general, as the result of the ingestion of an unfamiliar substance. The young men and women we studied have almost certainly absorbed cultural messages that stress the powers of drugs such as tobacco and alcohol to produce strong emotional effects of their own accord. When they attend parties and use tobacco, they are likely to attribute a wide range of their experiences to the effects of tobacco, even though not all of these effects have a chemical origin.

In American society, lapses in agency can become particularly troubling. Such lapses call into question the status of a key ethical precept, the capacity of humans for free choice, an important part of the foundation of Western ethical systems. Yet at the same time, we now live in a society in which the sense of not being able to fully control one's desires is central to the functioning of our economic system. This cultural contradiction is, it could be argued, disguised as an individual problem of appetite, in the most extreme cases as what we call addiction. In this sense American society, and Western societies in general, are like many other human social groups which exploit lapses in the sense of agency to encourage or discourage certain behaviors. Because of the central role of consumption in our economy, it should come as no surprise that in our society such social technologies are often harnessed to the promotion of consumerism by fostering the idea that our appetites are so powerful that they can easily escape our control.

Acknowledgments

Several of the participants at the conference "The Encultured Brain" offered helpful comments on this paper. Comments from Claudia Strauss, Rachel Brezis, and Michael Jindra were especially helpful, although many of the other participants offered useful suggestions. In particular I want to thank Greg Downey and Daniel Lende, not only for organizing the conference, but for careful and constructive comments that helped me to improve this paper. I have not followed all of the suggestions I have received, which will account in part for the paper's remaining imperfections.

Notes

1. See Kockelman (2007) for an ambitious (and terminologically fecund) synthesis of the ways in which speakers can exploit the resources of language at the same time that they are limited in what they can do by a range of contextual factors.

2. My formulation here has the virtue of being straightforward, but it could be argued that it disguises a number of complexities concerning the sense of agency. To take one example, once one begins to look closely one recognizes that there are probably distinguishable varieties of the sense of agency: one's sense of agency can

be fairly conscious and articulable or more embodied and not brought fully to awareness. I would admit the complexity of the sense of agency while still maintaining that for the present argument my fairly simple formulation of sense of agency is sufficient, and that indeed to treat the idea in its full complexity would require heading off in different directions. At base, my concern here is more with the point that human beings have a sense of agency rather than questions about how that sense of agency is constituted. Readers who would like to explore the sense of agency in greater depth should first consult Marcel (2003).

3. Marcel (2003) raises objections to the notion of authorship of action on the grounds that this approach entails the assumption that one's actions correspond to one's conscious intentions. While this may be the way some have used the conception of authorship, there is no necessary connection of this sort. For example, I could certainly admit that I authored a comment that was taken as an insult while at the same time denying that the insult was my intention.

4. Another related issue that must be set aside in the present chapter is the question, "are human beings in fact agents, capable of consciously willing their actions?" My analysis here focuses on the sense of agency, not on the ultimate question of the character of human agency. Readers who are interested in the latter question should consult, on two sides of the debate, Wegner (2003) and Nahmias (2005).

5. This view is a variation of what is now widely known as simulation theory. See Goldman (2008).

6. Another term often used by philosophers and cognitive scientists who study the sense of agency, that being "sense of ownership" (see, for example, Gallagher, 2000), would seem to be relevant to what I am calling a lapse in the sense of agency. Typically I claim responsibility for what my body does, but not always. Suppose, for example, someone grabs my hand and moves it. In such a case I would acknowledge that it was my hand that had moved but not that I was responsible for the movement. Unfortunately, the term seems to me to be used in contradictory ways in the literature, even by theorists of considerable stature such as Marcel (2003) and Gallagher (2000). Specifically, Gallagher says that I may own an action for which I deny responsibility, while Marcel (and others) seem to say that I may "disown" certain of my acts; that is, I may deny responsibility for them. The problem here may be with my reading, but in this situation, I cannot call upon the concept of ownership to further clarify my own position.

7. Greg Downey points out (personal communication) that group interactions such as an inspiring speech are another example of this, for in such a case both speaker and audience may have a sense of being carried away by something transcendent (hence "inspiring.").

8. An extensive literature on craving (see, for example, Rock & Kambouropoulos, 2008; Elster, 2000) overlaps to some extent with the sort of inquiry I am pursuing

here. Appetitive behavior is widely thought to be associated with the dopaminergic system (see, for example, Lende, 2005; Foxall, 2008; Collins & Woods, 2008). Although I do not explore that connection here, nothing in this analysis is incompatible with an association to the dopaminergic system.

9. For some of these smokers, these routines develop into a form of compulsive behavior we call addiction. However, most of the subjects in our sample had not yet developed a regular pattern of cigarette smoking, and therefore should not be considered addicted to tobacco. In a few cases, I will quote from interviews with students who are regular smokers, and in all cases I will designate these persons as more experienced smokers.

10. The reasoning that would link a human mirror neuron system to automatic imitation is complex. The evidence is strong for a mirror neuron system in the premotor cortex of monkeys such as macaques (Gallese, Fadiga, Fogassi, & Rizzolatti, 1996). That is, there are neurons in this area that become active both when the individual is performing certain goal-directed actions and when the individual observes these actions being performed by a conspecific. In the early research, mirror properties were observed only with certain kinds of goal-directed movements, especially involving manual grasping of objects. Subsequent work has produced strong indirect evidence that mirror neurons are present in humans and may be involved in much broader behavioral patterns, such as language and empathy. Nevertheless, one cannot assume that any human imitation behavior is based in a mirror neuron system (Pascherie & Dokic, 2006). At this point, a relationship between mirror neurons, imitation, and cigarette smoking in groups can only be mentioned as an intriguing possibility.

11. Randall Collins (2004) refers to this as "situational stratification, " and points out that in contemporary society this capacity to dominate in social situations is a form of stratification that deserves to be considered along with economic stratification as a significant social force.

12. Not all technologies for creating lapses in agency are equally open to improvisation. For example, ritual is generally less so than play. See the discussion in Stromberg (2009, p. 101ff.).

13. The felicitous formulation of "authorship" of action comes from Humphrey and Laidlaw (1994). More generally, Humphrey and Laidlaw develop an understanding of ritual action that is similar in some ways to what I am saying here about lapses in the sense of agency.

14. Some would dispute Seligman and Kirmayer's (2008) broad definition of dissociation, which is based on the assumption of a continuity between pathological and non-pathological forms of dissociation. In a taxometric analysis of reports of dissociative experiences (on the Dissociative Experiences Scale) from a large sample

of individuals, Waller, Putnam, and Carlson (1996) presented results suggesting that dissociation is of two types, normal and pathological. Obviously, in this situation, my own account of the neural basis of dissociation must be taken as tentative.

15. Yet another concept that overlaps with the lapse in the sense of agency is hypnosis. Psychologists Erik Woody and Henry Szechtman (2007, p. 242) write: "a true hypnotic response necessarily involves an alteration in the sense of agency, in which one's own will or volition is not experienced as the origin of the response."

References

Acosta, M. C., Eissenberg, T., Nichter, M., Nichter, M., & Balster, R. L. (2008). Characterizing early cigarette use episodes in novice smokers. *Addictive Behaviors, 33,* 106–121.

Bourdieu, P. (1977). *Outline of a theory of practice.* Cambridge: Cambridge University Press.

Colder, C. R., Flay, B. R., Segawa, E., Hedeker, D. & TERN members. (2008). Trajectories of smoking among freshmen college students with prior smoking history and risk for future smoking: Data from the University Project Tobacco Etiology Research Network (UpTERN) study. *Addiction, 103,* 1534–1543.

Collins, G. T., & Woods, J. H. (2008). Narrowing in on compulsions: Dopamine receptor functions. *Experimental and Clinical Psychopharmacology, 16,* 498–502.

Collins, R. (2004). *Interaction ritual chains.* Princeton, NJ: Princeton University Press.

David, N., Newen, A., & Vogeley, K. (2008). The "sense of agency" and its underlying cognitive and neural mechanisms. *Consciousness and Cognition, 17,* 523–534.

Downey, G. (2010). "Practice without theory": A neuroanthropological perspective on embodied learning. *Journal of the Royal Anthropological Institute, 16*(Suppl. 1), S22–S40.

Durkheim, E. (1995). *The elementary forms of religious life* (K. E. Fields, Trans.). New York: Free Press. (Original work published 1912)

Elster, J. (2000). *Strong feelings: Emotion, addiction, and human behavior.* Cambridge, MA: MIT Press.

Emmons, K. M., Wechsler, H., Dowdall, G., & Abraham, M. (1998). Predictors of smoking among US college students. *American Journal of Public Health, 88,* 104–107.

Foxall, G. R. (2008). Reward, emotion, and consumer choice: From neuroeconomics to neurophilosophy. *Journal of Consumer Behaviour, 7,* 368–396.

Gallagher, S. (2000). Philosophical conceptions of the self: Implications for cognitive science. *Trends in Cognitive Sciences, 4,* 14–21.

Gallese, V., Fadiga, L., Fogassi, L., & Rizzolatti, G. (1996). Action recognition in the premotor cortex. *Brain, 119,* 593–609.

Goldman, A. I. (2005). Imitation, mind-reading and simulation. In S. Hurley & N. Chater, *Perspectives on imitation: From neuroscience to social science* (Vol. 2, pp. 79–94). Cambridge, MA: MIT Press.

Goldman, A. I. (2008). *Simulating minds: The philosophy, psychology, and neuroscience of mindreading.* Oxford: Oxford University Press.

Hatfield, E., Cacioppo, J. T., & Rapson, R. L. (1992). Primitive emotional contagion. In M. S. Clark (Ed.), *Review of personality and social psychology* (Vol. 14, pp. 151–177). Newbury Park, CA: Sage Publications.

Hatfield, E., Cacioppo, J. T., & Rapson, R. L. (1994). *Emotional contagion.* Cambridge: Cambridge University Press.

Hollan, D. (2000). Constructivist models of mind, contemporary psychoanalysis, and the development of culture theory. *American Anthropologist, 102,* 538–550.

Humphrey, C., & Laidlaw, J. (1994). *The archetypal actions of ritual: A theory of ritual illustrated by the Jain rite of worship.* Oxford: Clarendon Press.

Keane, W. (2007). *Christian moderns: Freedom and fetish in the mission encounter.* Berkeley: University of California Press.

Kirschner, S., & Tomasello, M. (2009). Joint drumming: Social context facilitates synchronization in preschool children. *Journal of Experimental Child Psychology, 102,* 299–314.

Kockelman, P. (2007). Agency: The relation between meaning, power and knowledge. *Current Anthropology, 48,* 375–401.

Lende, D. (2005). Wanting and drug use: A biocultural approach to the analysis of addiction. *Ethos, 33,* 100–124.

Marcel, A. (2003). The sense of agency: Awareness and ownership of action. In J. Roessler & N. Eilan (Eds.), *Agency and self-awareness* (pp. 48–93). Oxford: Clarendon Press.

Nahmias, E. (2005). Agency, authorship, and illusion. *Consciousness and Cognition, 14,* 771–785.

Neumann, R., & Strack, F. (2000). "Mood contagion": The automatic transfer of mood between persons. *Journal of Personality and Social Psychology, 79,* 211–223.

Pascherie, E., & Dokic, J. (2006). From mirror neurons to joint actions. *Cognitive Systems Research, 7,* 101–112.

Patterson, F., Lerman, C., Kaufmann, V. G., Neuner, G. A., & Audrain-McGovern, J. (2004). Cigarette smoking practices among American college students: Review and future directions. *Journal of American College Health, 52*, 203–210.

Rigotti, N. A., Moran, S. E., & Wechsler, H. (2005). US college students' exposure to tobacco promotions: Prevalence and association with tobacco use. *American Journal of Public Health, 95*, 138–144.

Rock, A. J., & Kambouropoulos, N. (2008). Conceptualizing craving: Extrapolations from consciousness studies. *North American Journal of Psychology, 10*, 127–146.

Seligman, R., & Kirmayer, L. (2008). Dissociative behavior and cultural neuroscience: Narrative, metaphor, and mechanism. *Culture, Medicine and Psychiatry, 32*, 31–64.

Silverstein, B., Kozlowski, L. T., & Schacher, S. (1977). Social life, cigarette smoking, and urinary pH. *Journal of Experimental Psychology. General, 106*, 20–23.

Strauss, C. (2007). Blaming for Columbine: Conceptions of agency in the contemporary United States. *Current Anthropology, 48*, 807–832.

Stromberg, P. (2008). Symbolic valorization in the culture of entertainment: The example of legal drug use. *Anthropological Theory, 8*, 430–448.

Stromberg, P. (2009). *Caught in play: How entertainment works on you*. Stanford, CA: Stanford University Press.

Stromberg, P., Nichter, M., & Nichter, M. (2007). Taking play seriously. *Culture, Medicine and Psychiatry, 31*, 1–24.

Synofzik, M., Vosgerau, G., & Newen, A. (2008). Beyond the comparator model: A multifactorial two-step account of agency. *Consciousness and Cognition, 17*, 219–239.

Taylor, C. (1985). The person. In M. Carrithers, S. Collins, & S. Lukes (Eds.), *The category of the person* (pp. 257–281). Cambridge: Cambridge University Press.

Tognoli, E., Lagarde, J., DeGuzman, G. C., & Scott Kelso, J. A. (2007). The phi complex as a neuromarker of human social coordination. *Proceedings of the National Academy of Sciences of the United States of America, 104*, 8190–8195.

Tomasello, M. (1999). *The cultural origins of human cognition*. Cambridge, MA: Harvard University Press.

Tomasello, M., Carpenter, M., Call, J., Behne, T., & Moll, H. (2005). Understanding and sharing intentions: The origins of cultural cognition. *Behavioral and Brain Sciences, 28*, 675–735.

Waller, N. G., Putnam, F. W., & Carlson, E. B. (1996). Types of dissociation and dissociative types: A taxometric analysis of dissociative experiences. *Psychological Methods, 1*, 300–321.

Wegner, D. M. (2003). The mind's best trick: How we experience conscious will. *Trends in Cognitive Sciences, 7*, 65–69.

Wetter, D. W., Kenford, S. L., Welsch, S. K., Smith, S. S.; Fouladi, R. T., Fiore, M. C., et al. (2004). Prevalence and predictors of transitions in smoking behavior among college students. *Health Psychology, 32*, 168–177.

Woody, E., & Szechtman, H. (2007). To see feelingly: Emotion, motivation and hypnosis. In G. A. Jamieson (Ed.), *Hypnosis and conscious states: The cognitive neuroscience perspective* (pp. 241–255). Oxford: Oxford University Press.

13   Addiction and Neuroanthropology

Daniel H. Lende

In Colombia, the adolescents I worked with often described the experience of addiction as *querer más y más*, "to want more and more." These young men and women agreed that wanting more, alongside *ansiedades, deseo,* and *ganas* (cravings, desire, and urges), were what made addiction so difficult. They saw drugs as similar to other *vicios*, or vices, like video games and pool—an activity that can take up time and energy, that hooks a person in. But drug use could go to much greater extremes, and when it did, it took adolescents into *la calle*, the street. That was too far. By abandoning family, drug users committed a social violation. Families and communities reacted strongly to those who got too caught up in these vicios, especially *la droga*, because they betrayed basic social values in Colombia— to protect family and friends and community. For those who became even more deeply involved, focusing more and more on drug use rather than on family and friends, the problem wasn't pleasure. It was that they wanted drugs more than they wanted to be involved with their family. The nature of their involvement was off, not the nature of their self. As my ethnographic research revealed, adolescents moved beyond the initial pharmacological and social pleasures of use into a pattern where drug use became marked by "wanting more and more" and increasingly became a habit that was hard to control.

This view of addiction stands in stark contrast to ideas in the United States. In the country that is the world's largest consumer of drugs, addiction is seen as a disease or moral failing, placing an emphasis on either biology or control in relation to self. Too much pleasure is seen as one of the main problems—rats pressing a lever until death, deviants using to get high, cheerleaders swept away by the temptations of prescription drugs. Scientifically, this disordered need for pleasure was once explained through tolerance and withdrawal, of drugs demanding ever more use to get the same pleasurable effect, and when they weren't used, creating physiological urges that drove people to try to get even a modicum of that pleasure back. More recently, the brain has become the key site of too much pleasure, in particular the notion of hard-wired pleasure circuits that can be hijacked by these immoral drugs.

The views surrounding addiction in Colombia were remarkably different during the decade that I did research on drug use in Bogotá, the capital city of seven million, and Popayán, a small city in the southwest. Significantly, Colombian views provided novel ways to link research on the brain with an on-the-ground assessment of what drug users actually did and experienced. This combination of neuroscience and ethnography revealed that addiction is a problem of involvement, not just of pleasure or of self. That decade showed me that addiction is profoundly neuroanthropological.

Chapter Overview

Rather than a brain-driven theory of addiction, I propose a neuroanthropological one in this chapter, with a focus on two key dynamics that lie at the core of addiction. First, many users report compulsive desire for drugs, urges that can drive them to relapse and lead to excess. Second, users emphasize the repetitive nature of drug use, using drugs over and over—it is a drug *habit*, as much as something they want or that gives them pleasure or a break from everyday life. Both intense desire and repetitive habits lie at the center of addiction. Understanding both dynamics requires seeing them as neurocultural processes, in which the neurobiology of addiction meets up with the subjective and cultural dynamics of drug use in inextricable ways. For example, incentive salience is mediated by brain-reward systems, in particular the mesolimbic dopamine system; but what counts as salient, and why drug use actually matters to users, is part of the same neurocultural operation. Similarly, as addiction progresses, there is a shift from ventral striatum to dorsal striatum activation, a shift that is associated with more habitual and automatic use. But understanding "the habit" requires looking at the chunking of series of actions, experiences, and cues derived from the patterns of everyday life of hardcore users. Chunking, or habitual patterns, are not content-free; rather, they work within the sociocultural dynamics that specific individuals both must go through (school or being on the street) and seek out (places to use drugs, with other drug users there). Together, the compulsive involvement that marks addiction, that can lead it to such destructive ends, is jointly defined by the neurocultural dynamics of desire and habit.

Addiction as a Problem

The forthcoming criteria for addiction in the fifth edition of the *Diagnostic and Statistical Manual* (DSM), the basic guide used to diagnosis substance abuse clinically in the United States, show the multifaceted nature of the problem (American Psychiatric Association, 2011). Two basic criteria focus on the pharmacology of drugs, on whether individuals demonstrate tolerance and withdrawal, using more drugs to achieve the same effect and demonstrating known withdrawal symptoms and often using drugs to get rid of those negative effects. A second set of criteria focuses on continued use despite known risks and consequences, from health outcomes to interpersonal problems. These symptoms highlight how addiction is linked to changes in how people evaluate and learn from their behavior. Another set of criteria focuses on use—for example, taking larger amounts and for longer than intended—and a persistent desire to use along with failed efforts at controlling use. These are the criteria linked to wanting drugs and to seeking them out. A final set focuses on how substance use comes to interfere with life's roles and activities; users become so caught up in use that major social roles and obligations are neglected and previously important activities are given up. Here the deep involvement with drugs, the habitual pattern of use and more use, predominates.

While once assumed sufficient to account for drug abuse, it is now clear that tolerance and withdrawal do not drive addictive behavior in themselves (Koob & Nestler, 1997). Certainly for some drugs, heroin for example, physical dependence and the negative effects of withdrawal are an important part of the clinical problem. But many drugs produce no physiological adaptations, particularly stimulants like cocaine; thus, the search for the basic causes of addiction needs to go beyond a focus on withdrawal.

Decision making, our ability to evaluate and control behavior, is definitely altered in substance abuse. A more extensive examination of how the brain instantiates decision making and the resulting vulnerabilities to drug use is beyond the scope of this chapter (Lende, 2007; Redish, Jensen, & Johnson, 2008; Schultz, 2011). But a full explanation for addiction is not to be found in deviations from rational choice, leaving out community dynamics, social meanings, and other important aspects of substance abuse. Moreover, this approach also treats addiction as solely an individual problem, in particular, by assuming that addiction is a brain disease that limits the mind's ability to lead a rational life. This type of approach does not explain what drives the actual behavior of addiction—the excessive, habitual use that ends up being so destructive for some people.

Rather, wanting more and more, and then developing a deep involvement with drug use, are the truly problematic parts of drug use. These are the DSM criteria related to using more and longer than intended, and becoming so involved with drugs that other life roles are neglected. As this chapter will discuss, neuroscience research now reveals major aspects of how each of these works in the brain. In conjunction, first-hand anthropological research shows why desire and involvement matter, what is often missing from the neurobiological account, and how the various brain circuits involved rely on subjective and sociocultural dynamics in a formative way.

Neuroscience and Addiction

Chemical imbalances and hard-wired pleasure circuits have been prominent public explanations advanced by some biologists for addiction. But the real story is more complex, even at the level of neurobiology. Addiction is not simply a chemistry experiment gone wrong, some poor sap in the "laboratory of the street" mixing the wrong substances inside his brain. The parts of the brain where addiction happens are not single, isolated circuits—rather, these areas handle emotion, memory, and choice, and are complexly interwoven to manage the inherent difficulty of being a social self in a dynamic world.

These brain areas, which start front and center and extend down to the basic core of the brain, orchestrate motor and sensory integration with action and bodily regulation. Addiction runs from the basal parts of the brain that handle regulation and activation of the body, through limbic circuits that handle emotions and evaluation of environmental information, to frontal cortices that perform higher-order cognition and control. In general, these areas deal with what to do when, and how much to do it. They handle decision making, emotional engagement, and self-control, and range in functions from reacting to positive and negative events both internally and externally to pursuing complex goals and negotiating social worlds. Given the number and complexity of the neural circuits involved in addiction, distilling what matters most is a challenge. Here I focus on the pursuit of drugs and the repetition of drug use as the most problematic parts of addiction, with a corresponding emphasis on subcortical structures such as the ventral and dorsal striatum and the ventral tegmental area, part of the mesolimbic pathway.

Wanting More and More: The Neuroanthropology of Involvement

From 1993 on, at exactly the time I started working in Colombia, a new line of research on addiction examined "wanting" as key to addiction. Robinson and Berridge (1993) proposed a novel theory of addiction, arguing that incentive salience—literally, determining which incentives or indicators for eventual reward are salient to the individual in a particular environment—was the core psychobiological process that went awry with continued drug use. Addicts, with overly sensitized mesolimbic dopamine systems, wanted drugs too much, engaging in compulsive seeking and consumption of drugs, even when the pleasure from use was no longer as powerful as it once had been.

In historical terms, the most important effect the theory of incentive salience had with regard to addiction studies was its attack on the main assumption at the heart of the models dominant among US researchers—that the condition is a problem of pleasure, whether that is understood morally (i.e., as not having the moral backbone to resist pleasure) or as a disease (i.e., drugs create a chemical imbalance, a "reward deficiency syndrome"). By using novel research to get at a core problem with addiction, Robinson and Berridge opened our eyes to *how* to think about addiction.

In the 1990s, Robinson and Berridge still relied on the basic idea of one chemical, one imbalance to explain addiction. In their model, incentive salience was mediated by dopamine, and the attribution of salience happened in only one part of the brain—the mesolimbic dopamine system. The problem with addiction, the imbalance it created, was that drugs caused sensitization of this "wanting" response, making addicts want drugs more and more and more.

It was a compelling narrative, matching well with parts of what I had seen as a counselor in Bogotá. Subsequently, a central part of my research in the 1990s examined how incentive salience as a theory related to adolescents' actual experiences with seeking out and using drugs. My epidemiological research showed that these ideas applied to humans, building on the animal research of Robinson and Berridge (Lende, 2005). In this research, a specific scale was developed to capture the experience of incentive salience toward drugs, using the combination of ethnographic research and neuroscience theory to capture the core elements of the experience of salience, and then structured and psychometric techniques to build the scale. The resulting scale had robust psychometric properties, and proved a significant predictor of addicted status in regression analysis. Incentive salience worked for people, not just for rats.

Ethnography then provided a better sense of incentive salience out in the real world. Users described wanting drugs, and how their attention would focus on getting and having drugs, and finally a sense of urgency and effort as they moved toward getting drugs.

It's like you are in prison and you want to see your family and you can't. And you want so much to see them and you get really worked up to get out of there, out of prison where you are in a cage and you can't take it, you start to sweat, to be restless, and no matter what, you have to. What I mean is, you pay whatever it is to get out of there. It's something like that, you can't even put up with yourself. You've got to satisfy those desires.

This ethnographic work went beyond just trying to show parallels between individual experiences and neuroscience research. This qualitative research can critically inform knowledge about incentive salience; for example, it revealed significant amounts of "wanting" during actual drug use itself (something later confirmed by laboratory research). It also demonstrated that the drug experience, while centrally about salience, was also informed by meaning and by social context, as I discuss below.

Incentive Salience and Addiction

Incentive salience, in retrospect, is more narrow than the broad explanatory power claimed by both my epidemiological work and the approach advocated by Robinson and Berridge. Incentive salience applies to specific parts of addictive behavior, but not necessarily to the overall package. This point is consistent with the idea that incentive salience works in conjunction with learning, pleasure, habit, and other neural dynamics to shape behavior. But acknowledging this complexity does not have the same rhetorical power as saying, "wanting more and more—that's addiction." Nevertheless, with the advancing research on addiction and incentive salience, we can hone in on how the dynamics of incentive salience can help drive addictive behavior and, from there, draw more general conclusions about what this research tells us about how the brain works and how that illuminates human behavior and anthropological research.

The Mediation of Salience

Incentive salience has provided an important addition to understanding how reward, or decision making more broadly, works in the brain. Traditionally, reward theory has focused on how stimuli in the environment shape animal response, focusing on both unconditional rewards, such as sweetened water given to laboratory rats, and conditional rewards, such as Pavlov's bell to signal dinner for his dogs. More recent work on reward has also focused on how signaling in the brain can provide a teaching signal that provides feedback to neural pathways to improve reward prediction, based on the differences between predicted rewards and actual rewards received (Schultz, 2011). Incentive salience broadens this view of decision making.

Rather than assume that "reward" is a unitary process, Robinson and Berridge argue that learning about and responding to positive environmental stimuli consists of several different processes that all feed into behavior. What looks like unified behavior at the organism level is actually mediated by different processes that function together. These different processes relate to the ecological problem animals face. In the wild, animals are not simply given rewards; they have to seek them out as well (Lende & Smith, 2002). Putting effort into activities that lead to rewards and following the cues that indicate the possibility of reward are important parts of this adaptive problem. Processed pellets and sweetened water might work in a cage environment, but the real world is more complex. At a minimum, an animal has to learn about stimuli, decide whether or not these stimuli are appealing, and decide how involved to get with the stimuli.

Berridge and Robinson have long proposed that there are three "dissociable components" in the overall reward process: "liking" (hedonic impact), "learning" (making predictive associations), and "wanting" (incentive salience) (Berridge, Robinson, & Aldridge, 2009). *Liking* references pleasure and other direct aspects of how much reward an activity or substance provides. *Learning* is about relating cues to rewards, for example, being on one side of the cage to receive a pellet. Incentive salience is what links together cues and rewards through motivation and action.

The Feel of Salience

Through incentive salience, cues—like the lever that needs to be pressed to receive the food pellet in another part of the cage—can become "fascinating" and "attractive" in themselves, shaping how animals allocate attention by becoming focused on these cues for reward (Berridge et al., 2009, p. 68). These stimuli can also trigger "wanting" for absent rewards: "Encounters with incentive stimuli can dynamically increase motivation to seek out rewards, and increase the vigor with which they are sought" (Berridge et al., 2009, p. 68).

Incentive salience, though glossed by Robinson and Berridge as "wanting," is not the same thing as conscious desire; it is "distinguishable from more cognitive forms of desire meant by the ordinary word, wanting, that involve declarative goals or explicit expectations of future outcomes, and which are largely mediated by cortical circuits" (Berridge et al., 2009, p. 68). Incentive salience is "mediated by more subcortically weighted neural systems that include mesolimbic dopamine projections [and do] not require elaborate cognitive expectations (Berridge et al., 2009, p. 68)." That urge for pizza people get, that sense that they want it now, that they just have to have it—that is incentive salience.

Salience in Addiction

Asking informants to describe a "typical day" proved an important way for me to get data on what individuals did and felt as they sought out and consumed drugs. This approach engaged users in providing the thick description of their own experiences, thoughts, and emotions, data that could be usefully linked to neurological and psychological theories. Particularly important was getting informants to go beyond the typical answers they gave to others, ones that were either socially acceptable or provided a quick explanation. Engaging them in actual description, with follow-up questions about emotions, thoughts, experiences, and contexts in different moments, helped produce a rich understanding of ongoing action.

In a prototypical day, heavy drug users generally described a sequence of feeling ambivalent about drug use, followed by a decision to use, and then going to seek out drugs. Consistently, individuals reported that they felt more desire to use, a greater urgency to use, once they had made the decision and were actively engaged in moving toward drugs. As one young person reported:

Before consuming, at the beginning like you repented, like you were thinking if you consumed or not, that was the dilemma. But after, once you were inside ... then you were more anxious to consume.

Heavy drug users also reported "wanting more and more" in the moment of drug use itself. Rather than just the drug high—that euphoria of pleasure often assumed to be the prime driver of drug use—users described wanting more and more, of having to have it all, as they continued using. It was this feeling, rather than the pleasant effects themselves, which often drove excessive use, of lighting one *basuco* cigarette after another, *"sin parar."*

I'd be using and each time I wanted more and more and to be in it each moment and like not waste any little thing.

Thus, incentive salience takes on specific forms in addictive behavior. The urge to use, the desire to seek out drugs, is a major way that incentive salience translates into pushing heavy drug users from an initial desire or temptation for drugs to actually having drugs in their hands. Furthermore, in the actual moment of use, that sense of "more and more" can drive users to great excess, further implicating salience as a core part of addiction.

But salience is not a complete explanation, as if the salience—the desire—exists without content or context, without the meaning that people provide to every situation and behavior they encounter. Anthropology insists that feelings like desire and involvement are subjective and cultural, as well as biological. In an ethnographic sense, the salience of the moment was that it transported Colombian adolescents away from their everyday life. It took them on a "*viaje*," a trip, or put them into a "*video.*" The desire is not something solely driven by a malfunctioning brain; even in a problem like addiction, the person matters. In the case of Colombian adolescents, they wanted to continue with the experience that drugs afforded them.

You get this desire to like continue on in the same video, in the same game, like to go back to feeling nothing, to not feeling any problem.

Incentive Salience, Brain, and Environment

Alongside being only one component of larger decision making, incentive salience is not some innately determined function provided by a specific circuit in the brain. First, incentive salience cannot be localized to one spot in the brain. No one circuit and no one neurotransmitter alone accounts for incentive salience. At present, research indicates that incentive salience can be mediated by both the nucleus accumbens and the ventral pallidum, forming part of the overall set of circuits that shape an individual's pursuit of goals: "Neurochemical 'wanting' mechanisms are more numerous and diverse in both neurochemical and neuroanatomical domains. ... In addition to opioid systems, dopamine and dopamine interactions with corticolimbic glutamate and other neurochemical systems activate incentive salience 'wanting' (Berridge et al., 2009, p. 69)."

In other words, there is no one dedicated pathway, no single brain module for salience. Rather, the attribution of salience—and what that actually means—depends on patterns of firing, individual experience, and social and environmental context. For example, salience can be paired with hedonics by firing in the nucleus accumbens, if both lever pressing and reward delivery are experienced together. However, physically separate out the hedonic reward—the site to get the rat pellet—from the place for pressing the lever, and dopamine signaling mediates the salience of the lever (Berridge et al., 2009). In this case, the lever pressing has become a "sign" for reward, and thus distinguishable from the end goal, getting the pellet. Humans themselves often construct environments like this, social structures that represent a series of steps before a final reward. For example, there can be a series of hoops to jump through before getting a job promotion.

Individual state and experience can also shape incentive salience signaling. Normally, extremely salty water is aversive to rats. However, deprive that rat of salt, and the animal has a large incentive signal to cues for salty water, indicating that water should be sought out while in this state (Tindell, Smith, Berridge, & Aldridge, 2009). Moreover, rats that were deprived of good mothering when young are more likely to signal salience to cues when juveniles or adults (Lomanowska, Lovic, Rankine, Mooney, et al., 2011). These animals are in a state where quick reaction to cues for reward might make a difference in getting the resources that matter. The environment did not provide these resources when young, and now they have developed into animals that react easily to possibilities and options that might lead to benefits.

Addiction and the Environment for Salience

As this research shows, the problem of incentive salience in addiction is not simply due to "drugs did this to him or her." Rather, signals for salience depend on the presence of cues, the structure of environments, and present and past states. In other words, when behavioral options are salient, animals will pursue them—that is what incentive salience does. Experimenters manipulate the testing environment, shaping the potential options available, the contexts animals occupy, and the timing and amount of any rewards. The environment that these animal models occupy is never natural. Neither is the human environment—the options individuals have, how those options are understood, and the positive and negative consequences they might have are generally set by human forces well beyond the power of any one individual.

Let me give a concrete example. Drug use is often associated with school problems among adolescents. Often these kids are seen as double losers, not taking care of school work and getting involved with drugs, and blame is placed directly on the boy and girl for failing at school and for using drugs. The explanation then generally goes into the kneejerk answer realm—if only they had stayed away from the pleasures of drugs, buckled down and really did well at school, then everything would have turned out all right.

However, from the adolescent's point of view, the structure and people at school are often set against him or her. Other students will easily do better, and in difficult cases, the teachers work actively against the success of the student. There are few options for long-term success at school, and thus jumping through hoops now has very little salience for these kids. Kids like this often have troubles at home as well, and thus the rewards of family and of school are not available. They are, in one sense, in a deprived state, and other options become salient—other options for fun, for success, for a sense of involvement.

These adolescents often tried drugs, and between the social relationships and the fun of passing the time, this type of activity became important to them. Given how modern societies approach drug use—often demonizing it and confining its use to marginalized places of the social map—drug cues and drug availability come packaged together in specific environments. When users are in these types of environment, few options lead away to that "better life" many others pursue in different social environments. There are few cues to school, few cues to family, few cues to successful jobs. If anything, these are often seen in a negative light in drug-using contexts because the people in these spaces have had negative experiences in all three areas. They simply are not salient, either in the present or in the past.

Incentive Salience and Anthropology

Incentive salience can have a broader relevance within anthropology. I have focused on the experience of salience within a particular behavioral domain, in this case heavy drug use, and on salience as mediated by the social environment, in this case, the combination of problems at school and at home that adolescent drug users in Colombia often faced. Building on these examples, incentive salience can be applied in two ways: (1) mediating the sense of involvement in cultural activities and meanings, of commitment as a subjective state rather than an abstract belief; and (2) signaling transitions between activities, of shifting attention and involvement from what matters throughout a day or a set of important activities.

Salience can mark the conjunction of events and contexts with an indication of overall importance and self-involvement.

Berridge et al. (2009) highlight how incentive salience is about a subjective sense of "this matters," rather than conscious desire. Incentive salience is the motivation, rather than appreciation, of want and desire. In this broader sense, incentive salience can be related to "passion" in the word's older meaning—the sense of being carried away, of something that can overwhelm one's conscious sense of self but which also marks whatever is happening as vital. In this way, salience is the passion to our cold calculation of costs and benefits—it is Tarde's "passionate interests," the signal for the amplification in the intensity of passions that plays such a role both in capitalist economics and consumption and also in our social relations (Latour & Lepinay, 2010). Dopamine also mediates effort, of how much work an animal will put into a task—something related to sense of self and to commitment (Salamone, Correa, Farrar, & Mingote, 2007). Thus, the concept of salience captures a neurological part of how we decide what is important and how much effort we put into it.

Our irrational wants are some of our most self-defining. Our vices, the things we truly pursue, where we put our effort—these things often define us, both to ourselves and to others. Saliency can contribute to this process. As Northoff and Hayes (2011, p. 1023) argue, "Saliency may provide a conceptual bridge from reward to the self: stimuli related to the self are highly relevant and important and therefore may recruit and allocate more sensory, affective, and cognitive resources than nonself-related ones." We are what we do, as it were, and the push toward certain types of doing—mediated by salience—plays an important role in defining a practice-oriented self. This is even more the case when these activities are tied to specific social contexts and to transitions in self-awareness and subjectivity—exactly the sort of thing that happens with psychoactive drugs.

But salience is not only about relevance to self. It can signal relevance or importance in the environment. Zacks, Kurby, Eisenberg, and Haroutunian (2011) point out that midbrain and striatal areas associated with dopamine release increased their activity during testing of human subjects engaged in short-term predictions regarding what would happen next. They write: "Those points where prediction was more difficult corresponded with subjective boundaries in the stream of experience. ... This [dopamine] activity could provide a global updating signal, cuing other brain systems that a significant new event has begun." Similarly, Redgrave and Gurney have long argued that dopamine signals represent shifts in attention, which an animal can use to reorient toward significant events and environmental opportunities (Redgrave & Gurney, 2006; Redgrave, Prescott, & Gurney, 1999).

Salience—a signal to seek and of importance—is one way this could happen. For example, hard-core drug users often remarked about how the end of school, when they knew they could go use, came with a heightened anticipation and desire, and then an urge to seek out drugs. As school ended—one social activity finished—incentive salience marked these individuals' transition to another social activity, that of going to find drugs and hang out with friends. More broadly, the sort of signaling provided by salience can contribute to the patterned behavior that we often view as institutional, ritualized, and cultural. Through providing impetus to what we seek and the urges and effort that mark daily life, salience contributes to the acquisition and instantiation of our passions and our customs. What we find salient arises from within as well as from without, through how institutional strictures and goals interact with individuals' signaling of salience. Schools provide the institutional setting for learning; students comply, but with some significant variation in internalizing and executing the preferred actions. In Colombia, almost all of the adolescents showed up. Many of them put in the effort and wanted good grades. Others, like kids not doing well at home and at school, found the first school bell marking the time they spent away from drugs.

But patterned behavior is not due to salience alone. Habits also play a role in what we do, when, and why.

Habits and Addiction

As I said earlier, the conjunction of "wanting more and more" and a handy neurobiological theory initially limited my explanatory frame. However, brain research on addiction has exploded over the past five years, with genuine advances taking us away from the simple story that dopamine is to blame. The good thing is that the rich ethnography of the project, the thick description of lives and experiences, still stands valid. New theories about the neurobiology of addiction can be tested and informed by real life.

In Colombia, adolescents spoke of days of drug use, or of the pattern of going to school and then going to use. They described ritually preparing drugs to use with friends, rolling the drugs into a cigarette just so and then passing it around. They also recounted lighting one *basuco* (crack) cigarette after another, driven in part by want but also, especially as drug use continued for extended periods, by an excess of repetition.

Over time, the habitual nature of addiction became clear—the repetitive aspect of use, separate from the description of wanting and the pleasure and meaning of use itself. This repetition of seeking out and using drugs is a core behavioral aspect of addiction. This habitual use is directly related to the neglect of other social roles that is a core diagnostic criterion for substance abuse, as well as the sense from both users and outsiders that addiction has a compulsive side to it, a sense of just doing it over and over.

Recently, neuroscience research on addiction has focused on the habitual aspects of addiction (Everitt & Robbins, 2005; Kalivas & O'Brien, 2008; Ostlund & Balleine, 2008). It is now clear that habits are mediated by neurobiological systems that are separate from the mesocorticolimbic dopamine system implicated in reward and wanting. That system runs through the nucleus accumbens, located in the ventral striatum area of the brain. But habits, particularly drug habits, involve the dorsal striatum. Different lines of research have established that as animals transition from initial learning to habitual responding, neural activation moves from the ventral striatum to the dorsal striatum (Graybiel, 2008). For drug users, the dorsal striatum mediates "the maintenance of drug seeking," rather than initial learning about use.

Graybiel (2008, p. 363) defines habits as "learned, repetitive, sequential, context-triggered behaviors, [which are] performed not in relation to a current or future goal but rather in relation to a previous goal and the antecedent behavior that most successfully led to achieving that goal." This definition highlights that habits are not as subject to patterns of reinforcement as initial learning (and this serves as a key way to distinguish types of behavior in the laboratory setting). Graybiel's definition also points to habits being driven by context, and that the sequence of learned behaviors is related to previous experience and behavior in that setting. For addiction, this indicates that if an addict is prompted to seek out a high (say, by salience signaling the transition to after school, and then increasing wanting for drugs), the habit takes that person through the steps of getting that high based on some previous context, whether on the street or with a stash at home.

The move from the anterior part of the striatum (learning) to the dorsal part of the striatum (habit) is more complicated than localization in the brain (learning here, habits there). "The transition from goal-directed to habitual control of behavior is highly dynamic and ... the early phrase of the habit learning process occurs even while behavior is still demonstrably goal-directed" (Tricomi, Balleine, & O'Doherty, 2009, p. 2230). The systems can work together in guiding behavior. Furthermore, it is behavior that ties together the ventral and dorsal striatum. Drug administration by the researchers alone (for example, through direct injections) is not enough to establish a habitual behavior (Kalivas & O'Brien, 2008). The animal actually has to learn to use drugs. That learning, which sets up a sequence of behavior and a goal for the behavior, is what can get turned into the habit. This research brings neuroscience closer to classic social science research that showed that marijuana users had to learn the expected effects from marijuana smoking (Becker, 1953), and indicates that functional use, the utilization of drugs to achieve specific experiences and outcomes, likely plays an operative role in driving the transition from casual to serious use (Lende et al., 2007; Müller & Schumann, 2011).

Once behavior is mediated by the dorsal striatum, neuroscience researchers have found that it becomes "relatively insensitive" to changes in the actual rewards that happen from the behavior (Ostlund & Balleine, 2008). At this point, even punishing the behavior generates "exceedingly slow" change. One way to explain this discrepancy is that the habitual behavior is cued by context and by activation in the dorsal striatum. Since the ventral striatum is not activated, there is little evaluation of the behavior in relation to actual outcomes and consequences (Graybiel, 2008). This disparity is one way that neuroscientists are attempting to account for the inability of many addicts to change their behavior, despite the often adverse consequences they face.

An important way that habitual behavior is examined in the laboratory is in relation to "extinction training." With extinction training, the animal is placed in a new context and allowed to press a lever that delivers no drugs until the lever pressing completely disappears—the lever press response is then extinguished. Extinction training has been a good way to test for the reinstatement of drug seeking by animal models. With research on habits, an intriguing result has emerged. Unless animals go through this extinction process, drug-seeking reappears in the appropriate context as long as the dorsal striatum is intact (Fuchs, Branham, & See, 2006). Continued drug-seeking happens even when other brain structures associated with drug use have been knocked out by, for example, using surgical lesions or drugs to stop salience and reward processing in the nucleus accumbens. However, there is nothing like "extinction training" in the life of Colombian adolescents. They can get drugs when they want them, and are often exposed to the contexts where they used in their everyday lives. This exposure means that habitual use can continue even if the initial learning dynamics have changed. Even if drugs no long provide the same pleasure because of tolerance, and even if the signaling of salience is blunted because of extreme levels of use, dorsal striatum activation should still be sufficient to carry habitual drug use along.

Overall, the salience model works well in conjunction with habit responding. The habit side can mediate the procedural part of seeking out and consuming drugs; the heightened desire for drugs can come from incentive salience. The urge to use itself likely comes from the dorsal striatum and the mesocorticolimbic system working together. As Everitt and Robbins (2005) write, a sense of "must do" could come from the dorsal striatum, different from desire itself. Indeed, this type of insight can better explain adolescents who told me how they went toward using drugs without really understanding why, without that hard craving—just an urge pushing them along.

I started to get that little taste and those little tickles around here [her stomach], all strange, and that anxiety that you want to … and I was going at once, I caught the bus, I stopped drinking and everything and went straight for 'la olla' [where drugs were bought and consumed]. … So I went along like that, I went by the 'olla' and I got down from the bus just like that. That's what the impulse was like, I got off without even wanting to, I got off and when I was off I was saying, two and that's it. And then I was on two and I just went along and went along and went along.

Training Habits

Research on habits in animal models has revealed a range of specific results that have bearing on addiction. First, there are specific ways to generate habits in the laboratory. In the laboratory, scientists have found that there are two main ways to create habits—lots of repetition, or "overtraining," and using a "variable interval schedule" of reinforcement, meaning that the reward only comes after an unpredictable period of time has passed since the last reward (Ostlund & Balleine, 2008). These laboratory results correspond well with my ethnographic research in Colombia. There, as in many places, drug and alcohol use happen on a variable interval, for example, using on the weekends, and then transitioning to using every day after school. At the level of daily use, the amount of repetition of use that happens over weeks is quite comparable to the "overtraining" seen in laboratory models.

The variable interval schedule raises an important point about how excessive drug use can be maintained. One of the great fallacies of drug addiction research is that addicts find little pleasure in their use, particularly at the most desperate of stages. While addicts can use just to maintain themselves and get away from the negative side of withdrawal (Koob & Le Moal, 2008), this type of approach misses how drug use can remain a rewarding activity. As my ethnographic research identified, it is the shift in subjective experience caused by drug use that is one of the key rewarding factors from use. In Colombia, that shift took users into a different subjective domain, often seen as being "*englobado*" (wrapped up in things) or as being in a "video." In both cases, this shift was different from everyday life—that was the reward. Even if the pharmacological kick was not quite what it used to be, the subjective shift tied to specific social contexts remained a potent reward for users. They might not have felt the same euphoria drugs once brought, but at least they exited the mundane or uncomfortable existence they inhabited without using drugs.

Moreover, the social interaction young users found in the drug use setting were actually one of the main motivators and rewards for people with deep involvement in that setting. Oftentimes, as individuals became more involved in a drug-using scene, their other social relationships suffered. For example, in Colombia, relationships with family often grew strained, and social stigma against drug users in the local community came to bear heavily on the adolescents. The daily interaction and the approval of use, as well as the sense of belonging to a group bonded together against outsiders, all provided powerful feedback to individuals. Even if drugs no longer provided the same pharmacological reward they once offered, the increased social interaction and alternative social network could become more important than ever for hard-core users.

Stress is known to increase habitual behavior. Faced with stress, more automatic responses can lead to maladaptive behavior because the habit may not match the behavior appropriate to the situation or because the individual may not be able to set aside a habitual response in favor of pursing a long-term goal (Dallman, 2010). Recently, stress has even been shown to increase the activation of habitual behaviors keyed to the dorsal striatum (Schwabe, Dickinson, & Wolf, 2011). For hard-core drug users, stress—alongside involvement in a social scene that cues drug use—can prompt greater use. Oftentimes, in the adolescents I knew, the deeper their involvement with street drug use, the greater the stress they faced, ranging from family disapproval to violence and stigma on the street. In this way, structural violence and inequality worked against these adolescents, heightening the consequences of drug use and worsening the stress associated with use, and thus directly increasing the chances of these young men and women continuing with their often destructive behavior.

Habits, Models, and Culture

Research on habitual behavior, in conjunction with neurological under-standings of ventral and dorsal striatum dynamics, highlights an important way of thinking about how the brain mediates habitual action—how it instantiates the practices and models of behavior that make up such a large part of our daily sociocultural repertoire. Graybiel (2008, pp. 361–362) makes a compelling argument that one basic way that the brain handles the transition from learning about potential behaviors and their associated cues and rewards is through "chunking."

Many of these repetitive behaviors, whether motor or cognitive, are built up in part through the action of basal ganglia–based neural circuits that can iteratively evaluate contexts and select actions, and can then form chunked representations of action sequences that can influence both cortical and subcortical brain structures ... These findings suggest that one result of habit learning is to build in the senso-rimotor striatum chunked, boundary-marked representations of the entire set of action steps that make up the behavioral trait ... As the behavior became repetitive, the neural activity took on the accentuated beginning and end patterns (neural exploitation). This shift could represent part of the process by which action sequences are chunked for representation as a result of habit learning: When they are packaged as a unit ready for expression, the boundaries of the unit are marked and the behav-ioral steps unfold from the first to the last boundary marker. (Graybiel, 2008, pp. 361, 377)

Graybiel's point about chunking is important for understanding how the transition from learning to take drugs to habitual use happens, and why that can become problematic with deepening social involvement. Graybiel (2008) proposes a two-step model—first, chunking happens, bringing together a representation of action sequences that draw on both internal and external dynamics (internal states, external cues, potential rewards, and so forth); second, once that chunked representation is in place, neural signaling marks the beginning and end, leading to a relatively automatic execution of the representation. In other words, the focus becomes on signaling what to do when, followed by the chunked habitual behavior. When that action sequence ends another chunked sequence or evaluation can happen.

In anthropological terms, the chunking of an action representation captures individual-environment interactions—it is those interactions that dramatically shape the chunking, rather than some innate process of the brain. For example, there is a dramatic difference between the drinking done at a refined dinner party and at an out-of-control frat party. The styles of drinking—what counts as an appropriate level—are dramatically different in both situations, and the signaling of when to start, how much to ingest, and when to stop are at opposite ends. A sip of wine after tasting a morsel of food and repeated returns to the keg of beer involve radically different levels of alcohol consumption; but they are the expected actions in each social situation. The chunking of action—of sipping versus chugging—is going to be different as a result of each situation.

One model for the transition from goal-oriented to habit-based behavior is based on a learning model that moves from a model-based system to a cache-based system (Daw, Niv, & Dayan, 2005; Daw, Gershman, Seymour, Dayan, et al., 2011; Graybiel, 2008). Through reward error learning, animals use the dopamine signal to match potential behaviors and environmental cues with specific social and physiological rewards—that is, they are building a model, a cognitive representation of the task structure and how best to accomplish the task. This type of learning is largely mediated through the mesocorticolimbic dopamine system. Once the model is in place, due both to the neural dynamics of learning and to efficiency of execution, the striatum takes over on executing the behavior using a "cache" system, which contains stored inflexible sequences that can be executed quickly. The cache system is used to identify the right situation for the model and then mediates the automatic execution of a particular behavior sequence.

As both research on cultural models and on habitual action indicates (D'Andrade, 1995; Graybiel, 2008), the underlying representation (the model) will generally fade into the background as it is learned. Once it is internalized most people will pick up on the surface details—on whether the host has brought a good bottle of wine, requiring more appreciative drinking, or opened the tap on the keg, declaring that it won't close until the keg is kicked. But in each case, the model of the behavior—sipping wine or chugging beer—will be the basis for the action sequence. This particular practice, socially encoded and neurally represented, is thus intimately neuroanthropological, constructed by socially sanctioned and culturally motivated patterns of action.

Chunked representations of action sequences, linked to particular social contexts, provides a dramatic way to think about cultural models and practice theory within anthropology. In psychological anthropology, cultural models are often seen as flexible, fill-in-the-blank guides, for example, a restaurant model for ordering food that can be used either in a fast-food place or an elegant restaurant (D'Andrade & Strauss, 1992). By having prototypical action sequences—chunked together by the brain—people can execute culturally patterned behavior effectively and rapidly. Graybiel's chunking and anthropology's cultural models are fairly close to one another as concepts. What is even more interesting is that the brain uses signals to emphasize the start and finish of a prototypical action chunk. In other words, the neural signals provide a start/stop sequence, with the ability to string together different chunked models or to repeat a particular type of action as necessary. But for addiction, this type of engrained practice can be problematic, since it can get started over and over, the sequence repeated as long as the start signal is provided. Social contexts and incentive salience together can provide that sort of start cue.

The larger point is that something like "practices" or "models" in the sense used in anthropology are handled by the brain, where a model of a behavior is created and then cached, and then applied in appropriate situations. This neural instantiation view of practices provides a good complement to work that examines how enculturation happens through training, skill acquisition, and other forms of patterned practice (Downey, 2010; Roepstorff, Niewöhner, & Beck, 2010). In acquiring culture, different parts of the brain take part in the overall process, meaning that enculturation into specific practices or types of social action is more complex than "what fires together, wires together." Specific social practices are literally learned, modeled, cached, and then used, in conjunction with ongoing feedback on how behavior matches that model or, where previous experience and specific social situations indicate, simply executed in an efficient fashion.

With drug use, this neuroanthropological approach offers a powerful way to tie together the habitual nature of serious drug use with the social nature of serious drug use, recognizing how use happens in specific social situations. Early research in anthropology and sociology highlighted this as a "subculture of drug use," and the habit-learning model offers a way to see how that subculture gets under the skin to guide drug behavior in those types of settings. But the anthropology of drug use has moved beyond the subculture concept, which assumes a homogeneity and unity that simply is not there on the ground (similar to how the "culture" concept has also changed, no longer emphasizing bounded, homogeneous groups). Rather, people come together in specific social contexts based on social relations, shared values, and specific practices, prominent among them drug use (Moore, 2004; Rhodes, 2009).

Not everyone is the same in this social setting; rather, they are distinguished by how often they come to this drug-use context, how much they use, what types of identifiers they use (e.g., clothing and language) to indicate approval of drug use and solidarity with others in the group, and so forth. Behaviors—how often, how much—and contexts—particular sites of use, ways of talking, types of clothing—work together to shape drug use, and ultimately, addiction. As involvement deepens, and drug use becomes more salient and more habitual, excessive drug use becomes the norm, bolstered by the sense of commitment and engagement with a social role and the rewards of participating with people in a joint, often pleasurable experience that generally stands in sharp contrast with the options and the rewards available in other parts of these adolescents' lives.

Drug use in real life remains a dynamic mix of habit and goal-oriented behavior. The habit can drive a great deal of excessive drug use, reinforced by the shifts in subjective experience that the combination of drug use and social contexts consistently generates. As individuals become more involved in these drug scenes, and incentive salience signaling increases, individuals seek out and use drugs more. That use then drives further habitual learning, leading to the repetition of a behavior. Addicts want more and more, and do more and more. "More and more" exists only in reference to life experiences and social contexts, with that reference signaled by salience and by cached models. Thus, addiction is both anthropological and neurological. Neuroanthropology helps us better understand and study how those two arenas come together to define the problem of addiction.

References

American Psychiatric Association. (2011). DSM-5 development proposed revisions: Substance use and addictive disorders. Retrieved from http://www.dsm5.org/proposedrevision/Pages/SubstanceUseandAddictiveDisorders.aspx

Becker, H. S. (1953). Becoming a marihuana user. *American Journal of Sociology, 59*(3), 235–242.

Berridge, K. C., Robinson, T. E., & Aldridge, J. W. (2009). Dissecting components of reward: "Liking", "wanting", and learning. *Current Opinion in Pharmacology, 9*(1), 65–73.

Dallman, M. F. (2010). Stress-induced obesity and the emotional nervous system. *Trends in Endocrinology and Metabolism, 21*(3), 159–165.

D'Andrade, R. G. (1995). *The development of cognitive anthropology*. Cambridge: Cambridge University Press.

D'Andrade, R. G., & Strauss, C. (Eds.). (1992). *Human motives and cultural models*. Cambridge: Cambridge University Press.

Daw, N. D., Niv, Y., & Dayan, P. (2005). Uncertainty-based competition between prefrontal and dorsolateral striatal systems for behavioral control. *Nature Neuroscience, 8,* 1704–1711.

Daw, N. D., Gershman, S. J., Seymour, B., Dayan, P., & Dolan, R. J. (2011). Model-based influences on humans' choices and striatal prediction errors. *Neuron, 69*(6), 1204–1215.

Downey, G. (2010). "Practice without theory": A neuroanthropological perspective on embodied learning. *Journal of the Royal Anthropological Institute, 16*(S1), S22–S40.

Everitt, B. J., & Robbins, T. W. (2005). Neural systems of reinforcement for drug addiction: From actions to habits to compulsion. *Nature Neuroscience, 8,* 1481–1489.

Fuchs, R. A., Branham, R. K., & See, R. E. (2006). Different neural substrates mediate cocaine seeking after abstinence versus extinction training: A critical role for the dorsolateral caudate-putamen. *Journal of Neuroscience, 26*(13), 3584–3588.

Graybiel, A. M. (2008). Habits, rituals, and the evaluative brain. *Annual Review of Neuroscience, 31,* 359–387.

Kalivas, P. W., & O'Brien, C. (2008). Drug addiction as a pathology of staged neuroplasticity. *Neuropsychopharmacology, 33,* 166–180.

Koob, G., & Le Moal, M. (2008). Addiction and the brain antireward system. *Annual Review of Psychology, 59,* 29–53.

Koob, G. F., & Nestler, E. J. (1997). The neurobiology of drug addiction. *Journal of Neuropsychiatry and Clinical Neurosciences, 9,* 482–497.

Latour, B., & Lepinay, V. A. (2010). *The science of passionate interests: An introduction to Gabriel Tarde's economic anthropology.* Chicago: Prickly Paradigm Press.

Lende, D. H. (2005). Wanting and drug use: A biocultural analysis of addiction. *Ethos, 33*(1), 100–124.

Lende, D. H. (2007). Evolution and modern behavioral problems. In W. Trevathan, E. O. Smith, & James J. McKenna (Eds), *Evolutionary medicine and health: New perspectives* (pp. 277–290). New York: Oxford University Press.

Lende, D. H., Leonard, T., Sterk, C. E., & Elifson, K. (2007). Functional methamphetamine use: The insiders' perspective. *Addiction Research and Theory, 15*(5), 465–477.

Lende, D. H., & Smith, E. O. (2002). Evolution meets biopsychosociality: An analysis of addictive behavior. *Addiction, 97*(4), 447–458.

Lomanowska, A. M., Lovic, V., Rankine, M. J., Mooney, S. J., Robinson, T. E., & Kraemer, G. W. (2011). Inadequate early social experience increases the incentive salience of reward-related cues in adulthood. *Behavioural Brain Research, 220*(1), 91–99.

Moore, D. (2004). Beyond "subculture" in the ethnography of illicit drug use. *Contemporary Drug Problems, 31*, 181–212.

Müller, C. P., & Schumann, G. (2011). Drugs as instruments: A new framework for non-addictive psychoactive drug use. *Behavioral and Brain Sciences, 34*(6), 293–310.

Northoff, G., & Hayes, D. J. (2011). Is the self nothing but reward? *Biological Psychiatry, 69*(11), 1019–1025.

Ostlund, S. B., & Balleine, B. W. (2008). On habits and addiction: An associative analysis of compulsive drug seeking. *Drug Discovery Today: Disease Models, 5*(4), 235–245.

Redish, A. D., Jensen, S., & Johnson, A. (2008). A unified framework for addiction: Vulnerabilities in the decision process. *Behavioral and Brain Sciences, 31*(4), 415–437.

Redgrave, P., & Gurney, K. (2006). The short-latency dopamine signal: A role in discovering novel actions? *Nature Reviews: Neuroscience, 7*, 967–975.

Redgrave, P., Prescott, T. J., & Gurney, K. (1999). Is the short-latency dopamine response too short to signal reward error? *Trends in Neurosciences, 22*(4), 146–151.

Rhodes, T. (2009). Risk environments and drug harms: A social science for harm reduction approach. *International Journal on Drug Policy, 20*(3), 193–201.

Robinson, T. E., & Berridge, K. C. (1993). The neural basis of drug craving: An incentive-sensitization theory of addiction. *Brain Research: Brain Research Reviews, 18*(3), 247–291.

Roepstorff, A., Niewöhner, J., & Beck, S. (2010). Enculturing brains through patterned practices. *Neural Networks, 23*(8-9), 1051–1059.

Salamone, J. D., Correa, M., Farrar, A., & Mingote, S. (2007). Effort-related functions of nucleus accumbens dopamine and associated forebrain circuits. *Psychopharmacology, 191*(3), 461–482.

Schultz, W. (2011). Potential vulnerabilities of neuronal reward, risk, and decision mechanisms to addictive drugs. *Neuron, 69*(4), 603–617.

Schwabe, L., Dickinson, A., & Wolf, O. T. (2011). Stress, habits, and drug addiction: A psychoneuroendocrinological perspective. *Experimental and Clinical Psychopharmacology, 19*(1), 53–63.

Tindell, A. J., Smith, K. S., Berridge, K. C., & Aldridge, J. W. (2009). Dynamic computation of incentive salience: "Wanting" what was never "liked." *Journal of Neuroscience, 29*(39), 12220–12228.

Tricomi, E., Balleine, B. W., & O'Doherty, J. P. (2009). A specific role for posterior dorsolateral striatum in human habit learning. *European Journal of Neuroscience, 29*, 2225–2232.

Zacks, J. M., Kurby, C. A., Eisenberg, M. L., & Haroutunian, N. (2011). Prediction error associated with the perceptual segmentation of naturalistic events. *Journal of Cognitive Neuroscience, 23*, 4057–4066.

14 Cultural Consonance, Consciousness, and Depression: Genetic Moderating Effects on the Psychological Mediators of Culture

William W. Dressler, Mauro C. Balieiro, and José Ernesto dos Santos

Paulo is a 32-year-old man living in a working-class neighborhood in Ribeirão Preto, Brazil. He works as a computer technician in a local soft-drink bottling plant. Married, with one child, Paulo scored slightly below average on a composite measure of cultural consonance at an initial interview. This indicates that, overall, he lives a life that corresponds reasonably well, although far from perfectly, to the shared cultural models that Brazilians use to talk about daily life in terms of their lifestyles, families, network of friends, and views of society. For example, when talking about the most important characteristics of a family, Brazilians value affect and communication within the family, as well as organization and structure. Paulo tended to see his family as possessing these characteristics, but perhaps not quite as strongly as other families. Similarly, in terms of lifestyle, Paulo owned his own home and had most of the major appliances Brazilians regard as essential to a good life. He engaged in leisure activities such as going out with friends and participating in sports at a local club. On a scale of depressive symptoms, Paulo scored exactly zero, not reporting a single symptom. At a follow-up interview two years later, basically nothing had changed in his life, and his expression of symptoms was again zero.

João is also 32 years old and married, with one child. He is a metal worker and lives in a different, but comparable, working-class neighborhood. João scored somewhat lower than Paulo on the measure of cultural consonance. He did not own his own home, and valued material goods were more scarce. Significantly, João engaged in virtually no leisure-time activities beyond resting alone and reading, eschewing the sociality that Brazilians cherish. He also expressed a more jaundiced view of his family in the terms Brazilians use in everyday conversation, especially seeing his family as lacking in organization and not having good communication. João talked in terms of not feeling in control of his life, that it didn't, in some significant way, "make sense." Finally, with respect to depressed affect, João expressed feelings of sadness, low self-esteem, fatigue, and helplessness, although not at quite the intensity required for a diagnosis of probable depression. His symptom score did, however, place him in the upper 30% of men in this community in terms of depressed affect.

Two years later, things had changed substantially for João. His cultural consonance in the domain of lifestyle had increased considerably, particularly in terms of engaging in valued leisure-time activities such as going out with friends, going to the city's biggest shopping center on the weekend just to pass some time, and going out to the movies. His perception of his family had changed as well, and he now saw them as more organized and more caring. He expressed virtually no symptoms of depression. We will argue that this increase in cultural consonance—or the increased correspondence of his own life with Brazilian cultural models of how life is to be lived—led to his reduced depressed affect.

There is, however, one more difference between Paulo and João, deep in the structure of their brains. Paulo and João, like all other human beings, have receptors on their brain nerve cells to which the neurotransmitter serotonin attaches. Serotonin is released by a neuron in order to transmit a signal across the synapse that separates one neuron from another, and the receptor is there to, as its name implies, receive the signal. But Paulo's gene producing that receptor differs from João's very slightly (literally in terms of one pair of bases that make up DNA). While each genetic variant produces a fully functioning receptor, João's receptor works a little differently. It tends to react more strongly to the signal than does Paulo's. As we will argue below, this may tend to amplify the signal. João's increase in cultural consonance over the two years would have been sufficient to reduce his depressed affect. But, we will argue, João's specific genotype helped to increase the impact of that change in cultural consonance, because of the cultural salience of the domains of life (lifestyle, the family) in which the changes occurred. We will argue, in other words, that depressed affect is a function of how culturally salient beliefs and behaviors intersect with the genetic substrate that structures the neural net and its functioning.

Introduction

Research on the causes of depression has focused on stressors emanating from the social environment (such as stressful life events); psychological mediators (such as perceived stress); and, with the mapping of the human genome, on specific genetic polymorphisms thought to influence risk

(Monroe & Reid, 2009). More recently, research has examined the synergistic effects of genes and stressors (Monroe & Reid, 2008). Of the variety of factors thought to influence mood, emotion, and psychological distress, cultural influences have been perhaps the most difficult to study, due to seemingly intractable theoretical and methodological problems; however, the recent development of the theory of cultural consonance offers a new direction in research on culture (Dressler, 2007). The concept and measurement of cultural consonance are embedded in a theory of culture emphasizing the sharing of meaning and knowledge within social groups. Cultural consonance captures the sense in which individuals incorporate shared meanings guiding belief and behavior into their own lives. It is, in short, a measure of the degree to which people actually live like the prototypes encoded in shared cultural models. Lower cultural consonance is, in turn, associated with greater depression, because to live at the margins of the social and cultural space defining one's existence can be a profoundly stressful experience (Dressler, Balieiro, Ribeiro, & dos Santos, 2007a).

Recently, we also reported on an interaction between cultural consonance and a genetic polymorphism in the brain serotonin system. The magnitude of the effect of cultural consonance on depression changes depending on the specific variant of the gene coding for one of the serotonin receptors (Dressler, Balieiro, Ribeiro, & dos Santos, 2009). In this chapter, we will examine these findings further, especially in terms of the way in which the psychological processes that mediate the link of cultural consonance and depression are in turn modified in the presence of a specific genetic variant. These results have implications for how the intersection of neurophysiology, culture, and consciousness influence individual human adaptation.

Theoretical Background

Cultural Consonance and Depression
Few would doubt that the cultural milieu impacts depression, but research in this area has been hampered by conceptual and methodological problems. Many researchers implicitly rely on general notions of culture as a system of shared beliefs and values that define societal traditions (Kuper, 1999). This general orientation does not address several fundamental conceptual problems in defining and studying culture: (a) it is based in a weak social ontology (i.e., a theory of the nature of social and cultural facts); (b) culture is conflated with conventional social-psychological constructs; (c) intracultural diversity is not taken into account; (d) culture is

defined exclusively at the aggregate or societal level, without explicitly linking culture to the individual; and, (e) methods for assessing cultural sharing at the aggregate level, or measuring culturally relevant variables at the individual level, have not been applied.

The theory of "cultural consonance" described by Dressler (2007) and colleagues (Dressler et al., 2007a; 2009) was developed to address these issues. As mentioned previously, cultural consonance is the degree to which individuals, in their own beliefs and behaviors, approximate the prototypes for belief and behavior encoded in shared cultural models. The theory of cultural consonance is derived from a cognitive theory of culture, in which culture is defined as that which one needs to know in order to function adequately in a given society (Goodenough, 1996). This knowledge is cognitively stored and structured as schematic models of specific cultural domains that are both shared and differentially distributed within society (D'Andrade, 1995). The meanings encoded in cultural models are rarely defined by what any given individual knows. Culture resides in part in individual minds, but it is ultimately defined by the knowledge distributed across minds. Culture is thus a term the referents of which are both individuals and social aggregates (D'Andrade, 1995; Dressler, 2007).

A cultural model is a blueprint or schematic, and how it is realized in practice will vary. Cultural consonance captures this sense of practice. Individuals may, in their own beliefs and behaviors, diverge from cultural prototypes for a variety of reasons. In many instances, the economic resources to put cultural motives into action may be lacking. Or individuals may, upon conscious reflection, choose a different course of action. Whatever the cause, there can be a gap between what cultural models describe and what individuals are actually doing and believing. Given that cultural models describe what is widely expected with respect to behavior and belief, when individuals fail to correspond in their own beliefs and behaviors to these expectations (low cultural consonance) there may be confusion, misunderstanding, and negative social sanction in mundane social interaction. Individuals may fail to receive the positive social feedback in social interaction that reassures them that they are accepted and valued members of society. Low cultural consonance can be, in short, a chronically stressful experience and it is a potent predictor of a variety of health outcomes (Dressler, 2007).

Measuring cultural consonance requires a two-stage, mixed-methods research design. The first stage consists of a cultural domain analysis in which the elements making up a particular domain and their semantic relationships within the domain are explored in detail (Borgatti, 1999). One dimension that structures many domains, and that is of particular relevance to the measurement of cultural consonance, is value or importance. For each domain in which cultural consonance will be measured, respondents rank or rate the importance of items, and cultural consensus is used to verify the sharing of that evaluation (Romney, Weller, & Batchelder, 1986). Cultural consensus analysis generates a "best cultural estimate" of the relative importance of items within that domain. Then, in a community survey, respondents' actual behavior or beliefs (depending on the nature of the domain) are ascertained, and these are linked to the consensus data. Higher scores on a measure of cultural consonance indicate that individuals endorse beliefs or report behaviors that are collectively rated as important in that society (the measurement of cultural consonance in presented in detail elsewhere; see Dressler, Borges, Balieiro, & dos Santos, 2005).

Cultural consonance has been measured in several domains, including lifestyles, social support, family life, food, and national identity. Higher cultural consonance in each of these domains was associated with lower arterial blood pressure, fewer reported depressive symptoms, lower perceived stress, and lower body mass, controlling for appropriate covariates (Dressler, 2007). We also discovered a tendency for individuals to be consistent in their cultural consonance across different domains, and a summary measure of "generalized cultural consonance" was a significant predictor of these outcomes (Dressler et al., 2007a). In a two-year longitudinal study, change in cultural consonance was a significant predictor of depressive symptoms at follow-up (Dressler, Balieiro, Ribeiro, & dos Santos, 2007b).

As noted above, we have hypothesized that low cultural consonance is a stressful experience and that this experience of stress mediates its effects. In a recent paper we examined this directly using longitudinal data from Brazil. Perceived stress was assessed using Cohen's Perceived Stress Scale, which focuses on perceived control and degree of threat in assessing perceived stress (Cohen, Kamarck, & Mermelstein, 1983). The mediating effects of perceived stress were assessed using the Baron-Kenny model (Baron & Kenny, 1986). When domain-specific measures of cultural consonance were examined, perceived stress partially mediated the effects of cultural consonance in family life on depression; however, when generalized cultural consonance was examined, there was little evidence of the mediating effect of perceived stress on depression (Balieiro, dos Santos, dos Santos, & Dressler, 2011).

Gene-Environment Interactions and Depression

The mapping of the human genome, coupled with increased interest in the neurotransmitter serotonin, has opened a new direction in research on depression. Two genetic polymorphisms in the serotonin system have been associated with depression. One polymorphism (5HTTLPR) is in the gene that codes for the serotonin transporter. Variants of this gene are characterized by a variable number of tandem repeats in the promoter region of the gene, resulting in one shorter and one longer allele. The short version of the allele has lower transcriptional efficiency of the promoter (Firk & Markus, 2007). At least fifteen serotonin receptors have been described with single nucleotide polymorphisms in the genes that code for the receptors. The 2A receptor (5HTR2A) has been of particular interest. One polymorphism in the 2A receptor (-1438 G/A) has been shown to be functional in a study of cell lines; mRNA activity was greater in cells with the AA variant (Parsons, D'Souza, Arranz, Kerwin, et al., 2004).

Gene-environment interactions involving environmental events and circumstances and these polymorphisms have been investigated in depression and other mood outcomes. Caspi, Sugden, Moffitt, Taylor, et al. (2003) found that the likelihood of depression increased more with increasing numbers of life events (such as loss of a job or death of a close family member) for persons homozygous for the short form of the 5HTTLPR gene than for persons heterozygous, or homozygous for the long form. Keltikangas-Järvinen, Puttonen, Kivimäki, Elovainio, et al. (2007) found an interaction of the 5HT2RA polymorphism and childhood temperament to predict adult hostility. Genotype thus appears to moderate the effect of stressful life experiences on mood.

A review paper by Monroe and Reid (2008) reports on twelve studies of these interactions. Four studies found the same interaction effect as Caspi et al. (2003), and five studies were "partial" replications (the gene-environment interaction was detected for a subgroup of the sample). More recently, Risch, Herrell, Lehner, Liang, et al. (2009) performed a meta-analysis of the interaction of life stress and the 5HTTLPR polymorphism in predicting depression in which they found no significant gene-environment interaction with data pooled across studies. However, in a meta-analysis it is assumed that all variables in different studies have been measured in the same way. In studies of gene-environment interaction, this assumption is questionable (Monroe & Reid, 2008). Therefore, the results of the meta-analysis must be viewed with caution.

These studies offer a promising, but only partially explored, direction in research. Individuals with specific genotypes may not be at risk of depressed affect, but they may be more vulnerable to the impact of negative events and circumstances in their social environments. There is, however, a glaring shortcoming in this literature. Research carried out thus far has taken a casual attitude to the measurement of environmental events and circumstances. As Monroe and Reid (2008, p. 950) note: "In general, recent research on life stress and 5HTTLPR in depression is unified only by the broadest abstraction of 'stress.'" They conclude that studies of gene-environment interaction should devote more attention to the nature and measurement of the environmental conditions that form half of the gene-environment equation.

In one sense, researchers appear to have taken the attitude that the impact of *anything* that happens to an individual will be moderated by genotype. This assumption ignores the literature emphasizing the importance of the social or collective meaning (not the subjective or personal meaning) of events and circumstances in determining their impact. A variety of studies suggest that stressful experiences are generated in the context of culturally defined central life goals; thus, if an event signifies the loss of an achieved goal (e.g., divorce) or on-going problems arise within a central cultural domain (e.g., chronic difficulties in the family), these events or circumstances are experienced as stressful and increase the risk of depression (Brown, 1974; Monroe & Reid, 2009). What, unfortunately, has limited much of this research is the lack of a well-articulated theory of culture and a well-defined set of methods for assessing shared meaning and linking culture and the individual.

Gene-Culture Interactions and Depression

We have carried out a preliminary study in which it was hypothesized that in Brazil the effect of cultural consonance in family life on depressive symptoms would be moderated by genotype (Dressler et al., 2009). The outcome was predicted precisely because there is greater cultural consensus—and hence cultural salience—in the cultural model of the family in Brazil than in other cultural domains studied. The interaction of cultural consonance and the -1438 G/A polymorphism for 5HTR2A was examined. The effect of cultural consonance in family life was enhanced in the presence of the AA variant of the polymorphism. Individuals with the AA variant were *more* depressed if cultural consonance in family life worsened over a two-year period than persons with either the GA or GG variants. Similarly, individuals with the AA variant were *less* depressed if cultural consonance in family life improved over a two-year period than persons with either the GA or GG variants. Roughly speaking, the AA variant of the -1438 G/A polymorphism appears to enhance the "throughput" of socially relevant affective information (Dressler et al., 2009, p. 96).

The study of gene-environment interaction in general represents a substantial step forward in the biocultural analysis of health outcomes such as depression. While researchers have argued for years that treating genetic endowment and social experience as independent influences on mood and emotion was incorrect, without the ability to specify genetic risk at the individual level, more complex synergistic models could not be directly investigated. Even with the mapping of the human genome and the technical capability of identifying variants of polymorphisms at the individual level, however, these models continue to be rather crude. As we noted above, many researchers seem to think that any measure of stressful experience in the social environment will interact with genetic background, with little regard to the theoretical or ethnographic relevance of that measure. We have shown that shared cultural meaning is important and that cultural consonance has a potent interaction with genes in predicting depression, yet this is a somewhat simple model as well. Without more nuanced, theoretically based, and ethnographically specific models, the full potential for understanding gene-environment interactions will not be realized.

The Genetic Moderation of the Psychological Mediators of Cultural Consonance

One potential step in this direction would be to integrate the model of how genes *moderate* the effects of cultural consonance (Dressler et al., 2009) with the model of how psychological factors *mediate* the effects of cultural consonance (Balieiro et al., 2011). (A *mediating variable* is literally an influence that stands in between a predictor and an outcome in a causal sequence; a *moderator variable* is a factor that changes the nature of the relationship between two variables, depending on the particular value assumed by the moderating factor.) This proposition, however, immediately poses the question: what are the mediators of cultural consonance? In our initial study of psychological mediators, we relied solely on a global measure of perceived stress. This model was appropriate for an initial step, because it directly evaluated an hypothesis that is consistent with both professional theories of stress appraisal and with Euroamerican folk theories of how stress works, and because the research utilized what is regarded as the most generally useful measure of overall stress appraisal: Cohen's Perceived Stress Scale. But perceived stress only partially mediated the effects of cultural consonance.

There is, however, more to the emotional and cognitive processing of social experience than the evaluation of psychological threat and potential control over stressors. Arguably the most influential theory of the psychological mediators of depression in recent years has been Aaron Beck's cognitive theory of depression (Beck, 2008). In Beck's theory, depression results from a consistent and dysfunctional self-schema that individuals develop. This self-schema organizes experience as an ongoing series of failures. The individual in essence defines him- or herself as incapable of achieving short-term goals or, especially, life goals. Such persons will consistently distort experience in terms of overgeneralizing (e.g., "I failed at one thing so I'll fail at everything"), selectively abstracting (i.e., recalling only the single failure in a set of achievements), and personalizing (i.e., identifying their own shortcomings as the cause of an unavoidable setback). Beck emphasizes the importance of early childhood experience in forming this schema, although the experience of loss later in life can also be important. However it is formed, an individual with such a self-schema will ruminate on these negative self-evaluations and be plagued by intrusive thoughts, with the eventual outcome being depressed affect. Beck argues that ultimately the schema becomes such a powerful cognitive organizing force that conscious rumination or negative thoughts are bypassed by an automatic (unconscious) interpretation of experience in terms of these dysfunctional beliefs and subsequent depressive episodes (Beck, 2008, pp. 971–972).

The utility of a cognitive-behavioral perspective in understanding the effects of cultural consonance seems clear. In our research, we have investigated cultural consonance in cultural domains that define the structure and organization of major life goals. In addition, these are life goals that are widely socially valued, as evidenced by cultural consensus regarding their importance. Therefore, difficulties in achieving consonance with those goals in one's behaviors and beliefs will lead both to negative self-evaluation and a lack of confirmation of social status in mundane social interaction, since one's basic social status is enhanced by one's public identity as culturally successful. A lack of cultural consonance could then both generate and reinforce a self-schema of failure and helplessness, which in turn could give rise to depressed affect. This chain of events describes a simple pathway from low cultural consonance to a negative self-schema to depression, with a negative self-schema as the mediator of the impact of cultural consonance on depression.

How, then, could genetic differences influencing neurophysiology affect this process? We have already presented evidence that genotype can alter the impact of cultural consonance on depressive symptoms. Does genotype also alter the mediators of cultural consonance, such that both the direct effect of cultural consonance on depressive symptoms and the intervening effect of a negative self-schema are enhanced in the presence of a specific genetic variant? Or, can genotype moderate the mediation process itself? Examining these questions with our data from Brazil will provide insight into the intersection of culture, individual psychological processes, neuro-physiology, and depression.

Cultural Consonance, Self-Schema, Genotype, and Depression in Brazil

Research was carried out in Ribeirão Preto, Brazil, a city of approximately 500,000 persons located in the north of the state of São Paulo. Ribeirão Preto is a provincial economic powerhouse, a center for manufacturing, finance, health, and education serving a rich agricultural region.

In the midst of this affluence, however, Ribeirão Preto exhibits the same degree of social and income-inequality characteristic of Brazil as a whole. Its *favelas* (or "shanty-towns") may be smaller than those of Rio de Janeiro or São Paulo, but the existence of poverty and its attendant problems of crime and drug trafficking are no less real.

The Research Site: Sampling

For all data collection, we sampled all segments of the socioeconomic continuum. In the cultural domain analysis, which relied on purposive sampling, key respondents were systematically selected to differ by age, gender, and socioeconomic status. In survey analysis, a stratified random sample was selected with four neighborhoods as strata. The first, a lower-class neighborhood, began as a *favela* or squatter settlement. The municipality moved the population to a public housing project after a few years, but the neighborhood retains its lower-class economic status as well as unsavory reputation for crime and drug trafficking. The second neighborhood was a *conjunto habitacional*, which is a neighborhood developed in partnership between the municipality and private developers. Residents qualify for low-cost mortgages by demonstrating stable employment. Over time, this *conjunto* has developed into a stable working-class community. The third neighborhood is an old and well-known middle-class neighborhood named for one of the founders of the city. While it is beginning to lose some of its status to newer neighborhoods, the neighborhood is still regarded as solidly middle-class. The fourth neighborhood is a new development close to the university favored by upper-middle-class professionals and university professors.

Cultural Domain Analyses

Cultural domain analysis is a systematic progression of interviews designed to discover the vocabulary (literally, the words) that individuals use in talking about an area of life of importance to them, and for exploring systematically the basic criteria (or dimensions of meaning) that structure the domain (Borgatti, 1999). Key cultural domains in a society are best identified through participant-observation and the analysis of cultural products, such as media. In this study, five cultural domains were examined: (a) lifestyle, which refers to the material goods and leisure-time activities associated with middle-class social status; (b) social support, which refers to the pattern of seeking help among family, friends, and co-workers during times of felt need; (c) family life, which describes the characteristics of a good Brazilian family; (d) national identity, or the characteristics that make a Brazilian, Brazilian; and, (e) food, in terms of the foods most culturally salient and their defining features (e.g., health, convenience). These domains were selected both because these topics arise spontaneously in conversation in mundane social settings, and because they represent variables of interest in the study of individual well-being.

These domains were first explored with free lists. Free lists are generated by a respondent literally listing the terms that are used in talking about a domain. Free lists are an economical way of generating the elements that compose specific domains. When a set of terms making up a domain was determined, constrained and unconstrained pile sorts were used to explore the major dimensions of meaning used in talking about the domains. In a pile sort, a respondent is asked to sort terms into different piles, putting similar terms in the same pile and using however many piles are needed to differentiate sets of terms. During and after the pile-sorting exercise, respondents were interviewed to determine their criteria for sorting terms. The pile-sort data were then analyzed with multidimensional scaling and cluster analysis, to display graphically the similarity and differences in meaning among the terms (Dressler et al., 2005).

The qualitative and quantitative analyses of the pile-sort data identify the dimensions of meaning individuals use to distinguish among terms. For example, in the domain of the family, certain characteristics (e.g., love, understanding, organization, being hard-working) were defined as more or less *important* for "having a family" (*para ter uma família*), while another dimension distinguished *structure* and *affect* within families. These dimensions of meaning were confirmed in focus group and individual interviews. Cultural consensus analysis was then used to evaluate the degree to which subjects agree overall on the meaning of terms along these dimensions (Romney et al., 1986). In cultural consensus analysis it is hypothesized that when two individuals respond to a set of questions, and their responses have some degree of similarity, they may be drawing on a single knowledge base—or culture—in responding to the questions. The correlations between respondents can then be used to estimate each respondent's understanding of that underlying culture, referred to as cultural competence (in analogy to "linguistic competence"; see Keesing, 1974). If all respondents' answers are highly correlated, then a strongly shared cultural model is likely operating in that domain. If the respondents' answers are less strongly correlated, then the sharing of that cultural model is weaker.

Using the cultural competence coefficients, an "answer key" can be generated for the questions, giving higher weight to those respondents who have higher cultural competence. The key is a "best cultural estimate" of how a reasonably culturally competent respondent would answer the questions. The answer key is not the same as the cultural model; rather, the answer key represents the result of a group of respondents using their shared cultural model to reason through a problem, in this case to answer a set of questions. Also, respondents are not asked questions about personal preferences or how they themselves do some particular thing. The emphasis instead is on the community or the aggregate.

In our data, a central evaluative dimension was important in structuring each domain. A sample of respondents was then asked to rate or rank the importance of each item in each domain. These data were used to test for cultural consensus, and the results were consistent with shared cultural models for each of these domains.[1]

At this point, the results of the cultural consensus analyses were used to develop measures of cultural consonance. Single measures of cultural consonance in each domain are justified by the substantial cultural consensus for each domain (three of the four domains show evidence of intracultural diversity and residual agreement, but taking this pattern into account makes no difference in the measurement of cultural consonance). For lifestyle, cultural consonance was straightforward: the lifestyle items owned, such as a house and a car, and the leisure activities reported, such as going out with friends or to the cinema, that were rated in the consensus model as at least "important" were counted. The number of items represents how closely individuals approximate the consensus rating of the elements in their own reported behaviors. Similarly, for social support, respondents ranked potential supporters in their (the supporters) importance in offering help for particular problems, and these rankings were matched to the consensus rankings.

For family life, a Likert-response scale was developed. Within the research group (the PI, project director and two research assistants), propositions regarding the family were generated for each of the elements of the model. These propositions were stated such that the respondent could agree or disagree regarding how well the proposition described his or her own family. An example is: "My family firmly confronts problems." Then, the items (a total of eighteen) in the scale were weighted by the rating of the importance of that particular element in the cultural consensus analysis. The scale has quite high internal consistency reliability (Cronbach's alpha = 0.89). A similar scale was developed for national identity (Dressler et al., 2005).

As noted above, we observed a tendency for individuals to be consistent across all four domain-specific measures of cultural consonance. These four cultural consonance variables load together on a single principal component that accounts for 46% of the variance shared among the measures.[2] There is, in other words, a single continuum along which individuals can be arrayed that represents "general cultural consonance," which we will use in subsequent analyses as our primary independent variable, for the sake of simplicity.

At this point, it's worth recapitulating what has been done. From asking Brazilians to first simply describe a particular cultural domain, through a series of exploratory analyses, and then ultimately a cultural consensus analysis, we discovered that Brazilians agree on the relative importance of a set of characteristics that define each domain. Based on those results, a social survey measure that phrases questions in terms generated by Brazilians, and that employs the relative cultural salience of each of the characteristics, was developed. A continuous, linked set of research operations connect the spontaneous speech of Brazilians to measures that order individuals along a continuum derived from the way they talk about each domain. These measures have high *emic validity*; that is, these measures all assess individual belief and behavior relative to shared Brazilian cultural meaning.

Survey Research

In the original research, we surveyed 271 individuals randomly selected from the four neighborhoods described above. After two years, 210 of these respondents agreed to be re-interviewed. Of these respondents, we have genotypes on 145. At the follow-up interviews, we collected psychological data that can be used to create a measure of dysfunctional beliefs, not all of which was collected at baseline. Therefore, for the purposes of this case study, we will use the follow-up participants with genetic data as a cross-sectional sample. As we have shown elsewhere (Dressler et al., 2009), no significant differences were found between the original sample and the follow-up sample with genetic data on either socio-demographic or study variables, suggesting that the follow-up pool can be viewed as a random subsample of the original survey.

The dependent variable in the research reported here is a Brazilian Portuguese version of the Center for Epidemiologic Studies Depression Scale (CES-D), translated and validated by da Silveira and Jorge (2000). The CES-D has acceptable reliability in this sample (Cronbach's alpha = 0.88).

For the purposes of this paper, three covariates are employed: age, gender, and socioeconomic status (SES). SES was assessed as a composite scale of household income (reported in numbers of minimum salaries per month), education (reported as years of education completed), and occupational status (a ranking of occupational prestige developed in Brazil).

Four variables were combined to create a measure of dysfunctional beliefs. The first is Cohen's Perceived Stress Scale (PSS; Cohen et al., 1983). Items for the scale assess the degree to which individuals feel themselves to be overloaded and to lack control over events and circumstances in their lives. The second is the Health Locus of Control Scale (HLOC; Coreil & Marshall, 1982). Items in this scale again emphasize the individual's sense of control over events and circumstances, but these items are phrased more explicitly in terms of being controlled by forces outside of oneself and, of course, concentrate on the domain of health. The third variable is the Sense of Coherence Scale (SOC), based on Antonovsky's (1993) concept. Items in this scale emphasize the degree to which individuals feel that their lives are predictable and stable and that, when unpredictable events occur, they have the resources to deal with the unforeseen. Finally, we employed a self-efficacy scale (SE; Sherer, Maddux, Mercadante, Prentice-Dunn, et al., 1982). Items in this scale emphasize the individual's sense that he or she can act in the world effectively and carry through on his or her own plans. These four scales were translated and back-translated by our research team, with questions about translation resolved through group discussion. The PSS and HLOC scales have been validated in previous research in Brazil (Dressler et al., 2007a). The SOC and SE were translated and included specifically for this study.

These scales were not intended to measure dysfunctional attitudes; nevertheless, individuals scoring high on the PSS, and low on the HLOC, SE, and SOC scales, are expressing an enduring sense of themselves as buffeted by the world around them, as being incapable of changing their lives, and as existing in a world that fails to live up to their expectations. The four scales load a single principal component that explains 50% of the variance among the scales.[3] For the purposes of this case study, this latent variable can be interpreted as assessing dysfunctional beliefs in Beck's sense of the term.

Individuals were genotyped for the –1438 G/A polymorphism using standard polymerase chain reaction analyses and restriction enzymes. Details of this analysis are reported elsewhere (Dressler et al., 2009).

Data Analysis and Results

Descriptive statistics for the sample are presented in table 14.1.

To test the mediation hypothesis, data analytic procedures proposed by Baron and Kenny (1986) were employed. The essential features of this

Table 14.1

Descriptive statistics for all variables used in the analysis (N = 144)

	Mean (or %)	Std. Dev.
Depression	11.80	9.57
Age	40.57	11.38
Socioeconomic status	0.10	0.99
Sex (% male)	38.00	—
Cultural consonance*	0.06	1.04
Cultural consonance in lifestyle	0.70	0.15
Cultural consonance in social support	0.51	0.15
Cultural consonance in family life	103.17	23.19
Cultural consonance in national identity	10.97	3.38
Dysfunctional beliefs*	0.00	0.99
Perceived stress scale	10.02	5.55
Sense of coherence	25.97	6.29
Health locus of control	27.81	4.83
Self-efficacy	34.48	5.92
2A receptor polymorphism (% AA)	27.08	—

*Cultural consonance and dysfunctional beliefs are principal component scores. The descriptive data for variables making up those composite scores are shown below the latent variable name.

approach are first to determine that the independent variable of interest (cultural consonance in this case) has an independent effect (net of covariates) on the dependent variable (depressive symptoms), using ordinary least squares regression analysis. It should also be associated with the mediating variable (dysfunctional beliefs), which should itself have an effect on the dependent variable. Then, if dysfunctional belief has a mediating effect, when controlling for dysfunctional beliefs, the effect of cultural consonance should be substantially reduced, perhaps even to zero in the case of a very strong mediating effect. Finally, the indirect effect of cultural consonance on depressive symptoms operating through dysfunctional beliefs is equal to the product of the path from cultural consonance to dysfunctional beliefs and from dysfunctional beliefs to depression. The sum of the indirect and direct effects should equal the magnitude of the original direct path from cultural consonance to depression.

To facilitate the presentation of the results, we first removed the effects of age, sex, and socioeconomic status on cultural consonance, dysfunctional beliefs, and depression (this enables us to simply present correlation coefficients and standardized repression coefficients in the path analysis that tests the mediation hypothesis). The residual correlations among these variables are shown in table 14.2. As required for the model, both cultural consonance and dysfunctional beliefs have substantial correlations with depression and with each other. The decomposition of these bivariate correlations into direct and indirect effects is shown in figure 14.1. As shown in figure 14.1, the original overall association of cultural consonance (–0.36) is reduced to one-third of its original size (–0.12) when dysfunctional beliefs are taken into account. Most of the effect of cultural consonance on depression in this analysis is in the indirect effect of higher

Table 14.2

Correlations among depression, dysfunctional beliefs, and cultural consonance

	Depression	Cultural consonance	Dysfunctional beliefs
Depression	—	—	—
Cultural consonance	–.357*	—	—
Dysfunctional beliefs	.714*	–.349*	—

Correlations are calculated using residual scores removing the associations of age, sex, and socioeconomic status with each variable.

* $p < .001$

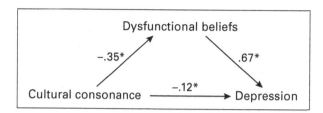

Figure 14.1
Coefficients testing the mediation hypothesis, full sample (standardized regression coefficients).

cultural consonance leading to lower dysfunctional beliefs (–0.35), and higher dysfunctional beliefs leading to greater depression (0.67). This indirect effect (–0.24, or the product of –0.35 and 0.67) and the remaining direct effect (–0.12) sum to equal the original overall effect (–0.36). So, looking at these results would lead to the conclusion that the Beck model is substantially correct, although cultural consonance has a small effect on depression that is left unmediated by dysfunctional beliefs.

We next tested for the genetic moderation of both cultural consonance and dysfunctional beliefs in this cross-sectional sample (our previous demonstration of moderation being in the longitudinal sample). Genotype was coded as a dichotomy (AA genotype versus either GA or GG), and a regression analysis was carried out with cross-products between this genotype dichotomy and cultural consonance and dysfunctional beliefs with depression as the dependent variable. Cultural consonance interacted significantly with genotype (t = 3.01, p < 0.01), but we found no significant dysfunctional beliefs x genotype interaction (t = 0.87, n.s.). Given this significant cultural consonance x gene interaction, dividing the sample by genotype and testing the mediation hypothesis *within* genotype is warranted.

Table 14.3 presents the within-genotype correlations among cultural consonance, dysfunctional beliefs, and depression for the AA genotype (lower-left triangle) and the GA and GG genotypes (upper-right triangle), respectively. As would be expected, the overall association of cultural consonance and depression is substantially larger for the AA genotype than for the GA and GG genotypes (hence the significance of the gene-culture interaction term in the regression analysis), while dysfunctional beliefs have a comparable association with depression regardless of genotype. Figure 14.2 presents the test of the mediation hypothesis for the AA genotype in the upper panel, and the GA and GG genotypes combined in the

Table 14.3
Correlations among depression, dysfunctional beliefs, and cultural consonance

	Depression	Cultural consonance	Dysfunctional beliefs
Depression	—	-.233[a]	.734[b]
Cultural consonance	-.658[b]	—	-.258[a]
Dysfunctional beliefs	.689[b]	-.600[b]	—

Correlations are calculated using residual scores removing the associations of age, sex, and socioeconomic status with each variable. AA genotype, lower-left triangle; GA/GG genotypes, upper-right triangle.
[a] $p < .01$
[b] $p < .001$

A – GA/GG genotypes

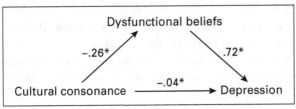

B – AA genotype

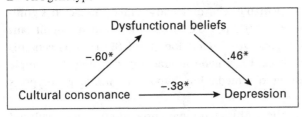

Figure 14.2
Coefficients testing the mediation hypothesis within genotypes (standardized regression coefficients).

lower panel. The key here is what happens to the direct effect of cultural consonance on depression. For the GA and GG genotypes, the direct effect drops to zero and the total effect of cultural consonance on depression is mediated by dysfunctional beliefs (–0.18). For the AA genotype, the direct effect of cultural consonance on depression is stronger (–0.38) than the indirect effect operating through dysfunctional attitudes (–0.28).

These results suggest that Beck's hypothesis—dysfunctional beliefs are a mediator between experience in the social environment and depression—is completely accurate if your genotype for the 2A receptor includes the GA or GG variants for the –1438 G/A polymorphism; if, on the other hand, your genotype is AA for that polymorphism, the Beck hypothesis, while still relevant, is inaccurate.

Discussion

There has been substantial interest in how genes that alter the functioning of the brain serotonin system might modify the impact of environmental events and circumstances on the risk of depression. As we argued earlier, this research has remained relatively crude in the sense that the environmental side of the gene-environment equation has not been very well specified. Our interest in this area has been two-fold. First, we argued (Dressler et al., 2009) that culture would be an important part of this process, in that low cultural consonance in domains that help to structure major life goals would be interpreted by oneself and by others as evidence of a kind of social and cultural failure, leading to self-doubt, dysfunctional beliefs, and depression. Second, as we have argued here, genotype and neurophysiology may alter more than the overall impact of cultural consonance, instead altering the psychological processes by which low cultural consonance is translated into depression.

There is a growing body of evidence that genes in the serotonin system, including especially the genes for the serotonin transporter and the 2A receptor, moderate stress reactivity. Recent research links the serotonin system to the sympathetic-adrenal-medullary (SAM) axis and the hypothalamic-pituitary-adrenal (HPA) axis, both of which govern the release of hormones such as epinephrine, cortisol, and aldosterone, all of which have been implicated in the stress process. Ultimately, genetic differences in the serotonin system have been linked to classic laboratory measures of stress reactivity, including acute changes in blood pressure and heart rate. It would appear that persons with certain genotypes are predisposed to react more strongly to the occurrence of a stressor (Alexander,

Kuepper, Schmitz, Osinsky, et al., 2009; Way & Taylor, 2010). This argument is consistent with our observation that for persons with the AA variant for the receptor gene, cultural consonance has a stronger impact on depression than for persons with either the GA or GG genotypes. Persons with the AA variant are more reactive to the stresses of low cultural consonance and respond with elevated depressive symptoms.

At the same time, the results observed here represent a challenge to current thinking in this area. Our results show that it is not just the reactivity of an individual that varies by genotype, but also the pathway from cultural consonance to depressive symptoms. For individuals with the GA or GG genotypes, depression appears to depend on the degree to which they consciously think about how life seems stacked against them, that they are unable to effect changes in their lives, and that life represents a struggle in which they are doomed to fail. The repeated evaluation of the world in terms of this negative self-schema then results in depressed affect. For individuals with the AA genotype, conscious rumination and negative thought intrusions are still important in mediating their depression, but cultural consonance also has a direct and substantial effect on depression. AA genotype individuals do not appear to need to engage in the conscious mental processing of problems in their lives; rather, the inability to achieve cultural consonance has a direct impact.

In other words, the nature of consciousness is implicated in this process. At the outset, we should emphasize that we do not need to take a strong theoretical stand regarding the nature of consciousness here. For our purposes, a simple distinction between cognitive and emotional processing that occurs with a reflexive awareness (or consciously) and mental processing that occurs without that reflexive awareness (or unconsciously) will do. When we interview respondents, we are inviting them to engage in this self-conscious reflection on their lives and to evaluate their lives in ways that can be used to create measures of cultural consonance, dysfunctional beliefs, and depression. For persons with the GA/GG genotypes, the reporting of low cultural consonance is associated with higher depression primarily because people reporting that pattern also report more dysfunctional beliefs. For persons with the AA genotype, persons reporting lower cultural consonance and higher depression may or may not report more dysfunctional beliefs; regardless of their reporting of dysfunctional beliefs, the link of low cultural consonance and higher depression persists.

Phrased in this way, the reporting of low cultural consonance may serve to "prime" the respondent to engage the negative self-schema of dysfunctional beliefs and in turn report more depressive symptoms. Priming, in the experimental literature, refers to an experimental intervention that is thought to engage a particular schema in the subject, which in turn leads the subject to perform in particular ways later in the experiment. Priming in different ways can lead to changes in performance. For example, an individual who sees the word "yellow" in a list of terms will more rapidly recall the word "banana" when asked to list fruit, presumably because the two words are linked in a schema. So, having to report low cultural consonance in various domains to an interviewer may engage that negative self-schema and lead to the reporting of more depressive symptoms.

When the individual has the AA variant of the 2A receptor for serotonin, however, that cognitive pathway is at least partially bypassed. Serotonin is important in the emotional processing of social stimuli. Experimental models often employ interventions (including tryptophan depletion or selective serotonin reuptake inhibitors) that alter circulating levels of serotonin. When, for example, individuals with lower levels of circulating serotonin are shown pictures of faces with ambiguous expressions, they are more likely to interpret those expressions as negative, hostile, or threatening. In the condition of higher circulating levels of serotonin, the same expressions are interpreted as neutral or benign (Harmer, 2008).

To the best of our knowledge, the 2A receptor polymorphism studied here has no effect on levels of circulating serotonin. But it has been shown to affect the functioning of the receptor. In the presence of the AA variant, when serotonin binds to the 2A receptor, greater messenger RNA activity results in the receptor, presumably leading to faster transmission of the nerve impulse. Furthermore, the psychoactive drugs lysergic acid diethylamide (LSD) and methylenedioxymethamphetamine (or "ecstasy") selectively bind to the 2A receptor, indicating that it is intimately involved in the processing of emotionally charged experience (Dressler et al., 2009).

For the person with the AA variant, therefore, this "upregulation" in the 2A receptor (i.e., greater activity in that receptor) in the presence of lower cultural consonance bypasses conscious negative self-evaluation and results directly in more depressive symptoms. Similarly, this upregulation in the presence of higher cultural consonance bypasses consciousness and results in fewer depressive symptoms. For persons with the GA/GG variants, that accompanying conscious rationale may be key in reporting depressed affect, because of the overall lower level of activity in the serotonin system.

This is especially important when considered in light of the fact that the AA variant does not increase the impact of dysfunctional beliefs; rather, the magnitude of the association of dysfunctional beliefs and depression is constant across the variants. It is only the impact of cultural consonance, and its mediation, that varies by genotype. This suggests a special set of associations among cultural salience, brain function, consciousness, and affect.

Much of the research in neuroscience emphasizes the importance of the amygdala, a neural structure associated with the limbic system, in the processing of emotions, especially fear and anger, and the storage of emotional memory (Ehrlich, Humeau, Grenier, Ciocchi, et al., 2009). Ohira (2011) suggests that the 2A serotonin receptor plays a role in reducing the inhibition of the amygdala, resulting in greater activity and presumably emotion processing. There are, however inconsistent associations among serotonergic polymorphisms, activity in the amygdala, and emotion, which have led some to suggest that the nature of these associations may vary depending on the larger—cultural—context within which stress and emotion processing occur (Ohira, 2011, p. 206). However, thus far, researchers suggesting this modifying influence of culture have relied on a relatively weak concept of culture, equating it exclusively with notions of individualism and collectivism (Chiao & Blizinsky, 2009).

The results presented here indeed highlight the importance of culture in this process, but in a somewhat different way than that suggested by other neuroscientists. Our results suggest that, in the presence of a polymorphism influencing the neural processing of emotion, an individual's relative success in personally enacting strongly shared collective representations of how life is to be lived bypasses conscious reflection and leads to either lower or higher depression, depending on the level of cultural consonance. It is, in other words, a direct neurocultural process, probably as a function of the high salience of the domains making up the measure of cultural consonance. Where the inhibition of this neural activity is greater, again based on a particular polymorphism, depression depends on a process of conscious reflection. These results suggest that the background of collective meaning, and then its realization in belief and behavior, can have a profound impact on mood and emotion outside of conventional psychological pathways.

Of course, many limitations and caveats must be applied to the study we have presented. We have a small sample size and an imperfect measure of dysfunctional beliefs. The data are cross-sectional, and we must assume that the various measures making up the indicator of dysfunctional beliefs and depression are associated beyond the semantic overlap in the items. The interpretation of our results, while plausible, pushes on the boundaries of understanding of the brain serotonin system. Nevertheless, despite these caveats, we have presented a model and an empirical test of that model that goes beyond the relatively crude examples of gene-environment interaction presented in much of the literature because it incorporates a more nuanced understanding of culture, consciousness, neurophysiology, and depression. The ideas offered here seem worthy of further exploration in future research.

Notes

1. Cultural consensus is evaluated by examining the latent structure of a correlation matrix measuring respondent agreement, and overall agreement is assessed by the ratio of the first-to-the-second eigenvalue for the matrix and by the mean (± s.d.) of the cultural competence coefficients. In these data, consensus was observed in each of the cultural domains, with the following eigenvalue ratios and average (± s.d.) cultural competence coefficients: family life, ratio = 7.42, avg. comp. = 0.82 (± 0.09); lifestyle, ratio = 6.59, avg. comp. = 0.71 (± 0.12); social support, ratio = 6.53, avg. comp. = 0.67 (± 0.14); national identity, ratio = 3.97, avg. comp. = 0.57 (± 0.19). Romney and colleagues (1986) recommend an eigenvalue ratio larger than 3.0 as indicative of sharing sufficient to make reasonable the inference that respondents are drawing on a single cultural model. Roughly speaking, mean competence should be larger than 0.50, and the range of competence coefficients should not include 0 or negative numbers.

2. The loadings on the principal component are as follows: cultural consonance in lifestyle, 0.781; cultural consonance in social support, 0.617; cultural consonance in family life, 0.513; cultural consonance in national identity, 0.765.

3. The principal component loadings are as follows: PSS, –0.789; HLOC, 0.575; SOC, 0.762; and, SE, 0.678.

References

Alexander, N., Kuepper, Y., Schmitz, A., Osinsky, R., Kozyra, E., Hennig, J. (2009). Gene-environment interactions predict cortisol responses after acute stress: Implications for the etiology of depression. *Psychoneuroendocrinology, 34,* 1294–1303.

Antonovsky, A. (1993). The structure and properties of the sense of coherence scale. *Social Science & Medicine, 36*(6), 725–733.

Balieiro, M. C., dos Santos, M. A., dos Santos, J. E., & Dressler, W. W. (2011). Does perceived stress mediate the effect of cultural consonance on depression? *Transcultural Psychiatry, 4*(4), 519–538.

Baron, R., & Kenny, D. (1986). The moderator-mediator variable distinction in social psychological research: Conceptual, strategic and statistical considerations. *Journal of Personality and Social Psychology, 51*(6), 1173–1182.

Beck, A. T. (2008). The evolution of the cognitive model of depression and its neurobiological correlates. *American Journal of Psychiatry, 165*(8), 969–977.

Borgatti, S. P. (1999). Elicitation techniques for cultural domain analysis. In J. J. Schensul, M. D. LeCompte, B. K. Nastasi, & S. P. Borgatti (Eds.), *Ethnographer's toolkit: Enhanced ethnographic methods* (pp. 115–151). Walnut Creek, CA: AltaMira.

Brown, G. W. (1974). Meaning and the measurement of life events. In B. S. Dohrenwend & B. P. Dohrenwend (Eds.). *Stressful life events: Their nature and effects* (pp. 217–244). New York: Wiley.

Caspi, A., Sugden, K., Moffitt, T. E., Taylor, A., Craig, I. W., Harrington, H., et al. (2003). Influence of life stress on depression: Moderation by a polymorphism in the 5-HTT gene. *Science, 301*, 386–389.

Chiao, J. Y., & Blizinsky, K. D. (2009). Culture-gene coevolution of individualism-collectivism and the serotonin transporter gene. *Proceedings of the Royal Society B: Biological Sciences, 277*, 529–537.

Cohen, S., Kamarck, T., & Mermelstein, R. (1983). A global measure of perceived stress. *Journal of Health and Social Behavior, 24*, 385–396.

Coreil, J., & Marshall, P. (1982). Locus of illness control: A cross-cultural study. *Human Organization, 41*, 131–138.

D'Andrade, R. G. (1995). *The development of cognitive anthropology*. Cambridge: Cambridge University Press.

Da Silveira, D. X., & Jorge, M. R. (2000). Escala de rastreamento populacional para depressão (CES-D) em populações clínica e não-clínica de adolescentes e adultos jovens [Scale for population tracking of depression in clinical and non-clinical groups of adolescents and young adults.]. In C. Gorenstein, L. H. S. G. Andrade, A. W. Zuardi (Eds.), *Escalas de avaliação clínica em psiquiatria and psicofarmacologia [Scales for clinical evaluation in psychiatry and psychopharmacology.]* (pp. 125–138). São Paulo, Brazil: Lemos-Editorial.

Dressler, W. W. (2007) Cultural consonance. In D. Bhugra & K. Bhui (Eds.), *Textbook of cultural psychiatry* (pp. 179–190). Cambridge: Cambridge University Press.

Dressler, W. W., Balieiro, M. C., Ribeiro, R. P., & dos Santos, J. E. (2007a). Cultural consonance and psychological distress: Examining the associations in multiple cultural domains. *Culture, Medicine and Psychiatry, 31*, 195–224.

Dressler, W. W., Balieiro, M. C., Ribeiro, R. P., & dos Santos, J. E. (2007b). A prospective study of cultural consonance and depressive symptoms in urban Brazil. *Social Science & Medicine, 65*, 2058–2069.

Dressler, W. W., Balieiro, M. C., Ribeiro, R. P., & dos Santos, J. E. (2009). Cultural consonance, a 5HT2A receptor polymorphism, and depressive symptoms: A longitudinal study of gene x culture interaction in urban Brazil. *American Journal of Human Biology, 21*, 91–97.

Dressler, W. W., Borges, C. D., Balieiro, M. C., & dos Santos, J. E. (2005). Measuring cultural consonance: Examples with special reference to measurement theory in anthropology. *Field Methods, 17*, 531–555.

Ehrlich, I., Humeau, Y., Grenier, F., Ciocchi, S., Herry, C., Lüthi, A. (2009). Amygdala inhibitory circuits and the control of fear memory. *Neuron, 62*, 757–771.

Firk, C., & Markus, C. R. (2007). Serotonin by stress interaction. *Journal of Psychopharmacology, 21*, 538–544.

Goodenough, W. H. (1996). Culture. In D. Levinson & M. Ember (Eds.), *Encyclopedia of cultural anthropology* (pp. 291–299). New York: Henry Holt.

Harmer, C. J. (2008). Serotonin and emotional processing. *Neuropharmacology, 55*, 1023–1028.

Keesing, R. (1974). Theories of culture. *Annual Review of Anthropology, 3*, 73–97.

Keltikangas-Järvinen, L., Puttonen, S., Kivimäki, M., Elovainio, M., Pulkki-Råback, L., Koivu, M., et al. (2008). Serotonin receptor genes 5HT1A and 5HT2A modify the relation between childhood temperament and adulthood hostility. *Genes, Brain & Behavior, 7*, 46–52.

Kuper, A. (1999). *Culture: The anthropologists' account.* Cambridge, MA: Harvard University Press.

Monroe, S. M., & Reid, M. W. (2008). Gene-environment interactions in depression research. *Psychological Science, 19*, 947–956.

Monroe, S. M., & Reid, M. W. (2009). Life stress and major depression. *Current Directions in Psychological Science, 18*, 68–72.

Ohira, H. (2011). Modulation of stress reactivity in brain and body by serotonin transporter promoter polymorphism. *Japanese Psychological Research, 53*, 193–210.

Parsons, M. J., D'Souza, U. M., Arranz, M. J., Kerwin, R. W., & Makoff, A. J. (2004). The -1438G/A polymorphism in the 5-hydroxytryptamine type 2A receptor gene affects promoter activity. *Biological Psychiatry, 56*, 406–410.

Risch, N., Herrell, R., Lehner, T., Liang, K.-Y., Eaves, L., Hoh, J., et al. (2009). Interaction between the serotonin transporter gene (5-HTTLPR), stressful life events, and the risk of depression: A meta-analysis. *Journal of the American Medical Association, 301*, 2462–2471.

Romney, A. K., Weller, S. C., & Batchelder, W. H. (1986). Culture as consensus: A theory of culture and informant accuracy. *American Anthropologist, 88*, 313–338.

Sherer, M., Maddux, J. E., Mercadante, B., Prentice-Dunn, S., Jacobs, B., & Rogers, R. W. (1982). The self-efficacy scale: Construction and validation. *Psychological Reports, 51*, 663–671.

Way, B. M., & Taylor, S. E. (2010). Social influences on health: Is serotonin a critical mediator? *Psychosomatic Medicine, 72*, 107–112.

IV Conclusion

15 The Encultured Brain—Toward the Future

Daniel H. Lende and Greg Downey

Neuroanthropology: The Here and Now

Neuroanthropology offers up a dual integration. First, neuroscience and anthropology have much to offer one another. As neuroscience has recognized neuroplasticity and moved away from old models of hard-wired circuits and computer algorithms, it has begun to embrace cross-cultural research methods and the understanding of human variation as both shaped by and shaping neural dynamics. Anthropology has sought to transcend its determinist strands, whether genetic or cultural, to address how practices, ideologies, and environments come together in the patterning of human behavior and societies around the world. Increasingly, anthropologists recognize that the body plays an important role in what people do and how they develop, but notions of embodiment on the cultural side have lacked the biological specificity to gain necessary empirical traction. Neuroanthropology offers one way to bring together two fields that are increasingly addressing similar questions.

The second integration is at the heart of anthropology itself. Anthropology is often described as the most scientific of the humanities, and the most humanistic of the sciences. However, a deep split remains between biological and cultural anthropologists within the field, even as many anthropologists and our organizations remain dedicated to the holistic ideal that is often used to distinguish anthropology from more narrow approaches to understanding ourselves. Neuroanthropology is an explicit effort to bridge that divide: to show that both neural processes and ethnographic insight matter in doing good research on human variation, whatever its shape and form. For outside readers, this internal debate within anthropology—and how neuroanthropology explicitly straddles it—might seem parochial and unimportant, a solution to petty bickering within departments. But bringing together scientific and interpretive approaches to ourselves is one of the great intellectual challenges we face; the friction within anthropology is an example in miniature of bigger challenges to our understanding of what it means to be human. We know that the brain makes a difference. We also know that our meaningful interpretation of the world makes a difference. The important part is to now bring the two together in a productive way. This biocultural synthesis is what neuroanthropology does.

Neural Diversity and Models of Culture: A Review of the Case Studies

Neural Diversity

Many of the case studies incorporate the concept of neural diversity in two ways: first, by positing that cultures can help create neural diversity, both across cultures and within them; and second, by asserting that neural diversity is a key starting point for thinking about neuroanthropological problems. The first point—that cultures can drive neural diversity—should be noncontroversial, given the increasing understanding of neural plasticity and the recognition that culture has a variety of ways to "get under the skin" (Downey, chapter 6, this volume; Dressler, Balieiro, & dos Santos, chapter 14, this volume; Worthman, 2009). Hay's chapter on memory and knowledge in Indonesia and the United States shows how knowledge traditions can lead to differential organization and use of memory across societies even to accomplish the same functions. Within cultures, forms of expertise, difficult social circumstances, social divisions like gender and race, and socioecological conditions can all work to shape neural patterns, as argued by Campbell, Downey, Lende, and others.

The second point—that understanding neural diversity sharpens anthropological thinking—is more subtle, but equally important. Here Brezis' chapter on autism stands out in showing how the condition becomes a prism to understand cultural development and, in particular, religious belief and practice. In particular, Brezis' research confounds Jesse Bering's influential theory that autistic individuals should lack an interpersonal relationship with God because they do not have an "existential theory of mind." Brezis found that autistic individuals themselves had different types of religious beliefs and that faith in God was not simply the product of a particular type of brain but a resource that individuals used in a variety of ways, including to address their own deficits. Similarly, Dressler's serotonin variants show that culture can work in diverging ways, sometimes through beliefs about self, other times more directly affecting how individuals react to their cultural fit. In more implicit fashion, many chapters also use ideas of neural diversity through, for example, a focus on what makes experts different, or how addicts and non-addicts vary in neural processing. As research continues, greater recognition of neural diversity as a fundamental part of human variation will surely become an even more substantive part of the neuroanthropological approach.

Multiple Approaches to Culture

Three different ways of using culture theory are apparent over the ten case studies. First is a cognitive approach to culture best typified by chapter 14 by Dressler, Balieiro, and dos Santos. In their specific approach, culture is comprised of areas of knowledge—what one needs to know in order to operate in a specific society. More broadly, a cognitive approach views culture as primarily symbolic, from cultural models that provide the cognitive template for behavior to the meanings people derive from the systems of symbols surrounding them. For Dressler and colleagues, a strategically narrow definition of culture permits measurement of culture as a group phenomenon rather than automatically reducing culture to a psychosocial phenomenon and helps them to quantify the cultural part of the environmental effects on the development of depression. This strategic choice in definition allows the team to tease out individuals who are cultural dissenters, who fail to match up to cultural ideals and patterns, rather than assuming every individual is equally embedded in their group. The subtlety allows the group to illustrate clearly how the neurocultural dynamics of depression are varied among sufferers, depending both on genetic differences and the individual's relationship to his or her group.

Downey's approach to culture in chapter 6 stands in contrast. His analysis of skill and embodiment highlights how culture can work through training, repetitive practices, and the cultivation of specific bodily and perceptive techniques. His nonsymbolic approach seeks to understand how culture can induce physiological and neurological variation. Whereas Dressler and colleagues use a narrow definition to support a fine-toothed methodology, Downey uses a strategically broad account to illustrate how developmental dynamics make a clear distinction between "biology" and "culture" untenable. The entanglement of these categories is what precisely demands a neuroanthropological suppleness.

Others are more agnostic about these two different views of culture, preferring to use them in conjunction. Chapter 5, by M. Cameron Hay, draws on medical traditions of knowledge, and thus represents a symbolic view; but the learning, memorization, and practice of this knowledge provides a more practice-oriented view of how culture gets under the skin.

Hay provides a tripartite model of cultural knowledge that captures these different levels and promises to be useful across many spheres of human activity, just like Dressler and colleagues' account of gene-beliefs-cultural consensus interaction and Downey's strategic blurring of culture and biology to discuss embodiment. Specifically, Hay describes how cultural forms of knowledge interact with the neurological organization of different types of memory in specific tasks and situations that demand knowledge. Similarly, Pettinen's work on learning Taijutsu in chapter 7 focuses primarily on the notions of habits and training, similar to Downey. However, Pettinen ties that embodied focus to an account of different social and historical traditions of understanding the mind and body and cultivating the senses, thus locating bodily practice in a longer-term symbolic and ideological stream of culture. In both Pettinen's and Hay's chapter, we see how cultural practices are the result of both neurologically diverse processes and larger scaled social and historical traditions; in both, forms of education must contend with initial diversity in individuals' habits, memories and abilities, instilling a degree of cultural uniformity.

Finally, some anthropologists approach culture as a system of mediators, rather than symbolic systems or enculturation techniques. Erin P. Finley's chapter on post-traumatic stress disorder, chapter 10, is the best example here. In Finley's account, forms of treatment, rituals of memory, webs of relationships and social support, and understandings of what trauma means all shape veterans' attempts to make sense of trauma and cope with PTSD, as well as doctors' and the military's attempts to address the extraordinary incidence of PTSD due to the wars in Iraq and Afghanistan. These mediators tie the different emotional elements—stress, horror, dislocation, and grief—together into a cohesive whole, capturing the lived reality of the condition and not just its clinical indicators. In similar fashion, other authors stress how culture can prime specific interpretations, or how social context limits the options available for people. This culture-as-mediation approach will be useful to neuroanthropologists, as it will permit researchers to make proposals about what neurocultural processes get tied together at specific places and times, how culture can bias neuropsychological events, and thus lead to different outcomes.

Neurocultural Processes, Models, and Vulnerabilities

A focus on *neurocultural processes* ties together the ten case studies. Neurocultural processes, like the term neuroanthropology, are straightforward. Processes are operations, components, or factors that shape the overall flow, development, or outcome of some phenomenon. As opposed to a simple cause-and-effect model, processes are considered formative rather than determinative. Identifying central processes—for example, the trajectory of stress, horror, dislocation, and grief in Finley's analysis of PTSD—was a central strategy of the chapters. But rather than treat these processes as isolated at some determined level, as older bio-psycho-social or gene-culture models did, these authors approached processes as inherently neuroanthropological in themselves. For example, Finley argues forcefully that horror, dislocation, and grief match better with both her ethnographic work and with an understanding of neuroscience than the typical psychiatric criteria of hyperarousal, reexperiencing, and avoidance. The traditional symptomology makes post-traumatic experiences into a largely psychobiological phenomenon, with the implication that if these three elements come together, then you get PTSD (in other words, they cause PTSD). In contrast, Finley emphasizes the inherently integrative dimensions of neurocultural processes, and argues furthermore that cultural mediators are what actually bring the other three elements together. The fact that these processes extend outside the individual, and are inflected by social interaction and symbolic interpretation, also provides points of entry methodologically, highlighting that all of the neurocultural action is not just in the synapses, but extends to other levels of complexity that are accessible even without imaging technology.

Similarly, Kathryn Bouskill's exploration of the role of humor in cancer recovery demonstrates how diverse forms of research, including a historical account of the emergence of cancer awareness movements as well as neuroimaging research on the effects of laughter, offer multiple ways of exploring the neurocultural process because different levels of phenomena are interacting at once. Humor only operates in its distinctive fashion within particular social circles of survivors—the same jokes cannot be shared with outsiders—and only at a historical moment in which cancer is no longer considered a personal burden to be carried in private. And "cancer world" itself is the result, not only of disease dynamics and psychological facts, but also of the technology of treatment available at this time. To collapse Bouskill's account into a simple discussion of stress and resilience would erase precisely the details we need to understand why this specific type of humor works for these groups of women.

Dressler and colleagues make their argument about neurocultural processes in even stronger terms. They write:

Our results suggest that, in the presence of a polymorphism influencing the neural processing of emotion, an individual's relative success in personally enacting strongly shared collective representations of how life is to be lived bypasses conscious reflection and leads to either lower or higher depression, depending on the level of cultural consonance. It is, in other words, a direct neurocultural process.

In this case, Dressler et al. explicitly mean that cultural consonance and serotonin function are the distinguishing elements, a *direct* neurocultural process without the same mediation by dysfunctional cognitions and beliefs as seen with other serotonin receptor polymorphism. But in the sense argued here, both models that Dressler and colleagues outline through their empirical analysis are distinguished by neurocultural processes. Dysfunctional beliefs are not in themselves simply a dimension of cognition, without culture or neurobiology, as other research on depression shows (Kleinman & Good, 1985 Bar-Haim, Lamy, Pergamin, Bakermans-Kranenburg, et al., 2007). Rather, they can also be considered neurocultural processes, manifesting themselves in cognition but mediated by neurobiology and culture as well.

Identifying neurocultural processes is thus a formative part of doing neuroanthropology. At times, these processes can remain somewhat nebulous, as in Lende's work in chapter 13 on the connections linking dopamine, incentive salience, what actually is salient in people's lives, and shifts in attention and desire that are a formative part of how salience works. At other times, neurocultural processes are embodied, reaching in dynamic ways from brain to body, out into the environment and back again, as revealed by Downey's analysis of the equilibrium system, in which the vestibular system meets with proprioception in feet, ankles, and wrists but also with constraints of proper technique and social coaching in the *capoiera roda* or on the balance beam.

A focus on neurocultural processes facilitates the proposal of specific models for the diverse phenomena the researchers examined through their case studies. These models bring together the neurocultural processes in explanatory frameworks. Hay examines knowledge traditions as neurocultural processes, a synthesis of cultural traditions, the neurological organization of memory and learning, and the specific demands and tasks that each knowledge tradition must meet. This model permits her to examine how healers in two extraordinarily different healing traditions, one located in rural Indonesia, the other in urban California, foster health and healing in patients. She proposes as next steps the examination of other medical traditions, as well as other types of expert knowledge, using this model.

Campbell, in his model of testosterone, social ecology, and the embodied experience of vitality in chapter 9, outlines how different studies can test various aspects of his model, demonstrating another way to approach using neuroanthropological accounts of neurocultural processes. Specifically, Campbell suggests that neuroanthropological research on embodiment needs to consider the tactile and experiencing body as we grapple with our neurophysiology, including rich physical sensations like thirst, hunger, sensual touch, sexual arousal, heat and cold, the fullness of a bladder or stomach. His investigation of the links between men's subjective sense of well-being and their endocrine profiles suggests a potential neurocultural process underlying a widespread association in a number of subsistence societies between semen, the spine, and the brain. At the same time that neurophysiology may help "explain" the association, these cultural models also help anthropologists, if they listen closely, to understand what it is like to live with a human nervous system in specific socioecological situations. Both biological and cultural materials provide complementary data to understand neurocultural processes.

Lende's opportunity for neuroanthropological research in chapter 13 also arises from close attention to ethnographic material. Specifically, the first-hand accounts provided to him by young people, users and former users struggling to stay off drugs in Colombia suggested that one of the competing neuropsychological models for understanding addiction hew closer to experiential dimensions of addiction than other models. But Lende's neuroanthropological research made him realize that increased salience of cues to "want" drugs alone did not explain the spiral of drug use. He recognized that multiple dynamics fed addiction, including a shift in the users' immediate environment due to changes in their patterns of interaction. Understanding social meaning and context allowed Lende to note how salient cues began to multiply, and the social conflicts that cued behavior became increasingly frequent: events like the end of a school day or a fight with a parent over drug use became cues to search out familiar scenes where drug taking was possible and socially supported. Environmental and biological effects were difficult to separate because changes in behavior exposed users to situations in which specific neurophysiological were cued with greater frequency; both environment and biology were moving together into a cycle of addiction.

An important implication of neurocultural processes appears in Stromberg's chapter on cigarette smoking, collective effervescence, and loss of agency. His ethnographic work in chapter 12 reveals that at parties, college students felt excited and sensed they got carried away with their smoking. Stromberg's analysis uses neurocultural processes to tie together two disparate social science traditions—that is, a focus on collective behavior and on agency—by examining how imitation and entrainment, pretend play, and emotional arousal can explain what these college students experienced and how they interpreted what had happened.

Stromberg, like Lende, notes the large gap between what actually happens—the neuroanthropology of the phenomenon—and how individuals and societies interpret such events. Generally, we lack direct insight into our own neuropsychology, and are not equipped by knowledge traditions to understand and act on what is happening. In the case of cigarettes, the folk understanding of their "power" bleeds over into public policy and even health research. This gap highlights "neuroanthropological vulnerabilities," a key idea for understanding how neuroanthropology can contribute to social analysis and applied work. In evolutionary medicine, "evolutionary mismatches" between our evolved physiology and modern environment have become a central analytical concept (Trevathan, Smith, & McKenna, 2007; Lende, 2007); neurocognitive and neuroeconomic analyses of decision making are similarly identifying key vulnerabilities that can arise from multiple, interacting systems in the brain (Redish, Jensen, & Johnson, 2008). The concept of neuroanthropological vulnerabilities utilizes a similar approach, focusing on the gap between neurocultural processes and social dynamics that can worsen outcomes for individuals and groups or obscure accurate assessment of the causes of a particular dynamic.

As one example, the effect of gender differences in the brain is a domain rich for neuroanthropological analysis (Fine, 2010). However, cultural ideologies of gender often utilize neurosexism to reinforce inequalities rather than confront entrenched patterns of socialization (see also Kraus, 2011). Similarly, mounting evidence makes apparent that poverty poisons the brain (Hackman & Farah, 2009). However, rather than focusing on the neurocultural processes that explain how that happens, some neuroscientists focus on designing pharmacological interventions even as continued ideas about the "culture of poverty" essentialize poverty and maintain the inequalities that set up the poisoning in the first place. The recognition of neuroanthropological vulnerabilities reveals that overly simple understandings of neurocultural processes, such as the effects of inequality or gender stereotypes on brain functioning, can foreclose any possibility of seeing all the opportunities at our disposal.

Moving against Essentialism, both Universal and Particular

Neuroanthropology emphasizes patterns of human variation around the world, with a focus on similarities and differences across and within societies. For this reason, its major intellectual opponent is essentialism. *Essentialism* is the idea that some particular essence, some defining attribute, makes a specific group or process irremediably different from other groups. Essentialism's two major forms are found in arguments about universal traits and human nature, and in defining specific cultures or social groups as irrevocably different (generally using an implicit "us" set against an explicit "them"). Oftentimes the two types of essentialism travel together, as seen in the "Great Chain of Being," wherein "higher" types of beings are believed to possess an essence that defines them as superior to "lower" forms (Lovejoy, 1976). This hierarchy is one important legacy that informs racism, and that underlies the use of the concept of "culture" to argue for the clash of opposing civilizations, or to define humans as having some essential trait that makes them fundamentally different from other primates.

This move against essentialism is foundational in modern anthropology. Franz Boas contrasted his historical particularism against prevailing notions of social evolution in the early 1900s, arguing for historically grounded analysis rather than assuming all societies passed through a series of evolutionary steps (Boas, 1920). Boas also used biological data to refute the racial essentialism of his day, demonstrating that immigrant children had similar cranial development to children born from US natives, thus showing how biology responds to environment and contradicting dominant ideas about racial purity, phrenology, and other essentialist views (Boas, 1912; Gravlee, Bernard, & Leonard, 2003). Boas' focus on acculturation and environment, on language and biology, on the importance of history and culture, even his early work on how culture informed psychophysics—these provided the legacy of holistic analysis in anthropology that focuses on how history, culture, and biology meet.

This holistic approach has broad relevance in scholarship across many domains today. Within the neurosciences, work on neuroconstructivism provides an empirical approach to understanding how experience, constraints, and development underlie the genesis of cognitive representations (Westermann, Mareschal, Johnson, Sirois et al., 2007). Research on neural reuse as a fundamental way to understand brain evolution, development, and function also offers the conceptual means to move beyond a hardwired or essentialist view (Anderson, 2010). The principle of neural reuse observes that, in spite of brain tissue specialization, circuits and regions get redeployed for new purposes over both evolutionary and developmental time, often without losing their original function. Together, neuroconstructivism and neural reuse offer ways to connect the increasing evidence for neuroplasticity—the activity-dependent and experience-mediated wiring and functionality of the brain (Wexler, 2011)—with human behavior, development, and culture.

Research from evolution and development also support the move toward a contingent and constructivist approach to our biology and brain, rather than an essentialist position (MacKinnon & Fuentes, chapter 3, this volume). Work on developmental systems theory (Oyama, Gray, & Griffiths, 2001); niche construction (Odling-Smee, Laland, & Feldman, 2003); multiple forms of inheritance (Jablonka & Lamb, 2005; Richerson & Boyd, 2006); and an extended synthesis (Pigliucci & Müller, 2010) move evolutionary theory away from a neo-Darwinian approach that emphasized only genes and adaptations as the dominant features of evolution. In cognitive science and psychology, similar intellectual changes are happening. Cognitive science is grappling directly with ideas of an extended mind, where cognition is understood to happen in relations that engage the material processes of the world, from bodily forms and actions to environmental structures and even processes like writing (Clark, 2008). Developmental approaches to "innateness" (Elman, Bates, Johnson, Karmiloff-Smith, et al., 1996) and experimental and theoretical work on embodied cognition (Barsalou, 2010) offer psychology an opportunity to move well beyond a view based solely on modules and algorithms installed in an isolated mind.

Similarly, recent work on language diversity and evolution provides ways to understand how language works without returning to notions of a "universal grammar." Rather, language is a "biocultural hybrid," in which a biological system meets with a particular language system, leading to linguistic diversity around the world (Evans & Levinson, 2009). Rather than being determined by evolution, language is better thought of as adapted to the brain (Changizi, 2011; Chater & Christiansen, 2010) and can evolve through social dynamics (Hruschka, Christiansen, Blythe, Croft, et al., 2009). The Sapir-Whorf hypothesis, that language directly shapes how we think, is also making a comeback after being sidelined by universal grammar views (Perlovsky, 2009; Regier & Kay, 2009). Appropriately enough, this revival of Sapir-Whorf is not in its most determinist form, in which language is believed to wholly determine thought (Boroditsky, 2011); rather, language, cultural experience, and brain interact in shaping thought.

A good example of this synthesis is Wierzbicka's cross-cultural work on emotion, language, and cognition (Wierzbicka, 1999). Across languages, similar words and concepts exist for fear, anger, and shame. These words are "overlapping (but not identical) in meaning," and Wierzbicka cautions against slotting these local concepts into our own native category of "emotion" (Wierzbicka, 1999, p. 276). The Western concept of "emotion" is not comparable across the languages she reviewed. Rather, the idea of "feeling" is closer to the concepts found in different societies: "'ordinary people' generally assume that the way one feels can be described and that one can tell other people how one feels" (p. 13).

Finally, philosophy also has significant strands that work against essentialist views. Lakoff and Johnson (1999) present the embodied mind as a powerful antidote to much of Western philosophy, which has emphasized reason and consciousness as the defining attributes of the mind. Lakoff and Johnson draw on cognitive science to show thought as inherently acting in the world rather than abstract, mostly unconscious, and largely metaphorical. More recently, Noë (2009) works against Descartes' famous dictum, "I think, therefore I am," and its more recent manifestations that locate thought and consciousness solely in the operations of the brain. Habits, language, body, and environment go into making up consciousness in processes that extend beyond the brain; in this way, we exist first, and then we think.

Another type of philosophical essentialism was seen in the rise of postmodernism, which made strong statements about humans' "infinite variability" and viewed ideologies as determining political structures and thus human agency, resulting in a sort of cultural essentialism. Such absolute relativity, where human nature is shaped solely by environment, also appears one-sided in a neuroanthropological light. A more grounded and nuanced view recognizes the power of culture and history without claiming they determine reality. As Barbara Herrnstein Smith (2011, p. 14) writes in "The Chimera of Relativism":

What has actually been said—as in the case of the figures I have mentioned, from Protagoras to Latour—are statements to the effect that human perceptions, interpretations, and judgments are not absolute, universal, or objective in the sense of being independent from all perspectives and/or invariant under all conditions; that what we take to be real, true, and good depends upon and varies with, among other things, our assumptions, expectations, categories, and existing beliefs as these are affected to one degree or another by, among other things, our particular experiences and situations, both past and ongoing; and that these in turn are affected to one degree or another by, among other things, our historically and otherwise particular social, cultural, and institutional environments, including the conceptual and verbal idioms in our communities.

Smith then demonstrates how this view does not mean "human perspectives vary *only* historically or culturally" or that "people from different eras or cultures have nothing in common." Rather than universals being opposed to cultural variability, Smith argues for an empirical relativism: "although various specieswide ('universal') cognitive capacities, traits, or tendencies may exist, they must, in their actual operations, interact continuously with our other more or less highly individuated traits and tendencies and also with the traces of our individual experiences in particular physical, social and cultural worlds" (Smith, 2011, p. 21).

Beyond Essentialism in Psychology

One important example of essentialism is the emphasis over many years in psychology on universal mechanisms located within the mind as a major approach to explanation and goal of research. Evolutionary psychology, for example, often posits a universal human psychology based on evolved modules that make up the mind. Behaviorism presents universal learning processes, ones that span animal and human models. Early cognitive science inherited this legacy, using algorithms and modules to provide an account of human cognition; Chomsky's universal grammar is a good example of this approach.

The critique of universalizing approaches to human psychology is important because neuroanthropology has its intellectual roots in psychological anthropology, which has long contrasted itself with mainstream psychology by emphasizing "the individual in context." That is, psychological anthropologists have countered the essentializing tradition by arguing that individual interactions with cultural environments characterized by fields of symbolic meaning and specific social roles and by concrete socioecological circumstances helped form internal psychology and individual behavior (Moore & Mathews, 2001). Similarly, cultural neuroscience has drawn on the research tradition of cultural psychology, psychological anthropology's sister discipline. Cultural psychology like psychological anthropology emphasizes the individual in context and can be distinguished from other approaches in psychology by its comparative approach. Furthermore, cultural psychologists look at psychological function across societies rather than in depth within a particular cultural setting and offer a greater focus on cognitive processes rather than cultural meanings (Kitayama & Cohen, 2007).

Cognitive science and anthropology can produce many positive collaborations (Bender & Beller, 2011; Bender, Hutchins, & Medin, 2010; Kitayama & Uskul, 2011; Medin, Bennis, & Chandler, 2010; Roepstorff, Niewöhnerc, & Beck, 2010). Nevertheless, neuroanthropology will need to actively work against some of the potential limitations that come from drawing on mainstream psychology and its general emphasis on content-free universal mechanisms. Psychology can impart an essentialism that gets in the way of understanding neurocultural processes. When cultural variation is only treated as a thin developmental gloss on underlying universal psychological traits, psychologists can underestimate the alienness of other ways of being human. As the examples in this volume show, both human psychological diversity and many neuropsychological processes depend on cultural meanings and social context. And as Henrich and colleagues (2010) have argued, basing models of universal traits on subject pools drawn from introductory psychology courses at US universities is fraught with empirical and theoretical problems.

Much of present-day psychology, for example, is built on a cognitive approach that has never considered either neuroplasticity or culture as foundational operating principles for the mind (Bender & Beller, 2011; Kitayama & Uskul, 2011). Neuroplasticity is at once grounded in neural operations and yet necessarily dependent upon environmental shaping, thus overcoming the closed-system, algorithmic approach used to such important effect over the previous decades (Wexler, 2011). A model of human nature as consisting entirely of encapsulated cognitive modules, operating according to universal mechanisms, is a radically different starting point for research than neuroanthropology's joint emphasis on neuroplasticity and human variation (Downey, chapter 6, this volume). Indeed, much of psychology is still largely premised on the assumption of universal mechanisms acting in specific individuals, thus closing off psychology to many of the insights anthropology has to offer. Positing substantive cultural variation in emotion, self, reasoning, and belief was simply not how experimental psychology and cognitive science progressed for many decades (though, of course, small groups of scholars did engage in these topics to great effect). Rather, culture and environment were aspects to be controlled and excluded in experimental design, the better to illustrate a mechanism assumed to operate solely within the individual and to work uniformly across all settings.

Second, much of psychology missed out on the "interpretive turn" that has been so productive in anthropology and in the humanities, an intellectual movement that foregrounded the way that humans' interpretations of their situation affected their behavior and perception (Geertz, 1973; Bender & Beller, 2011). The notion of meaning—of systems of symbols, ideologies of power, and rich histories of interpretation—has found little purchase in modern psychology (Geertz, 1973; Moore & Mathews, 2001). The guiding assumption often seemed to be: memory is memory, but the content of that memory is ephemeral. However, content can directly shape memory, from the enculturation of skill to the ways in which reinterpretation and application in varying contexts affect what and how we recall (Hay, chapter 5, this volume). Bringing the interpretive turn to neuroanthropology means recognizing that psychological processes generally do not happen in laboratories to subjects, but to individuals with biographies and biases in situations shot through with power, shaped by negotiation, and overshadowed by significance.

Third, psychology has one serious lacuna at odds with both anthropology and neuroplasticity: decades of investigation have built research paradigms based on a very narrow range of human variation, and that slice is dramatically different from what research in other societies reveals. This selective sampling of humanity is, simply put, WEIRD—Western, educated, industrialized, rich, and democratic (Henrich, Heine, & Norenzayan, 2010). Or, cast in another light, we might call them MYOPICS—materialist, young, self-obsessed, pleasure-seeking, isolated, consumerist, and sedentary.

The fact that WEIRD people are the outliers in so many key domains of the behavioral sciences may render them one of the worst subpopulations one could study for generalizing about Homo sapiens. … WEIRD people, from this perspective, grow up in, and adapt to, a rather atypical environment vis-à-vis that of most of human history. It should not be surprising that their psychological world is unusual as well. (Henrich et al., 2010, p. 79–80)

This sampling bias makes it difficult to draw on large areas of research in psychology, which focus on a narrow range of human variation and make generalizations about universal psychological processes. In contrast, the neuroanthropology project explicitly recognizes that brains will vary across time and space, and draws on cultural anthropology and ethnography as one of its formative strands. The data from psychology can sometimes simply be too weird, too unusual, to be useful in either trying to generalize from "the West to the rest" or as a useful starting point for understanding local variation in one specific time or place. In contrast, neuroanthropologists are likely to focus on studying subjects who are more unusual when compared to much of the intended audience of their research—men in subsistence economies, drug-addicted youth in Colombia, high-functioning autistic individuals in Israel—but our awareness of these subjects' peculiarity allow us to better trace out the envelope of human potentiality.

Addressing Essentialism in Anthropology and Culture

When anthropologists and others talk about other cultures or social groups, they can fall prey to another form of essentialism; as opposed to psychological essentialism (where the mind is treated as everywhere the same), cultural essentialism attributes some particular set of traits or characteristics to a group, thereby homogenizing them. Early views of culture held to this type of view. Cultures were defined as timeless, bounded entities that could be judged dispassionately by outside Western observers. However, the reality is that patterns of diffusion and globalization had already affected local communities for many, many years, and the outsider's view was hardly objective. Practice views of culture, examinations of globalization, and analyses of older biases in anthropology have all contributed to overcoming the "timeless, bounded" view of culture. The critical practice of reflexivity has emerged in recent decades to combat assumptions of researcher objectivity in anthropology, and presents a way not merely to recognize researcher bias but also to help others understand the potential distortions of research when they use data and read analyses.

Nevertheless, tendencies toward essentialism remain in anthropology, and more broadly in the social sciences, when addressing questions of "difference." For example, an emphasis on collecting data on "race, gender, and ethnicity," treating each of these as a category that defines a group, can be a form of essentialism. While the focus might help get at variation in a "population," treating race, gender, and ethnicity as a categorical filter can reduce the reality of race, gender, and ethnicity to an essentialized difference. What it means to be member of a racial or ethnic group, or a man or woman, is not simply to be an exemplar of a category, but is itself liable to shifting meaning and impact across our lives and contexts. Similar to critiques raised at many points in this volume about measures of "culture" that treat it solely as another exogenous variable in a data set, so too with race, gender, and ethnicity (Dressler, Oths, & Gravlee, 2005; Fine, 2010; Gravlee & Sweet, 2008).

Similarly, notions of cultural difference—of the "clash of civilizations"— can severely essentialize people across nationalities, religious background, linguistic identity, socioeconomic status, migration, and other important dimensions of the patterning of human variation (Ong, 1999; Todorov, 2010). In this "civilizational" approach, cultural or ideological difference is assumed to work rather like an industrial press, stamping down on people and forming them into near exact copies. Given the diversity demonstrated in this book, and neuroanthropology's approach to examining how brain, culture, and environment come together to induce patterns of variation, the hydraulic press metaphor of culture and the clash of civilizational behemoths are both to be avoided. For example, as Dressler, Balieiro, and dos Stantos's cultural consensus research methods highlights, not every individual feels equally at home in their own culture.

Beyond Essentialism

Patterns of variation can be perceived by their similarity or their difference; we must guard against essentializing both our perceptions of sameness and of difference. Both psychology and anthropology have engaged in essentialism in the past. Overcoming those legacies can increase the impact of our research. Essentialism is still used in potent combination today all around us—some group is described in popular or academic accounts as different, some core aspect of their character marks them as distinctive, as not being essentially the same as us. This social process can leverage ideologies against groups. Cleaving to either universalist or particularist essentialism is not a defense against this kind of co-optation.

Historically, biological and environmental explanations have been used to explain why other people are different, and then that ideology of difference has been used to justify mistreatment in many historical periods. Slavery and racism are powerful examples of how ideologies of biological inferiority have been employed to justify exploitation and violence. More recently, ideas about genetic difference have been coopted to promote or justify the special treatment of people with differences assumed to be "innate" because they are biological (Fine, 2010). Clearly, brain scans are being used in this way—as signs of biological differences that are taken to be unchangeable and which can then be used to justify institutional intervention against these supposedly unchangeable (i.e., essential) traits. For example, brain-imaging research on abnormalities in individuals with psychopathy suggests to some observers that detectable difference in brain function may be admissible in criminal courts, whether to increase a sentence or exonerate a person for insanity remains unclear (Koenigs, Baskin-Sommers, Zeier, et al., 2011; Motzkin, Newman, Kiehl, et al., 2011).

At the same time, notions of "cultural difference" have also been used to justify discrimination, exclusion, and violence. "Culture," in popular terms, often means whatever makes other people different; the folk use of the term can sometimes assume all difference is "cultural" because human nature is assumed to be universal. The idea goes something like, "It's their culture that makes them different; otherwise, they'd be just like us." This perspective has often been used to justify assimilation or even genocide.

Notions of cultural purity can be as pernicious as those of biological purity. Overall, then, both neuroscience and anthropological research have been used to reinforce and legitimize racial, ethnic, gender, sexual, age, and religious stereotypes and hierarchies.

But neuroanthropology can disrupt these folk models that allow Westerners to divide the world, and thus justify how some people are treated in much worse fashion than others. Good research does matter. Strong data to refute biased and manipulated studies are an important aspect of how we need to approach attempts to appropriate our ideas. At the same time, as scientists facing political and social dilemmas, we need to think seriously about how we frame our results, present our ideas in public, and engage in public outreach. It would be extremely naïve to assume that just doing good research will overcome historical legacies of prejudice and inequality, that finding good data that hew closer to the truth than previous results will somehow make others see the light. We must anticipate that our own data and arguments will be taken by others to justify their ideological and political arguments, even if diametrically opposed to our interpretations. For example, to argue that addiction is the conjunction of "bad brain biology" and "bad social environments" would be a simplification of Lende's research. Nevertheless, excessive salience and ingrained habits, coupled with social contexts that are deviant and cue drug-seeking behavior, could be easily taken by others to mean that addicts deserve harsh treatments or social isolation (Buchman, Illes, & Reiner, 2011; Garriott, 2011). Overall, we need to take responsibility for our ideas, and to ensure that out in the public realm, our data and the overall approach to that data are properly represented, in part by finding ways to present scientific work to a broad audience.

Potential Divisions within an Overall Synthesis

This edited volume has been written largely by anthropologists who have drawn on neuroscience, social and biological theory, and ethnography to understand contemporary problems at the individual-environment interface. One point all authors agree on is that an integrative effort leads to better understanding of problems ranging from agency to addiction. However, limitations exist in that integrative effort, and likely divisions are sure to emerge in this new arena.

One of the easiest divisions that could arise is between neuroscience-oriented research and anthropology-oriented practitioners. The work in this book has come from the side of anthropology and as a result often addresses anthropological concerns—questions of enculturation, society, religion, and the development of self and meaning. Meanwhile, work in cultural and social neuroscience often focuses on the neural mechanisms implicated in human cultural life. While these two approaches can and do mutually inform each other, we cannot over optimistically assume that integration will come naturally. Different methodologies, historical legacies of research, and core theoretical problems will generally keep anthropology and neuroscience separate, and make bridging that gap difficult for researchers in an individual or group fashion. Bridging the divide is certainly possible, as this volume and similar work in cultural neuroscience makes apparent. More difficult to overcome, however, are institutional barriers that force researchers into narrow specializations while often setting bounds on what counts as "good research." Neuroscientists and anthropologists (and sociologists and English scholars and cognitive scientists ...) generally find themselves in different departments, often housed in split sections within an overall university architecture, and dependent upon different sources of funding for their research. Publishing in top journals in specialized fields pushes a disciplinary focus on researchers that can often limit integrative tendencies, and certainly cut off the opportunity to actively learn about other perspectives. Yet exploration and training are absolutely essential to develop scholars who can do effective work across these traditional boundaries. Unfortunately, the rigors of the tenure and promotion processes often push this type of exploration later into a research career, rather than encouraging collaboration as an active part of scholarly development.

Nevertheless, small groups of scholars are developing these interstitial areas of academic life, because integration addresses fundamental questions we want to answer. But here, too, one can anticipate splits. One major division will likely emerge between research on "within-culture" dynamics against "cross-culture" dynamics. Take the study of emotion. Examining how grief is expressed across cultures is not the same question as how the expression, meaning, and practice of grief happen in one particular place and time. While each research project can inform the other, they address different theoretical problems, since cross-cultural variation is not the same as enculturation. For example, with addiction, the vast majority of existing research focuses on "within-culture" dynamics—how extreme drug use functions in a particular place or time—with the assumption that this research bears on the universal phenomenon of "addiction." However, actually studying the cross-cultural patterns of heavy drug use requires not only a different theoretical approach but also distinctive data-gathering techniques, which at a minimum means data from different sites and a theoretical framework that can lead to comparability. The effort to produce comparability often sacrifices exactly the sensitivity to peculiar dynamics that people interested in understanding a specific time and place know are necessary—history, ethnographic depth, a sense of local context, and so forth. Rather than talking past each other, recreating former divisions between universalists and relativists, the general and the particular, we encourage researchers to understand that their research problems and methodologies are by necessity different, and that each group can learn from the other.

That same point can be made of the likely divisions between sciences and the humanities, and quantitative as opposed to qualitative methods. We hope that with neuroanthropology, these potential divisions are more about flavoring than either/or splits or outright antagonism. The case studies demonstrate the importance of qualitative work, which requires an interpretive approach; however, qualitative work can be used as a basis for developing quantitative measures, test hypotheses about theory of mind, and see if laboratory results can pass the "real world" test. Similarly, quantitative work permits the evaluation of multiple strands of data simultaneously, which can help efforts to test what sets of factors explain an outcome of interest. Science, particularly emerging research about neuroplasticity and cultural variation, provides an all-important foundation for developing ideas about particular neurocultural processes. In other words, successful neuroanthropological research needs to continually navigate these basic academic divides, tacking back and forth between fundamentally different research approaches to advance upwind toward greater holistic understanding.

A final major split is already happening between people who favor rational or cognitive approaches and those who favor embodied or practice-oriented approaches. This split is very old in Western intellectual history, going back to Plato and Aristotle. Are you interested in abstract forms and rationality? Or practicality and bodies? Is the brain-culture interface a Platonic ideal—a meshing of optimality and consonance? Or is it tilted toward Aristotle's flesh-and-blood approach, focused on training, emotion, and failure? This split might be one that is harder to reconcile. The rationalist approach to synergy is already gaining ground, including new work on neuroeconomics and "teaching errors" by the dopamine system (Glimcher, Fehr, Camerer, & Poldrack, 2009) or studies of how people match up with overarching cultural ideals and the stress that results when they don't. Another strand of work embraces embodiment, which can bridge cognition, language, and anthropology (Clark, 2008) and highlights alternative approaches to culture that stress bodily practice, skill acquisition, and sensory perception.

This split recreates itself across many intellectual debates: selfish genes or life cycles, the invisible hand or irrational buying, universal grammar or generative rules, universal human rights or relativism. We urge researchers to recognize that their own intellectual leanings—toward a rational, universal type of explanation or a more relative, everyday life explanation—is grounded both in personal histories that reach back to individual upbringing and education, as well as in intellectual histories dating back thousands of years. Given Western intellectual history, neither of these orientations is likely to disappear soon. What we would like to suggest here is that neither should go away. Both reveal important dimensions of our lives, and provide important footing for how to think about our lives.

Future Developments

This volume has documented many of the present trends in neuroanthropology. However, not every researcher could be included, and new developments were already happening even as this book was being written and published. Below we cover some of the developments we believe will continue. We focus on work happening in neuroscience and anthropology. Other exciting developments involve a synthesis of related social sciences with brain research, such as sociology and economics (Camerer, Loewenstein, & Prelec, 2005; Franks, 2010; Glimcher et al., 2009; Rose, 2007; Slingerland & Collard, 2011; Smail, 2007; Sutton, Harris, Keil, & Barnier, 2010; Turner, 2006; Vrecko, 2010; Westen, 2008). For questions of space, we will leave these non-anthropological advances to one side to focus on the core intersection examined in this volume.

Within anthropology, there are several developments we expect:

1. continued work on neuroanthropology that comes more from the biological side, including the neural bases of behaviors using imaging technology and comparative primate research, the examination of phenomena like stress cross-culturally, and sophisticated examinations of human development that focus on biological factors but use a neuroanthropological orientation (Aiello, 2010; Arbib, 2011; Burbank, 2011; Flinn, 2008; Gettler, McDade, Feranil, & Kuzawa, 2011; Matthews, Paukner, & Suomi, 2010; Rilling & Sanfey, 2011; Worthman, 2009);

2. a hopeful combination of work in neuroanthropology and neuroarcheology, both of which draw on ideas of neural mechanisms and plasticity and how those interact with a particular space and time of human living (de Beaune, Coolidge, & Wynn, 2009; Malafouris, 2010; Renfrew, Firth, & Malafouris, 2009);

3. increased synergy with psychological and medical anthropology, which will address mental and social problems at the individual-culture interface the fields share in common (Antelius, 2008; Desjarlais & Jason Throop, 2011; Hinton & Good, 2009; Kohrt & Harper, 2008; Kirmayer, 2009; Luhrman, Nusbaum, & Thisted, (2010); Martin, 2010; Ortega, 2009; Raybeck & Ngo, 2011); and

4. increased collaboration with cultural anthropology and cultural studies that examine the influence of neuroscience and popular ideas about the brain more broadly, including in public debates and popular discussions (Buchman et al., 2011; Campbell, 2009; Dumit, 2003; Langlitz, 2010; Rees, 2010).

On the neuroscience side, we expect cross-fertilization to happen with cultural neuroscience and with critical neuroscience. Cultural neuroscience is really the sister discipline of neuroanthropology, with a focus on how culture patterns the brain using field studies and international research (Chiao et al., 2010; Bender et al., 2010; Domínguez Duque, Turner, Lewis, & Egan, 2010; Kitayama & Uskul, 2011; Northoff, 2010; Roepstorff et al., 2010; Seligman & Brown, 2010). Critical neuroscience uses imaging research and critical analysis to challenge how brain science is being used by society and by neuroscientists themselves, and to challenge the abuses that already exist (Choudhury, Nagel, & Slaby, 2009; Choudhury & Slaby, 2011; Fine, 2010; Kraus, 2011; Slaby, 2010).

Two potential developments should also be on any list of priorities. First, field studies that jointly use anthropologists and neuroscientists (or cognitive scientists more generally) are necessary to create a genuine synthesis. As imaging and biomarker technologies become increasingly portable and field ready (McDade, Williams, & Snodgrass, 2007), the opportunities to do work in cross-cultural settings as teams has become feasible. Work that brings together imaging, other types of quantitative data, ethnography, and a comparative framework should really begin to address many of the questions raised by researchers in this volume. Second, neuroanthropology, and even cultural neuroscience, have not drawn much on the work in social neuroscience, which has taken its inspiration more from social psychology than from sociology or anthropology

(Todorov, Fiske, & Prentice, 2011). However, neuroanthropology and social neuroscience are both focused on understanding the proximate determinants of behavior, from those that reside primarily in the brain, to those in interactions between people and with features of a historic social environment.

Applied work, policy development, and other types of engaged research already mark many of the entries in this volume. This area should be a vibrant arena for future research, in particular as research focuses on the combination of neurocultural processes and neuroanthropological vulnerabilities. Anthropology has often been split by a theoretical/applied distinction; neuroscience is similar in its divide between bench and applied researchers. However, by addressing fundamental questions about human behavior and variation, neuroanthropology positions itself to address applied issues, since these questions often bear on human problems rather than overarching theories of culture or neural function. Moreover, given its grounding in ethnography and community-based approaches, neuroanthropology develops the resources and insights necessary to have pragmatic implications.

Conclusion

Societies are complex phenomena; so are the brains that undergird them. We cannot do effective research by bracketing the complexity of either, as the interactions between parts of brains and facets of social life are just as complex. A plethora of intersections exist where valuable research can be done; we look forward to observing how neuroanthropology unfolds and takes advantage of these opportunities. This new field addresses two interrelated phenomena—the patterning of human variation and diverse neurocultural dynamics. At present, most prior research has focused either on single societies or on cross-cultural work using small samples, Western assumptions, and very broad and often nebulous assessments of culture and lived experience. Understanding patterns of human variation using samples from multiple sites, with methods that assess cross-cultural differences as well as similarities and both direct and indirect appraisals of neural function, represents a core area of research for the future. Similarly, in-depth work on human behavior, experience, and skill in particular times and places will provide a greater understanding of specific neurocultural processes and how neuroanthropological vulnerabilities come to be. As the case studies in this volume have shown, research on balance, agency, humor, memory, and post-traumatic stress disorder requires a depth and rigor of data that neither neuroscience nor anthropology can achieve on its own. With models that focus on specific neurocultural processes, the research in this volume provides a way to test ideas and apply theory to other arenas. Overall, by combining neuroscience and anthropology, and using both a comparative framework and in-depth analysis, neuroanthropology will continue to develop as a robust research endeavor.

References

Aiello, L. C. (Ed.) (2010). Working memory: Beyond language and symbolism [Special issue]. *Current Anthropology, 51* (Suppl. 1).

Anderson, M. L. (2010). Neural reuse: A fundamental organizational principle of the brain. *Behavioral and Brain Sciences, 33,* 245–266.

Antelius, E. (2008). The meaning of the present: Hope and foreclosure in narrations about people with severe brain damage. *Medical Anthropology Quarterly, 21*(3), 324–342.

Arbib, M. A. (2011). From mirror neurons to complex imitation in the evolution of language and tool use. *Annual Review of Anthropology, 40,* 257–273.

Bar-Haim, Y., Lamy, D., Pergamin, L., Bakermans-Kranenburg, M. J., & van Ijzendoorn, M. H. (2007). Threat-related attentional bias in anxious and nonanxious individuals: A meta-analytic study. *Psychological Bulletin, 133*(1), 1–24.

Barsalou, L. W. (2010). Grounded cognition: Past, present, and future. *Topics in Cognitive Science, 2,* 716–724.

Bender, A., & Beller, S. (2011). The cultural constitution of cognition: Taking the anthropological perspective. *Frontiers in Psychology, 2,* 67.

Bender, A., Hutchins, E., & Medin, D. (2010). Anthropology in cognitive science. *Topics in Cognitive Science, 2*(3), 374–385.

Boas, F. (1912). Changes in the bodily form of descendants of immigrants. *American Anthropologist, 14,* 530–562.

Boas, F. (1920). The methods of ethnology. *American Anthropologist, 22*(4), 311–321.

Boroditsky, L. (2011, February). How language shapes thought: The languages we speak affect our perceptions of the world. Scientific American. Retrieved from http://www.scientificamerican.com/article.cfm?id=how-language-shapes-thought

Buchman, D., Illes, J., & Reiner, P. (2011). The paradox of addiction neuroscience. *Neuroethics, 4*(2), 65–77.

Burbank, V. K. (2011). *An ethnography of stress: The social determinants of health in aboriginal Australia.* New York: Palgrave MacMillan.

Camerer, C., Loewenstein, G., & Prelec, D. (2005). Neuroeconomics: How neuroscience can inform economics. *Journal of Economic Literature, 43*(1), 9–64.

Campbell, N. D. (2009). Toward a critical neuroscience of "addiction". *Biosocieties, 5*, 89–104.

Changizi, M. (2011). *Harnessed: How language and music mimicked nature and transformed ape to man.* Dallas: BenBella Books.

Chater, N., & Christiansen, M. H. (2010). Language evolution as cultural evolution: How language is shaped by the brain. *Wiley Interdisciplinary Reviews: Cognitive Science, 1*, 623–628.

Chiao, J. Y., Hariri, A. R., Harada, T., Mano, Y., Sadato, N., Parrish, T. B., et al. (2010). Theory and methods in cultural neuroscience. *Social Cognitive and Affective Neuroscience, 5*(2–3), 356–361.

Choudhury, S., Nagel, S. K., & Slaby, J. (2009). Critical neuroscience: Linking neuroscience and society through critical practice. *Biosocieties, 4*(1), 61–77.

Choudhury, S. & Slaby, J. (Eds.) (2011). *Critical neuroscience: A handbook of the cultural and social contexts of neuroscience.* Hoboken, NJ: Wiley-Blackwell.

Clark, A. (2008). *Supersizing the mind: Embodiment, action, and cognitive extension.* Oxford: Oxford University Press.

De Beaune, S. A., Coolidge, F. L., & Wynn, T. (Eds.) (2009). *Cognitive archaeology and human evolution.* Cambridge: Cambridge University Press.

Desjarlais, R., & Throop, J. C. (2011). Phenomenological approaches in anthropology. *Annual Review of Anthropology, 40*, 87–102.

Domínguez Duque, J. F., Turner, R., Lewis, E. D., and Egan, G. (2010). Neuroanthropology: A humanistic science for the study of the culture-brain nexus. *Social Cognitive and Affective Neuroscience, 5*(2–3), 138–147.

Dressler, W. W., Oths, K. S., & Gravlee, C. C. (2005). Race and ethnicity in public health research: Models to explain health disparities. *Annual Review of Anthropology, 34*, 231–252.

Dumit, J. (2003). *Picturing personhood: Brain scans and biomedical identity.* Princeton, NJ: Princeton University Press.

Elman, J. L., Bates, E. A., Johnson, M. H., Karmiloff-Smith, A., Parisi, D., & Plunkett, K. (1996). *Rethinking innateness: A connectionist perspective on development.* Cambridge, MA: MIT Press.

Evans, N., & Levinson, S. C. (2009). The myth of language universals: Language diversity and its importance for cognitive science. *Behavioral and Brain Sciences, 32*(5), 429–492.

Fine, C. (2010). *Delusions of gender: How our minds, society, and neurosexism create difference*. New York: W.W. Norton.

Flinn, M. V. (2008). Why words can hurt us: Social relationships, stress, and health. In E. O. Smith, W. Trevathan, & J. McKenna (Eds.), *Evolutionary medicine and health: New Perspectives* (pp. 247–258). Oxford: Oxford University Press.

Franks, D. D. (2010). *Neurosociology: The nexus between neuroscience and social psychology*. New York: Springer.

Garriott, W. (2011). *Policing methamphetamine: Narcopolitics in rural America*. New York: NYU Press.

Geertz, C. (1973). *The interpretation of culture*. New York: Basic Books.

Gettler, L. T., McDade, T. W., Feranil, A. B., & Kuzawa, C. W. (2011). Longitudinal evidence that fatherhood decreases testosterone in human males. *Proceedings of the National Academy of Sciences, 108*(39), 16194–16199.

Glimcher, P. W., Fehr, E., Camerer, C., & Poldrack, R. A. (Eds.). (2009). *Neuroeconomics: Decision making and the brain*. London: Academic Press.

Gravlee, C. C., Bernard, H. R., & Leonard, W. R. (2003). Heredity, environment, and cranial form: A reanalysis of Boas's immigrant data. *American Anthropologist, 105,* 125–138.

Gravlee, C. C., & Sweet, E. (2008). Race, ethnicity, and racism in medical anthropology, 1977–2002. *Medical Anthropology Quarterly, 22*(1), 27–51.

Hackman, D. A., & Farah, M. J. (2009). Socioeconomic status and the developing brain. *Trends in Cognitive Science, 13*(2), 65–73.

Henrich, J., Heine, S. J., & Norenzayan, A. (2010). The weirdest people in the world? *Behavioral and Brain Sciences, 33,* 61–83.

Hinton, D., & Good, B. (Eds.). (2009). *Culture and panic disorder*. Stanford, CA: Stanford University Press.

Hruschka, D. J., Christiansen, M. H., Blythe, R. A., Croft, W., Heggarty, P., Mufwene, S. S., et al. (2009). Building social cognitive models of language change. *Trends in Cognitive Sciences, 13*(11), 464–469.

Jablonka, E., & Lamb, M. J. (2005). *Evolution in four dimensions: Genetic, epigenetic, behavioral, and symbolic variation in the history of life*. Cambridge, MA: MIT Press.

Kirmayer, L. J. (2009). Nightmares, neurophenomenology, and the cultural logic of trauma. *Culture, Medicine and Psychiatry, 33*(2), 323–331.

Kitayama, S., & Cohen, D. (Eds.). (2007). *Handbook of cultural psychology*. New York: Guilford Press.

Kitayama, S., & Uskul, A. K. (2011). Culture, mind, and the brain: Current evidence and future directions. *Annual Review of Psychology, 62*, 419–449.

Kleinman, A., & Good, B. (Eds.). (1985). *Culture and depression: Studies in the anthropology and cross-cultural psychiatry of affect and disorder.* Berkeley, CA: University of California Press.

Koenigs, M., Baskin-Sommers, A., Zeier, J., & Newman, J. P. (2011) Investigating the neural correlates of psychopathy: A critical review. *Molecular Psychiatry, 16*(8), 792–799.

Kohrt, B. A., & Harper, I. (2008). Navigating diagnoses: Understanding mind-body relations, mental health, and stigma in Nepal. *Culture, Medicine and Psychiatry, 32*(4), 462–491.

Kraus, C. (2011). Critical studies of the sexed brain: A critique of what and for whom? *Neuroethics*, 1–13.

Lakoff, G., & Johnson, M. (1999). *Philosophy in the flesh: The embodied mind and its challenge to Western thought.* New York, NY: Basic Books.

Langlitz, N. (2010). The persistence of the subjective in neuropsychopharmacology: Observations of contemporary hallucinogen research. *History of the Human Sciences, 23*(1), 37–57.

Lende, D. H. (2007). Evolution and modern behavioral problems. In W. Trevathan, E.O. Smith, & J. J. McKenna (Eds.), *Evolutionary medicine and health: New perspectives* (pp. 277–290). New York: Oxford University Press.

Lovejoy, A. O. (1976). *The great chain of being.* Cambridge, MA: Harvard University Press.

Luhrman, T., Nusbaum, H., & Thisted, R. (2010). The absorption hypothesis: Learning to hear God in evangelical Christianity. *American Anthropologist, 112*(1), 66–78.

Malafouris, L. (2010). Metaplasticity and the principles of neuroarchaeology. *Journal of Anthropological Sciences, 88*, 49–72.

Martin, E. (2010). Self-making and the brain. *Subjectivity, 3*(4), 366–381.

Matthews, L. J., Paukner, A., & Suomi, S. J. (2010). Can traditions emerge from the interaction of stimulus enhancement and reinforcement learning? An experimental model. *American Anthropologist, 112*(2), 257–269.

McDade, T., Williams, S., & Snodgrass, J. (2007). What a drop can do: Dried blood spots as a minimally invasive method for integrating biomarkers into population-based research. *Demography, 44*(4), 899–925.

Medin, D., Bennis, W., & Chandler, M. (2010). Culture and the home-field disadvantage. *Perspectives on Psychological Science, 5*(6), 708–713.

Moore, C. C., & Mathews, H. F. (Eds.). (2001). *The psychology of cultural experience*. Cambridge, MA: Cambridge University Press.

Motzkin, J. C., Newman, J. P., Kiehl, K. A., & Koenigs, M. (2011). Reduced prefrontal connectivity in psychopathy. *Journal of Neuroscience, 31*(48), 17348–17357.

Noë, A. (2009). *Out of our heads: Why you are not your brain, and other lessons from the biology of consciousness*. New York: Hill and Wang.

Northoff, G. (2010). Humans, brains and their environments: Marriage between neuroscience and anthropology? *Neuron, 65*(6), 748–751.

Odling-Smee, J. F., Laland, K. N., & Feldman, M. W. (2003). *Niche construction: The neglected process in evolution*. Princeton, NJ: Princeton University Press.

Ong, A. (1999). *Flexible citizenship: The cultural logics of transnationality*. Durham, NC: Duke University Press.

Ortega, F. (2009). The cerebral subject and the challenge of neurodiversity. *Biosocieties, 4*(4), 425–445.

Oyama, S., Gray, R. D., & Griffiths, P. E. (Eds.). (2001). *Cycles of contingency: Developmental systems and evolution*. Cambridge, MA: MIT Press.

Perlovsky, L. (2009). Language and emotions: Emotional Sapir-Whorf hypothesis. *Neural Networks, 22*(5–6), 518–526.

Pigliucci, M., & Müller, G. B. (Eds.). (2010). *Evolution: The extended synthesis*. Cambridge, MA: MIT Press.

Raybeck, D., & Ngo, P. Y. L. (2011). Behavior and the brain: Mediation of acquired skills. *Cross-Cultural Research, 45*(2), 178–207.

Redish, A. D., Jensen, S., & Johnson, A. (2008). A unified framework for addiction: Vulnerabilities in the decision process. *Behavioral and Brain Sciences, 31*, 415–437.

Rees, T. (2010). Being neurologically human today: Life and science and adult cerebral plasticity. *American Ethnologist, 37*(1), 150–166.

Regier, T., & Kay, P. (2009). Language, thought, and color: Whorf was half right. *Trends in Cognitive Sciences, 13*(10), 439–446.

Renfrew, C., Firth, C., & Malafouris, L. (2009). *The sapient mind: Archaeology meets neuroscience*. New York: Oxford University Press.

Richerson, P. J., & Boyd, R. (2006). *Not by genes alone: How culture transformed human evolution*. Chicago: University of Chicago Press.

Rilling, J. K. (2008). Neuroscientific approaches and applications within anthropology. *Yearbook of Physical Anthropology, 51*, 2–32.

Rilling, J. K., & Sanfey, A. G. (2011). The neuroscience of social decision-making. *Annual Review of Psychology, 62*, 23–48.

Roepstorff, A., Niewöhnerc, J., & Beck, S. (2010). Enculturing brains through patterned practices. *Neural Networks, 23*, 1051–1059.

Rose, N. (2007). *The politics of life itself: Biomedicine, power, and subjectivity in the twenty-first century*. Princeton, NJ: Princeton University Press.

Seligman, R., & Brown, R. A. (2010). Theory and method at the intersection of anthropology and cultural neuroscience. *Social Cognitive and Affective Neuroscience, 5*(2–3), 130–137.

Slaby, J. (2010). Steps towards a critical neuroscience. *Phenomenology and the Cognitive Sciences, 9*, 397–416.

Slingerland, E., & Collard, M. (Eds.). (2011). *Creating consilience: Integrating the sciences and the humanities*. Oxford: Oxford University Press.

Smail, D. L. (2007). *On deep history and the brain*. Berkeley, CA: University of California Press.

Smith, B. H. (2011). The chimera of relativism: A tragicomedy. *Common Knowledge, 17*, 13–26.

Sutton, J., Harris, C., Keil, P., & Barnier, A. (2010). The psychology of memory, extended cognition, and socially distributed remembering. *Phenomenology and the Cognitive Sciences, 9*(4), 521–560.

Todorov, T. (2010). *The fear of barbarians: Beyond the clash of civilizations* (A. Brown, Trans.). Chicago: University of Chicago Press.

Todorov, A., Fiske, S., & Prentice, D. (Eds.). (2011). *Social neuroscience: Toward understanding the underpinnings of the social mind*. New York: Oxford University Press.

Trevathan, W., Smith, E.O., & McKenna, J. J. (Eds.) (2007). *Evolutionary medicine and health: New perspectives*. New York: Oxford University Press.

Turner, M. (Ed.) (2006). *The artful mind: Cognitive science and the riddle of human creativity*. New York: Oxford University Press.

Vrecko, S. (2010). Neuroscience, power and culture: An introduction. *History of the Human Sciences, 23*(1), 1–10.

Westen, D. (2008). *The political brain: The role of emotion in deciding the fate of the nation*. New York: Public Affairs.

Westermann, G., Mareschal, D., Johnson, M. H., Sirois, S., Spratling, M. W., & Thomas, M. S. (2007). Neuroconstructivism. *Developmental Science, 10*, 75–83.

Wexler, B. E. (2011). Neuroplasticity: Biological evolution's contribution to cultural evolution. In S. Han & E. Pöppel (Eds.), *Culture and neural frames of cognition and communication*. Heidelberg: Springer.

Wierzbicka, A. (1999). *Emotions across languages and cultures: Diversity and universals.* Cambridge, MA: Cambridge University Press.

Worthman, C. M. (2009). Habits of the heart: Life history and the developmental neuroendocrinology of emotion. *American Journal of Human Biology, 21*(6), 772–778.

Contributors

Mauro C. Balieiro Paulista University (Brazil)

Kathryn Bouskill Emory University

Rachel S. Brezis University of California Los Angeles

Benjamin Campbell University of Wisconsin-Milwaukee

José Ernesto dos Santos University of São Paulo (Brazil)

Greg Downey Macquarie University (Australia)

William W. Dressler University of Alabama

Erin P. Finley South Texas Veterans Health Care System and University of Texas

Agustín Fuentes University of Notre Dame

M. Cameron Hay Miami University

Daniel H. Lende University of South Florida

Katherine C. MacKinnon Saint Louis University

Katja Pettinen Purdue University

Peter G. Stromberg University of Tulsa

Index

Printed in the United States
by Baker & Taylor Publisher Services